RADIA
IN ASTROPHYSICS

RADIATIVE PROCESSES IN ASTROPHYSICS

GEORGE B. RYBICKI
ALAN P. LIGHTMAN
Harvard-Smithsonian Center for Astrophysics

WILEY-VCH Verlag GmbH & Co. KGaA

Library of Congress Card No.: Applied for

British Library Cataloging-in-Publication Data:
A catalogue record for this book is available from the British Library

Bibliographic information published by
Die Deutsche Bibliothek
Die Deutsche Bibliothek lists this publication in the Deutsche Nationalbibliografie; detailed bibliographic data is available in the
Internet at <http://dnb.ddb.de>.

Printed in the Federal Republic of Germany
Printed on acid-free paper

Cover Picture Calculated 6 cm continuum emission of a protoplanetary disk illuminated by a nearby hot star
(courtesy of H.W. Yorke, NASA-JPL/California Institute of Technology and S. Richling, Institut d'Astrophysique de Paris)
Printing Strauss GmbH, Mörlenbach
Bookbinding Großbuchbinderei J. Schäffer GmbH & Co. KG, Grünstadt

ISBN-13: 978-0-471-82759-7
ISBN-10: 0-471-82759-2

To **Verena** ***and*** **Jean**

PREFACE

This book grew out of a course of the same title which each of us taught for several years in the Harvard astronomy department. We felt a need for a book on the subject of radiative processes emphasizing the physics rather than simply giving a collection of formulas.

The range of material within the scope of the title is immense; to cover a reasonable portion of it has required us to go only deeply enough into each area to give the student a feeling for the basic results. It is perhaps inevitable in a broad survey such as this that inadequate coverage is given to certain subjects. In these cases the references at the end of each chapter can be consulted for further information.

The material contained in the book is about right for a one-term course for seniors or first-year graduate students of astronomy, astrophysics, and related physics courses. It may also serve as a reference for workers in the field. The book is designed for those with a reasonably good physics background, including introductory quantum mechanics, intermediate electromagnetic theory, special relativity, and some statistical mechanics. To make the book more self-contained we have included brief reviews of most of the prerequisite material. For readers whose preparation is less than ideal this gives an opportunity to bolster their background by studying the material again in the context of a definite physical application.

A very important and integral part of the book is the set of problems at the end of each chapter and their solutions at the end of the book. Besides their usual role in affording self-tests of understanding, the problems and solutions present important results that are used in the main text and also contain most of the astrophysical applications.

We owe a debt of gratitude to our teaching assistants over the years, Robert Moore, Robert Leach, and Wayne Roberge, and to students whose penetrating questions helped shape this book. We thank Ethan Vishniac for his help in preparing the index. We also want to thank Joan Verity for her excellence and flexibility in typing the manuscript.

George B. Rybicki
Alan P. Lightman

Cambridge, Massachusetts
May 1979

CONTENTS

RADIATIVE PROCESSES IN ASTROPHYSICS

1

FUNDAMENTALS OF RADIATIVE TRANSFER

1.1 THE ELECTROMAGNETIC SPECTRUM; ELEMENTARY PROPERTIES OF RADIATION

Electromagnetic radiation can be decomposed into a *spectrum* of constituent components by a prism, grating, or other devices, as was discovered quite early (Newton, 1672, with visible light). The spectrum corresponds to waves of various wavelengths and frequencies, related by $\lambda\nu = c$, where ν is the frequency of the wave, λ is its wavelength, and $c = 3.00 \times 10^{10}$ cm s^{-1} is the free space velocity of light. (For waves not traveling in a vacuum, c is replaced by the appropriate velocity of the wave in the medium.) We can divide the spectrum up into various regions, as is done in Figure 1.1. For convenience we have given the energy $E = h\nu$ and temperature $T = E/k$ associated with each wavelength. Here h is Planck's constant $= 6.625 \times 10^{-27}$ erg s, and k is Boltzmann's constant $= 1.38 \times 10^{-16}$ erg K^{-1}. This chart will prove to be quite useful in converting units or in getting a quick view of the relevant magnitude of quantities in a given portion of the spectrum. The boundaries between different regions are somewhat arbitrary, but conform to accepted usage.

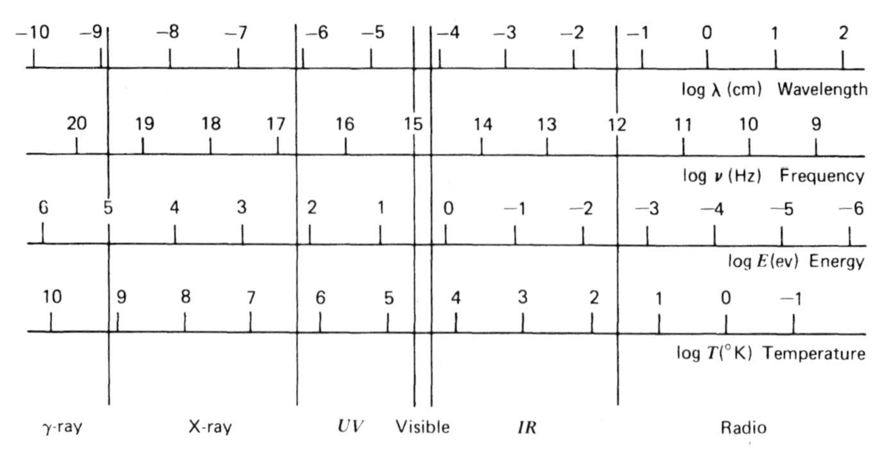

Figure 1.1 The electromagnetic spectrum.

1.2 RADIATIVE FLUX

Macroscopic Description of the Propagation of Radiation

When the scale of a system greatly exceeds the wavelength of radiation (e.g., light shining through a keyhole), we can consider radiation to travel in straight lines (called rays) in free space or homogeneous media—from this fact a substantial theory (transfer theory) can be erected. The detailed justification of this assumption is considered at the end of Chapter 2. One of the most primitive concepts is that of *energy flux*: consider an element of area dA exposed to radiation for a time dt. The amount of energy passing through the element should be proportional to $dA\,dt$, and we write it as $F\,dA\,dt$. The energy flux F is usually measured in erg s^{-1} cm^{-2}. Note that F can depend on the orientation of the element.

Flux from an Isotropic Source—the Inverse Square Law

A source of radiation is called *isotropic* if it emits energy equally in all directions. An example would be a spherically symmetric, isolated star. If we put imaginary spherical surfaces S_1 and S at radii r_1 and r, respectively, about the source, we know by conservation of energy that the total energy passing through S_1 must be the same as that passing through S. (We assume no energy losses or gains between S_1 and S.) Thus

$$F(r_1)\cdot 4\pi r_1^2 = F(r)\cdot 4\pi r^2,$$

or

$$F(r)=\frac{F(r_1)r_1^2}{r^2}.$$

If we regard the sphere S_1 as fixed, then

$$F=\frac{\text{constant}}{r^2}. \tag{1.1}$$

This is merely a statement of conservation of energy.

1.3 THE SPECIFIC INTENSITY AND ITS MOMENTS

Definition of Specific Intensity or Brightness

The flux is a measure of the energy carried *by all rays* passing through a given area. A considerably more detailed description of radiation is to give the energy carried along by *individual rays*. The first point to realize, however, is that a single ray carries essentially no energy, so that we need to consider the energy carried by sets of rays, which differ infinitesimally from the given ray. The appropriate definition is the following: Construct an area dA normal to the direction of the given ray and consider all rays passing through dA whose direction is within a solid angle $d\Omega$ of the given ray (see Fig. 1.2). The energy crossing dA in time dt and in frequency range $d\nu$ is then defined by the relation

$$dE=I_\nu\, dA\, dt\, d\Omega\, d\nu, \tag{1.2}$$

where I_ν is the *specific intensity* or *brightness*. The specific intensity has the

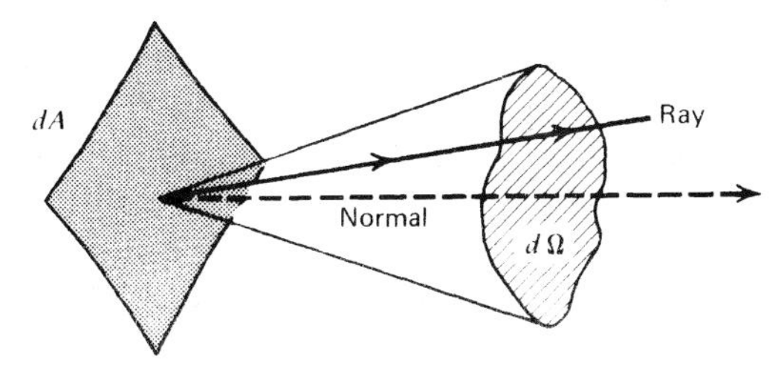

Figure 1.2 Geometry for normally incident rays.

dimensions

$$I_\nu(\nu,\Omega) = \text{energy (time)}^{-1}\,(\text{area})^{-1}\,(\text{solid angle})^{-1}\,(\text{frequency})^{-1}$$
$$= \text{ergs s}^{-1}\,\text{cm}^{-2}\,\text{ster}^{-1}\,\text{Hz}^{-1}.$$

Note that I_ν depends on location in space, on direction, and on frequency.

Net Flux and Momentum Flux

Suppose now that we have a radiation field (rays in all directions) and construct a small element of area dA at some arbitrary orientation $\mathbf{n}$ (see Fig. 1.3). Then the differential amount of flux from the solid angle $d\Omega$ is (reduced by the lowered effective area $\cos\theta\, dA$)

$$dF_\nu(\text{erg s}^{-1}\,\text{cm}^{-2}\,\text{Hz}^{-1}) = I_\nu \cos\theta\, d\Omega. \tag{1.3a}$$

The *net flux* in the direction $\mathbf{n}$, $F_\nu(\mathbf{n})$ is obtained by integrating dF over all solid angles:

$$F_\nu = \int I_\nu \cos\theta\, d\Omega. \tag{1.3b}$$

Note that if I_ν is an isotropic radiation field (not a function of angle), then the net flux is *zero*, since $\int \cos\theta\, d\Omega = 0$. That is, there is just as much energy crossing dA in the $\mathbf{n}$ direction as the $-\mathbf{n}$ direction.

To get the flux of momentum normal to dA (momentum per unit time per unit area = pressure), remember that the momentum of a photon is E/c. Then the momentum flux along the ray at angle θ is dF_ν/c. To get

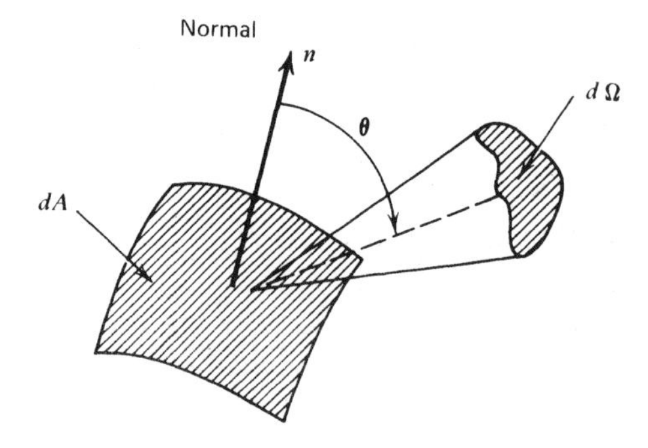

Figure 1.3 Geometry for obliquely incident rays.

the component of momentum flux normal to dA, we multiply by another factor of $\cos\theta$. Integrating, we then obtain

$$p_\nu(\text{dynes cm}^{-2}\,\text{Hz}^{-1}) = \frac{1}{c}\int I_\nu \cos^2\theta \, d\Omega. \tag{1.4}$$

Note that F_ν and p_ν are *moments* (multiplications by powers of $\cos\theta$ and integration over $d\Omega$) of the intensity I_ν. Of course, we can always integrate over frequency to obtain the total (integrated) flux and the like.

$$F(\text{erg s}^{-1}\,\text{cm}^{-2}) = \int F_\nu \, d\nu \tag{1.5a}$$

$$p(\text{dynes cm}^{-2}) = \int p_\nu \, d\nu \tag{1.5b}$$

$$I(\text{erg s}^{-1}\,\text{cm}^{-2}\,\text{ster}^{-1}) = \int I_\nu \, d\nu \tag{1.5c}$$

Radiative Energy Density

The specific energy density u_ν is defined as the energy per unit volume per unit frequency range. To determine this it is convenient to consider first the energy density per unit solid angle $u_\nu(\Omega)$ by $dE = u_\nu(\Omega)\,dV\,d\Omega\,d\nu$ where dV is a volume element. Consider a cylinder about a ray of length ct (Fig. 1.4). Since the volume of the cylinder is $dA\,c\,dt$,

$$dE = u_\nu(\Omega)\,dA\,c\,dt\,d\Omega\,d\nu.$$

Radiation travels at velocity c, so that in time dt all the radiation in the cylinder will pass out of it:

$$dE = I_\nu\,dA\,d\Omega\,dt\,d\nu.$$

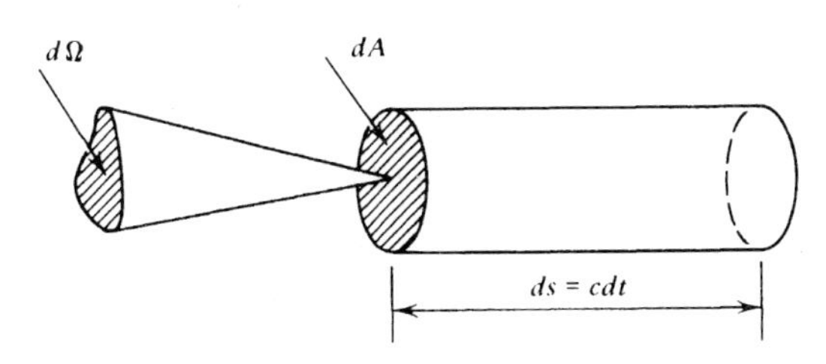

***Figure 1.4** Electromagnetic energy in a cylinder.*

Equating the above two expressions yields

$$u_\nu(\Omega) = \frac{I_\nu}{c}. \tag{1.6}$$

Integrating over all solid angles we have

$$u_\nu = \int u_\nu(\Omega)\, d\Omega = \frac{1}{c}\int I_\nu\, d\Omega,$$

or

$$u_\nu = \frac{4\pi}{c} J_\nu, \tag{1.7}$$

where we have defined the *mean intensity* J_ν:

$$J_\nu = \frac{1}{4\pi}\int I_\nu\, d\Omega. \tag{1.8}$$

The total radiation density (erg cm^{-3}) is simply obtained by integrating u_ν over all frequencies

$$u = \int u_\nu\, d\nu = \frac{4\pi}{c}\int J_\nu\, d\nu. \tag{1.9}$$

Radiation Pressure in an Enclosure Containing an Isotropic Radiation Field

Consider a reflecting enclosure containing an isotropic radiation field. Each photon transfers *twice* its normal component of momentum on reflection. Thus we have the relation

$$p_\nu = \frac{2}{c}\int I_\nu \cos^2\theta\, d\Omega.$$

This agrees with our previous formula, Eq. (1.4), since here we integrate only over 2π steradians. Now, by isotropy, $I_\nu = J_\nu$ so

$$p = \frac{2}{c}\int J_\nu\, d\nu \int \cos^2\theta\, d\Omega.$$

The angular integration yields

$$p = \tfrac{1}{3} u. \tag{1.10}$$

The radiation pressure of an isotropic radiation field is one-third the energy density. This result will be useful in discussing the thermodynamics of blackbody radiation.

Constancy of Specific Intensity Along Rays in Free Space

Consider any ray L and any two points along the ray. Construct areas dA_1 and dA_2 normal to the ray at these points. We now make use of the fact that energy is conserved. Consider the energy carried by that set of rays passing through both dA_1 and dA_2 (see Fig. 1.5). This can be expressed in two ways:

$$dE_1 = I_{\nu_1} dA_1 \, dt \, d\Omega_1 \, d\nu_1 = dE_2 = I_{\nu_2} dA_2 \, dt \, d\Omega_2 \, d\nu_2.$$

Here $d\Omega_1$ is the solid angle subtended by dA_2 at dA_1 and so forth. Since $d\Omega_1 = dA_2/R^2$, $d\Omega_2 = dA_1/R^2$ and $d\nu_1 = d\nu_2$, we have

$$I_{\nu_1} = I_{\nu_2}.$$

Thus the intensity is constant along a ray:

$$I_\nu = \text{constant}. \tag{1.11}$$

Another way of stating the above result is by the differential relation

$$\frac{dI_\nu}{ds} = 0, \tag{1.12}$$

where ds is a differential element of length along the ray.

Proof of the Inverse Square Law for a Uniformly Bright Sphere

To show that there is no conflict between the constancy of specific intensity and the inverse square law, let us calculate the flux at an arbitrary

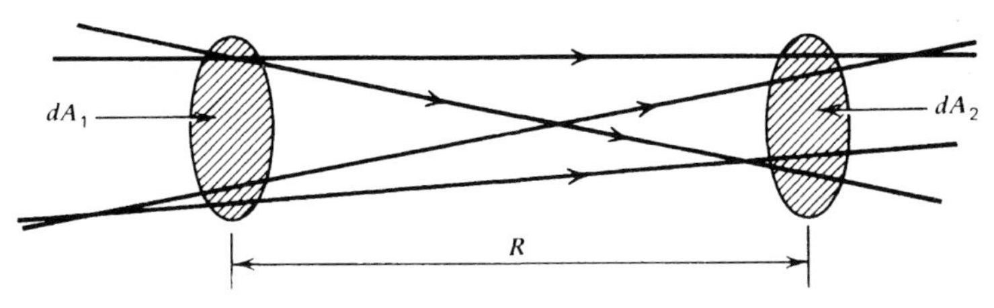

Figure 1.5 Constancy of intensity along rays.

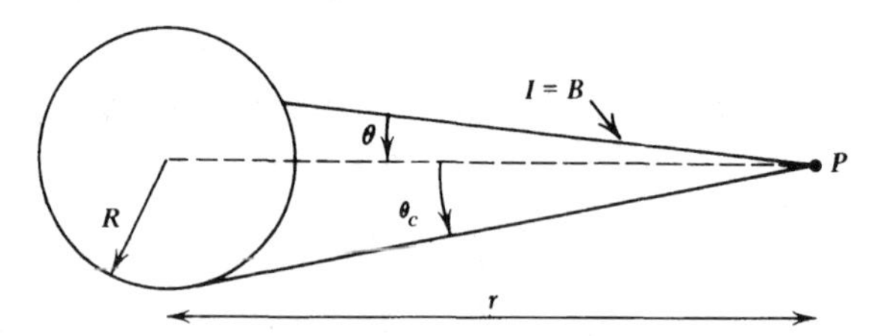

Figure 1.6 Flux from a uniformly bright sphere.

distance from a sphere of uniform brightness B (that is, all rays leaving the sphere have the same brightness). Such a sphere is clearly an isotropic source. At P, the specific intensity is B if the ray intersects the sphere and zero otherwise (see Fig. 1.6). Then,

$$F=\int I\cos\theta\, d\Omega=B\int_0^{2\pi}d\phi\int_0^{\theta_c}\sin\theta\cos\theta\, d\theta,$$

where $\theta_c=\sin^{-1}R/r$ is the angle at which a ray from P is tangent to the sphere. It follows that

$$F=\pi B(1-\cos^2\theta_c)=\pi B\sin^2\theta_c$$

or

$$F=\pi B\left(\frac{R}{r}\right)^2. \tag{1.13}$$

Thus the specific intensity is constant, but the solid angle subtended by the given object decreases in such a way that the inverse square law is recovered.

A useful result is obtained by setting $r=R$:

$$F=\pi B. \tag{1.14}$$

That is, the flux at a surface of uniform brightness B is simply πB.

1.4 RADIATIVE TRANSFER

If a ray passes through matter, energy may be added or subtracted from it by emission or absorption, and the specific intensity will not in general remain constant. "Scattering" of photons into and out of the beam can also affect the intensity, and is treated later in §1.7 and 1.8.

Emission

The spontaneous *emission coefficient* j is defined as the energy emitted per unit time per unit solid angle and per unit volume:

$$dE = j\,dV\,d\Omega\,dt.$$

A monochromatic emission coefficient can be similarly defined so that

$$dE = j_\nu\,dV\,d\Omega\,dt\,d\nu, \tag{1.15}$$

where j_ν has units of erg cm^{-3} s^{-1} ster^{-1} Hz^{-1}.

In general, the emission coefficient depends on the direction into which emission takes place. For an *isotropic* emitter, or for a distribution of randomly oriented emitters, we can write

$$j_\nu = \frac{1}{4\pi} P_\nu, \tag{1.16}$$

where P_ν is the radiated power per unit volume per unit frequency. Sometimes the spontaneous emission is defined by the (angle integrated) *emissivity* ϵ_ν, defined as the energy emitted spontaneously per unit frequency per unit time per unit mass, with units of erg gm^{-1} s^{-1} Hz^{-1}. If the emission is isotropic, then

$$dE = \epsilon_\nu \rho\,dV\,dt\,d\nu \frac{d\Omega}{4\pi}, \tag{1.17}$$

where ρ is the mass density of the emitting medium and the last factor takes into account the *fraction* of energy radiated into $d\Omega$. Comparing the above two expressions for dE, we have the relation between ϵ_ν and j_ν:

$$j_\nu = \frac{\epsilon_\nu \rho}{4\pi}, \tag{1.18}$$

holding for isotropic emission. In going a distance ds, a beam of cross section dA travels through a volume $dV = dA\,ds$. Thus the intensity added to the beam by spontaneous emission is:

$$dI_\nu = j_\nu\,ds. \tag{1.19}$$

Absorption

We define the *absorption coefficient*, α_ν(cm^{-1}) by the following equation, representing the loss of intensity in a beam as it travels a distance ds (by convention, α_ν positive for energy taken out of beam):

$$dI_\nu = -\alpha_\nu I_\nu \, ds. \tag{1.20}$$

This phenomenological law can be understood in terms of a microscopic model in which particles with density n (number per unit volume) each present an effective absorbing area, or *cross section*, of magnitude $\sigma_\nu(\mathrm{cm}^2)$. These absorbers are assumed to be distributed at random. Let us consider the effect of these absorbers on radiation through dA within solid angle $d\Omega$ (see Fig. 1.7). The number of absorbers in the element equals $n\,dA\,ds$. The total absorbing area presented by absorbers equals $n\sigma_\nu\,dA\,ds$. The *energy* absorbed out of the beam is

$$-dI_\nu \, dA \, d\Omega \, dt \, d\nu = I_\nu (n\sigma_\nu \, dA \, ds) \, d\Omega \, dt \, d\nu;$$

thus

$$dI_\nu = -n\sigma_\nu I_\nu \, ds,$$

which is precisely the above phenomenological law (1.20), where

$$\alpha_\nu = n\sigma_\nu. \tag{1.21}$$

Often α_ν is written as

$$\alpha_\nu = \rho\kappa_\nu, \tag{1.22}$$

where ρ is the mass density and $\kappa_\nu(\mathrm{cm}^2\ \mathrm{g}^{-1})$ is called the *mass absorption coefficient*; κ_ν is also sometimes called the *opacity* coefficient.

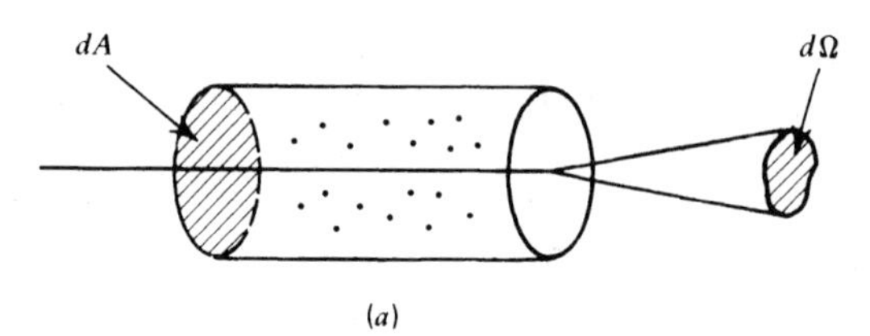

Figure 1.7a Ray passing through a medium of absorbers.

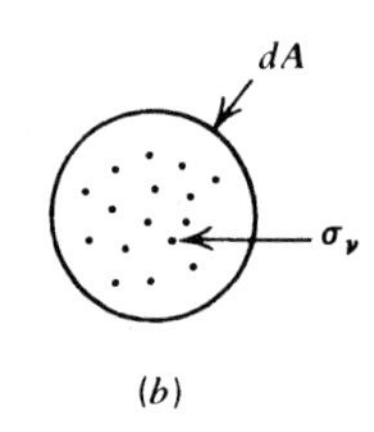

Figure 1.7b Cross sectional view of 7a.

There are some conditions of validity for this microscopic picture: The most important are that (1) the linear scale of the cross section must be small in comparison to the mean interparticle distance d. Thus $\sigma_\nu^{1/2} \ll d \sim n^{-1/3}$, from which follows $\alpha_\nu d \ll 1$ and (2) the absorbers are independent and randomly distributed. Fortunately, these conditions are almost always met for astrophysical problems.

As is shown in §1.6, we consider "absorption" to include both "true absorption" and stimulated emission, because both are proportional to the intensity of the incoming beam (unlike spontaneous emission). Thus the *net absorption* may be positive or negative, depending on whether "true absorption" or stimulated emission dominates. Although this combination may seem artificial, it will prove convenient and obviate the need for a quantum mechanical addition to our classical formulas later on.

The Radiative Transfer Equation

We can now incorporate the effects of emission and absorption into a single equation giving the variation of specific intensity along a ray. From the above expressions for emission and absorption, we have the combined expression

$$\frac{dI_\nu}{ds} = -\alpha_\nu I_\nu + j_\nu. \tag{1.23}$$

The transfer equation provides a useful formalism within which to solve for the intensity in an emitting and absorbing medium. It incorporates most of the macroscopic aspects of radiation into one equation, relating them to two coefficients α_ν and j_ν. A primary task in later chapters of this book is to find forms for these coefficients corresponding to particular physical processes.

Once α_ν and j_ν are known it is relatively easy to solve the transfer equation for the specific intensity. When scattering is present, solution of the radiative transfer equation is more difficult, because emission into $d\Omega$ depends on I_ν in solid angles $d\Omega'$, integrated over the latter (scattering from $d\Omega'$ into $d\Omega$). The transfer equation then becomes an integrodifferential equation, which generally must be solved partly by numerical techniques. (See §1.7 and 1.8.)

A formal solution to the complete radiative transfer equation will be given shortly. Here, we can give solutions to two simple limiting cases:

1—Emission Only: $\alpha_\nu = 0$. In this case, we have

$$\frac{dI_\nu}{ds} = j_\nu,$$

which has the solution

$$I_\nu(s) = I_\nu(s_0) + \int_{s_0}^{s} j_\nu(s')\, ds'. \tag{1.24}$$

The increase in brightness is equal to the emission coefficient integrated along the line of sight.

2—Absorption Only: $j_\nu = 0$. In this case, we have

$$\frac{dI_\nu}{ds} = -\alpha_\nu I_\nu,$$

which has the solution

$$I_\nu(s) = I_\nu(s_0)\exp\left[-\int_{s_0}^{s} \alpha_\nu(s')\, ds'\right]. \tag{1.25}$$

The brightness decreases along the ray by the exponential of the absorption coefficient integrated along the line of sight.

Optical Depth and Source Function

The transfer equation takes a particularly simple form if, instead of s, we use another variable τ_ν called the *optical depth*, defined by

$$d\tau_\nu = \alpha_\nu\, ds,$$

or

$$\tau_\nu(s) = \int_{s_0}^{s} \alpha_\nu(s')\, ds'. \tag{1.26}$$

The optical depth defined above is measured along the path of a traveling ray; occasionally, τ_ν is measured backward along the ray and a minus sign appears in Eq. (1.26). In plane-parallel media, a standard optical depth is sometimes used to measure distance normal to the surface, so that ds is replaced by dz and $\tau_\nu = \tau_\nu(z)$. We shall distinguish between these two definitions, where appropriate. The point s_0 is arbitrary; it sets the zero point for the optical depth scale.

A medium is said to be *optically thick* or *opaque* when τ_ν, integrated along a typical path through the medium, satisfies $\tau_\nu > 1$. When $\tau_\nu < 1$, the medium is said to be *optically thin* or *transparent*. Essentially, an optically thin medium is one in which the typical photon of frequency ν can traverse the medium without being absorbed, whereas an optically thick medium is one in which the average photon of frequency ν cannot traverse the entire medium without being absorbed.

The transfer equation can now be written, after dividing by α_ν,

$$\frac{dI_\nu}{d\tau_\nu} = -I_\nu + S_\nu, \tag{1.27}$$

where the *source function* S_ν is defined as the ratio of the emission coefficient to the absorption coefficient:

$$S_\nu \equiv \frac{j_\nu}{\alpha_\nu}. \tag{1.28}$$

The source function S_ν is often a simpler physical quantity than the emission coefficient. Also, the optical depth scale reveals more clearly the important intervals along a ray as far as radiation is concerned. For these reasons the variables τ_ν and S_ν are often used instead of α_ν and j_ν.

We can now formally solve the equation of radiative transfer, by regarding all quantities as functions of the optical depth τ_ν instead of s. Multiply the equation by the integrating factor e^{τ_ν} and define the quantities $\mathcal{I} \equiv I_\nu e^{\tau_\nu}$, $\mathcal{S} \equiv S_\nu e^{\tau_\nu}$. Then the equation becomes

$$\frac{d\mathcal{I}}{d\tau_\nu} = \mathcal{S},$$

with the solution

$$\mathcal{I}(\tau_\nu) = \mathcal{I}(0) + \int_0^{\tau_\nu} \mathcal{S}(\tau_\nu')\,d\tau_\nu'.$$

Rewriting the solution in terms of I_ν and S_ν, we have the *formal solution of the transfer equation*:

$$I_\nu(\tau_\nu) = I_\nu(0)e^{-\tau_\nu} + \int_0^{\tau_\nu} e^{-(\tau_\nu - \tau_\nu')} S_\nu(\tau_\nu')\,d\tau_\nu'. \tag{1.29}$$

Since τ_ν is just the dimensionless e-folding factor for absorption, the above equation is easily interpreted as the sum of two terms: the initial intensity diminished by absorption plus the integrated source diminished by absorption. As an example consider a *constant* source function S_ν. Then Eq. (1.29) gives the solution

$$\begin{aligned} I_\nu(\tau_\nu) &= I_\nu(0)e^{-\tau_\nu} + S_\nu(1-e^{-\tau_\nu}) \\ &= S_\nu + e^{-\tau_\nu}(I_\nu(0) - S_\nu). \end{aligned} \tag{1.30}$$

As $\tau_\nu \to \infty$, Eq. (1.30) shows that $I_\nu \to S_\nu$. We remind the reader that when scattering is present, S_ν contains a contribution from I_ν, so that it is not possible to specify S_ν a priori. This case is treated in detail in §1.7 and 1.8.

We conclude this section with a result of use later, which provides a simple physical interpretation of the source function and the transfer equation. From the transfer equation we see that if $I_\nu > S_\nu$ then $dI_\nu / d\tau_\nu < 0$ and I_ν tends to decrease along the ray. If $I_\nu < S_\nu$, then I_ν tends to increase along the ray. Thus the source function is the quantity that the specific intensity tries to approach, and does approach if given sufficient optical depth. In this respect the transfer equation describes a "relaxation" process.

Mean Free Path

A useful concept, which describes absorption in an equivalent way, is that of the *mean free path* of radiation (or photons). This is defined as the average distance a photon can travel through an absorbing material without being absorbed. It may be easily related to the absorption coefficient of a homogeneous material. From the exponential absorption law (1.25), the probability of a photon traveling at least an optical depth τ_ν is simply $e^{-\tau_\nu}$. The *mean* optical depth traveled is thus equal to unity:

$$\langle \tau_\nu \rangle \equiv \int_0^\infty \tau_\nu e^{-\tau_\nu} d\tau_\nu = 1.$$

The mean physical distance traveled in a homogeneous medium is defined as the *mean free path* l_ν and is determined by $\langle \tau_\nu \rangle = \alpha_\nu l_\nu = 1$ or

$$l_\nu = \frac{1}{\alpha_\nu} = \frac{1}{n\sigma_\nu}. \tag{1.31}$$

Thus the mean free path l_ν is simply the reciprocal of the absorption coefficient for homogenous material.

We can define a *local mean path* at a point in an inhomogeneous material as the mean free path that would result if the photon traveled through a large homogenous region of the same properties. Thus at any point we have $l_\nu = 1/\alpha_\nu$.

Radiation Force

If a medium absorbs radiation, then the radiation exerts a force on the medium, because radiation carries momentum. We can first define a *radiation flux vector*

$$\mathbf{F}_\nu = \int I_\nu \mathbf{n}\, d\Omega, \tag{1.32}$$

where $\mathbf{n}$ is a unit vector along the direction of the ray. Remember that a photon has momentum E/c, so that the vector momentum per unit area per unit time per unit path length absorbed by the medium is

$$\mathfrak{F} = \frac{1}{c}\int \alpha_\nu \mathbf{F}_\nu\, d\nu. \tag{1.33}$$

Since $dA\, ds = dV$, $\mathfrak{F}$ is the force per unit volume imparted onto the medium by the radiation field. We note that the force per unit mass of material is given by $\mathbf{f} = \mathfrak{F}/\rho$ or

$$\mathbf{f} = \frac{1}{c}\int \kappa_\nu \mathbf{F}_\nu\, d\nu. \tag{1.34}$$

Equations (1.33) and (1.34) assume that the absorption coefficient is isotropic. They also assume that no momentum is imparted by the emission of radiation, as is true for isotropic emission.

1.5 THERMAL RADIATION

Thermal radiation is radiation emitted by matter in thermal equilibrium.

Blackbody Radiation

To investigate thermal radiation, it is necessary to consider first of all *blackbody radiation*, radiation which is itself in thermal equilibrium.

To obtain such radiation we keep an enclosure at temperature T and do not let radiation in or out until equilibrium has been achieved. If we are careful, we can open a small hole in the side of the container and measure the radiation inside without disturbing equilibrium. Now, using some general thermodynamic arguments plus the fact that photons are massless, we can derive several important properties of blackbody radiation.

Since photons are massless, they can be created and destroyed in arbitrary numbers by the walls of the container (for practical purposes,

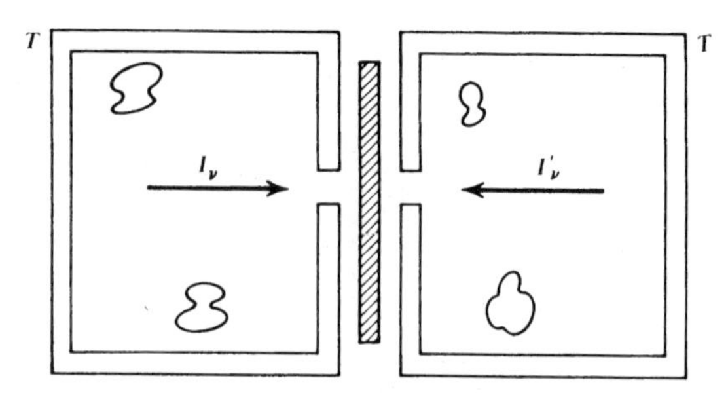

***Figure 1.8** Two containers at temperature T, separated by a filter.*

there is negligible *self-interaction* between photons). Thus there is no conservation law of photon number (unlike particle number for baryons), and we expect that the number of photons will adjust itself in equilibrium at temperature T.

An important property of I_ν is that it is independent of the properties of the enclosure and depends only on the temperature. To prove this, consider joining the container to another container of arbitrary shape and placing a filter between the two, which passes a single frequency ν but no others (Fig. 1.8). If $I_\nu \neq I'_\nu$, energy will flow spontaneously between the two enclosures. Since these are at the same temperature, this violates the second law of thermodynamics. Therefore, we have the relations

$$I_\nu = \text{universal function of } T \text{ and } \nu \equiv B_\nu(T). \tag{1.35}$$

I_ν thus must be independent of the shape of the container. A corollary is that it is also isotropic; $I_\nu \neq I_\nu(\Omega)$. The function $B_\nu(T)$ is called the Planck function. Its form is discussed presently.

Kirchhoff's Law for Thermal Emission

Now consider an element of some thermally emitting material at temperature T, so that its emission depends solely on its temperature and internal properties. Put this into the opening of a blackbody enclosure at the same temperature T (Fig. 1.9). Let the source function of the material be S_ν. If $S_\nu > B_\nu$, then $I_\nu > B_\nu$, and if $S_\nu < B_\nu$, then $I_\nu < B_\nu$, by the discussion after Eq. (1.30). But the presence of the material cannot alter the radiation, since the new configuration is also a blackbody enclosure at temperature T. Thus we have the relations

$$S_\nu = B_\nu(T), \tag{1.36}$$

$$j_\nu = \alpha_\nu B_\nu(T). \tag{1.37}$$

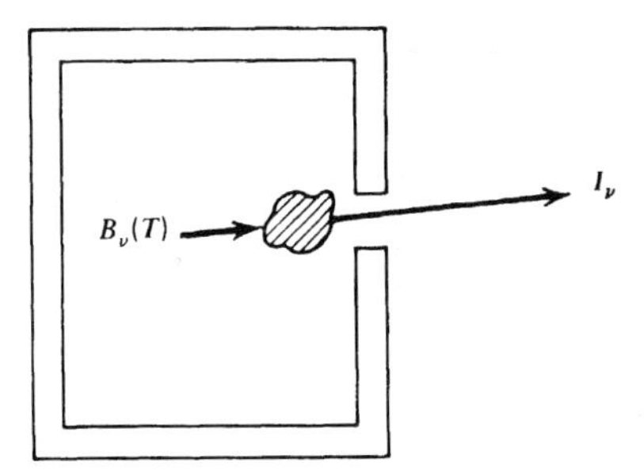

***Figure 1.9** Thermal emitter placed in the opening of a blackbody enclosure.*

Relation (1.37), called *Kirchhoff's law*, is an expression between α_ν and j_ν and the temperature of the matter T. The transfer equation for thermal radiation is, then, [cf. Eq. (1.23)],

$$\frac{dI_\nu}{ds} = -\alpha_\nu I_\nu + \alpha_\nu B_\nu(T),$$

or

$$\frac{dI_\nu}{d\tau_\nu} = -I_\nu + B_\nu(T). \tag{1.38}$$

Since $S_\nu = B_\nu$ throughout a blackbody enclosure, we have that $I_\nu = B_\nu$ throughout. Blackbody radiation is homogeneous and isotropic, so that $p = \frac{1}{3}u$.

At this point it is well to draw the distinction between *blackbody radiation*, where $I_\nu = B_\nu$, and *thermal radiation*, where $S_\nu = B_\nu$. Thermal radiation becomes blackbody radiation only for optically thick media.

Thermodynamics of Blackbody Radiation

Blackbody radiation, like any system in the thermodynamic equilibrium, can be treated by thermodynamic methods. Let us make a blackbody enclosure with a piston, so that work may be done on or extracted from the radiation (Fig. 1.10). Now by the first law of thermodynamics, we have

$$dQ = dU + p\,dV, \tag{1.39}$$

where Q is heat and U is total energy. By the second law of thermodynamics,

$$dS = \frac{dQ}{T},$$

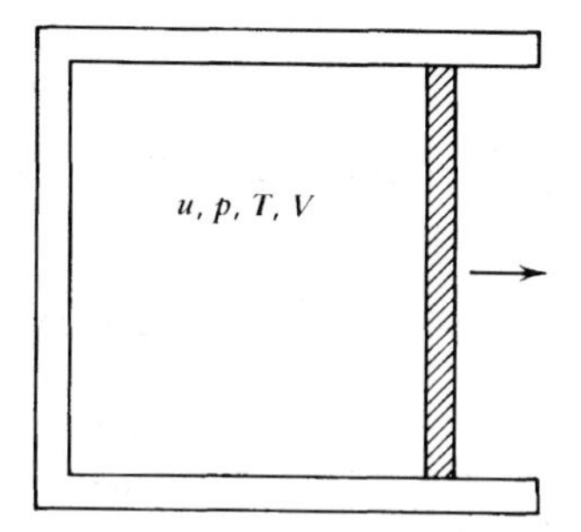

Figure 1.10 Blackbody enclosure with a piston on one side.

where $S \equiv$ entropy. But $U = uV$, and $p = u/3$, and u depends only on T since $u = (4\pi/c)\int J_\nu \, d\nu$ and $J_\nu = B_\nu(T)$. Thus we have

$$dS = \frac{V}{T}\frac{du}{dT}dT + \frac{u}{T}dV + \frac{1}{3}\frac{u}{T}dV,$$
$$= \frac{V}{T}\frac{du}{dT}dT + \frac{4u}{3T}dV$$

Since dS is a perfect differential,

$$\left(\frac{\partial S}{\partial T}\right)_V = \frac{V}{T}\frac{du}{dT} \qquad \left(\frac{\partial S}{\partial V}\right)_T = \frac{4u}{3T}. \tag{1.40}$$

Thus we obtain

$$\frac{\partial^2 S}{\partial T \partial V} = \frac{1}{T}\frac{du}{dT} = -\frac{4u}{3T^2} + \frac{4}{3T}\frac{du}{dT},$$

so that

$$\frac{du}{dT} = \frac{4u}{T}, \qquad \frac{du}{u} = 4\frac{dT}{T},$$
$$\log u = 4\log T + \log a,$$

where $\log a$ is a constant of integration. Thus we obtain the *Stefan–Boltzmann* law

$$u(T) = aT^4. \tag{1.41}$$

This may be related to the Planck function, since $I_\nu = J_\nu$ for isotropic radiation [cf. Eqs. (1.7)],

$$u=\frac{4\pi}{c}\int B_\nu(T)\,d\nu=\frac{4\pi}{c}B(T),$$

where the integrated Planck function is defined by

$$B(T)=\int B_\nu(T)\,d\nu=\frac{ac}{4\pi}T^4. \qquad (1.42)$$

The emergent flux from an isotropically emitting surface (such as a blackbody) is $\pi\times$brightness [see Eq. (1.14)], so that

$$F=\int F_\nu\,d\nu=\pi\int B_\nu\,d\nu=\pi B(T).$$

This leads to another form of the *Stefan–Boltzmann* law,

$$F=\sigma T^4, \qquad (1.43)$$

where

$$\sigma\equiv\frac{ac}{4}=5.67\times10^{-5}\text{ erg cm}^{-2}\text{ deg}^{-4}\text{ s}^{-1}, \qquad (1.44a)$$

$$a=\frac{4\sigma}{c}=7.56\times10^{-15}\text{ erg cm}^{-3}\text{ deg}^{-4}. \qquad (1.44b)$$

The constants a and σ cannot be determined by macroscopic thermodynamic arguments, but they are derived below. It is easily shown (Problem 1.6) that the entropy of blackbody radiation, S, is given by

$$S=\tfrac{4}{3}aT^3V, \qquad (1.45)$$

so that the law of adiabatic expansion for blackbody radiation is

$$TV^{1/3}=\text{constant, or} \qquad (1.46a)$$

$$pV^{4/3}=\text{constant}. \qquad (1.46b)$$

Equations (1.46) are the familiar adiabatic laws $pV^\gamma=$constant, with $\gamma=4/3$.

The Planck Spectrum

We now give a derivation of the Planck function. This derivation falls into two main parts: first, we derive the density of photon states in a blackbody enclosure; second the average energy per photon state is evaluated.

Consider a photon of frequency ν propagating in direction $\mathbf{n}$ inside a box. The wave vector of the photon is $\mathbf{k}=(2\pi/\lambda)\mathbf{n}=(2\pi\nu/c)\mathbf{n}$. If each dimension of the box L_x, L_y and L_z is much longer than a wavelength, then the photon can be represented by some sort of standing wave in the box. The number of nodes in the wave in each direction x,y,z is, for example, $n_x=k_xL_x/2\pi$, since there is one node for each integral number of wavelengths in given orthogonal directions. Now, the wave can be said to have changed states in a distinguishable manner when the number of nodes in a given direction changes by one or more. If $n_i \gg 1$, we can thus write the number of node changes in a wave number interval as, for example,

$$\Delta n_x = \frac{L_x \Delta k_x}{2\pi}.$$

Thus the number of states in the three-dimensional wave vector element $\Delta k_x \Delta k_y \Delta k_z \equiv d^3k$ is

$$\Delta N = \Delta n_x \Delta n_y \Delta n_z = \frac{L_x L_y L_z d^3k}{(2\pi)^3}.$$

Now, using the fact that $L_x L_y L_z = V$ (the volume of the container) and using the fact that photons have two independent polarizations (two states per wave vector $\mathbf{k}$), we can see that the number of states per unit volume per unit three-dimensional wave number is $2/(2\pi)^3$.

Now, since

$$d^3k = k^2\, dk\, d\Omega = \frac{(2\pi)^3 \nu^2\, d\nu\, d\Omega}{c^3},$$

we find the density of states (the number of states per solid angle per volume per frequency) to be

$$\rho_s = \frac{2\nu^2}{c^3}. \qquad (1.47)$$

Next we ask what is the average energy of each state. We know from quantum theory that each photon of frequency ν has energy $h\nu$, so we focus on a single frequency ν and ask what is the average energy of the state having frequency ν. Each state may contain n photons of energy $h\nu$, where $n=0,1,2,\ldots$. Thus the energy may be $E_n = nh\nu$. According to statistical mechanics, the probability of a state of energy E_n is proportional to $e^{-\beta E_n}$ where $\beta=(kT)^{-1}$ and $k=$Boltzmann's constant$=1.38\times10^{-16}$ erg deg^{-1}. Therefore, the average energy is:

$$\bar{E}=\frac{\sum_{n=0}^{\infty} E_n e^{-\beta E_n}}{\sum_{n=0}^{\infty} e^{-\beta E_n}} = -\frac{\partial}{\partial\beta}\ln\left(\sum_{n=0}^{\infty} e^{-\beta E_n}\right).$$

By the formula for the sum of a geometric series,

$$\sum_{n=0}^{\infty} e^{-\beta E_n} = \sum_{n=0}^{\infty} e^{-nh\nu\beta} = (1-e^{-\beta h\nu})^{-1}.$$

Thus we have the result:

$$\bar{E}=\frac{h\nu e^{-\beta h\nu}}{1-e^{-\beta h\nu}} = \frac{h\nu}{\exp(h\nu/kT)-1}. \tag{1.48}$$

Since $h\nu$ is the energy of one photon of frequency ν, Eq. (1.48) says that the average number of photons of frequency ν, n_ν, the "occupation number", is

$$n_\nu = \left[\exp\left(\frac{h\nu}{kT}\right)-1\right]^{-1}. \tag{1.49}$$

Equation (1.48) is the standard expression for Bose–Einstein statistics with a limitless number of particles (chemical potential$=0$). The energy per solid angle per volume per frequency is the product of $\bar{E}$ and the density of states, Eq. (1.47). However, this can also be written in terms of $u_\nu(\Omega)$, introduced in §1.3. Thus we have:

$$u_\nu(\Omega)\,dV\,d\nu\,d\Omega = \left(\frac{2\nu^2}{c^3}\right)\frac{h\nu}{\exp(h\nu/kT)-1}\,dV\,d\nu\,d\Omega,$$

$$u_\nu(\Omega) = \frac{2h\nu^3/c^3}{\exp(h\nu/kT)-1}. \tag{1.50}$$

Equation (1.6) gives the relation between $u_\nu(\Omega)$ and I_ν; here we have $I_\nu = B_\nu$

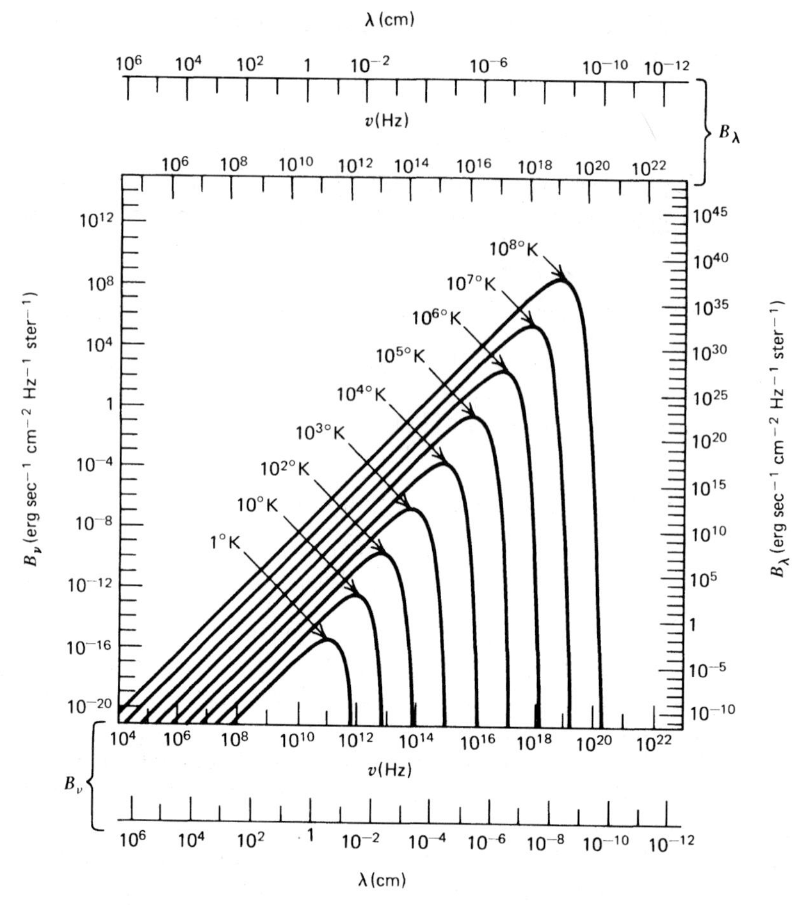

Figure 1.11 Spectrum of blackbody radiation at various temperatures (taken from Kraus, J. D. 1966, Radio Astronomy, McGraw-Hill Book Company)

so that

$$B_\nu(T) = \frac{2h\nu^3/c^2}{\exp(h\nu/kT)-1}. \tag{1.51}$$

Equation (1.51) expresses the *Planck law*.

If we express the Planck law per unit wavelength interval instead of per unit frequency we have

$$B_\lambda(T) = \frac{2hc^2/\lambda^5}{\exp(hc/\lambda kT)-1}. \tag{1.52}$$

$$B_{\tilde\nu}(T) = \frac{2hc\tilde\nu^3}{\exp(hc\tilde\nu/kT)-1}$$

A plot of B_ν and B_λ versus ν and λ for a range of values of T ($1K \leqslant T \leqslant 10^8 K$) is given in Fig. 1.11.

Properties of the Planck Law

The form of $B_\nu(T)$ just derived [Eq. (1.51)] is one of the most important results for radiation processes. We now give a number of properties and consequences of this law:

a—$h\nu \ll kT$: The Rayleigh–Jeans Law. In this case the exponential can be expanded

$$\exp\left(\frac{h\nu}{kT}\right) - 1 = \frac{h\nu}{kT} + \cdots$$

so that for $h\nu \ll kT$, we have the *Rayleigh–Jeans law*:

$$I_\nu^{RJ}(T) = \frac{2\nu^2}{c^2} kT. \tag{1.53}$$

Notice that this result does not contain Planck's constant. It was originally derived by assuming that $\bar{E} = kT$, the classical equipartition value for the energy of an electromagnetic wave.

The Rayleigh–Jeans law applies at low frequencies (in the radio region it almost always applies). It shows up as the straight-line part of the $\log B_\nu - \log \nu$ plot in Fig. 1.11.

Note that if Eq. (1.53) applied to all frequencies, the total amount of energy $\propto \int \nu^2 d\nu$ would diverge. This is known as the *ultraviolet catastrophe*. For $h\nu \gg kT$, the discrete quantum nature of photons must be taken into account.

b—$h\nu \gg kT$: Wien Law. In this limit the term unity in the denominator can be dropped in comparison with $\exp(h\nu/kT)$, so we have the *Wien law*:

$$I_\nu^W(T) = \frac{2h\nu^3}{c^2} \exp\left(\frac{-h\nu}{kT}\right). \tag{1.54}$$

This form was first proposed by Wien on the basis of rather ad hoc arguments. The brightness of a blackbody decreases very rapidly with frequency once the maximum is reached. Note the steep portions of the curves in Fig. 1.11 associated with the Wien law.

c—Monotonicity with Temperature. Of two blackbody curves, the one with higher temperature lies entirely above the other. To prove this we note

$$\frac{\partial B_\nu(T)}{\partial T}=\frac{2h^2\nu^4}{c^2kT^2}\frac{\exp(h\nu/kT)}{\left[\exp(h\nu/kT)-1\right]^2} \tag{1.55}$$

is positive. At any frequency the effect of increasing temperature is to increase $B_\nu(T)$. Also note $B_\nu\to 0$ as $T\to 0$ and $B_\nu\to\infty$ as $T\to\infty$.

d—Wien Displacement Law. The frequency $\nu_{\max}$ at which the peak of $B_\nu(T)$ occurs can be found by solving

$$\left.\frac{\partial B_\nu}{\partial \nu}\right|_{\nu=\nu_{\max}}=0.$$

Letting $x\equiv h\nu_{\max}/kT$, this is equivalent to solving $x=3(1-e^{-x})$, which has the approximate root $x=2.82$, so that

$$h\nu_{\max}=2.82\ kT, \tag{1.56a}$$

or

$$\frac{\nu_{\max}}{T}=5.88\times 10^{10}\ \text{Hz deg}^{-1}. \tag{1.56b}$$

Thus the peak frequency of the blackbody law shifts linearly with temperature; this is known as the *Wien displacement law*.

Similarly, a wavelength $\lambda_{\max}$ at which the maximum of $B_\lambda(T)$ occurs can be found by solving

$$\left.\frac{\partial B_\lambda}{\partial \lambda}\right|_{\lambda=\lambda_{\max}}=0.$$

Letting $y=hc/(\lambda_{\max}kT)$, this is equivalent to solving $y=5(1-e^{-y})$, which has the approximate root $y=4.97$, so that

$$\lambda_{\max}T=0.290\ \text{cm deg}. \tag{1.57}$$

This is also known as the Wien displacement law.

Equations (1.56) and (1.57) are very reasonable. By dimensional analysis, one could have argued that the blackbody radiation spectrum should peak at energy $\sim kT$, since kT is the only quantity with dimensions of energy one can form from k, T, h, c.

One should be careful to note that the peaks of B_ν and B_λ do not occur at the same places in wavelength or frequency; that is, $\lambda_{max}\nu_{max} \neq c$. As an example, if $T=7300$ K the peak of B_ν is at $\lambda=.7$ microns (red), while the peak of B_λ is at $\lambda=.4$ microns (blue). The Wien displacement law gives a convenient way of characterizing the frequency range for which the Rayleigh–Jeans law is valid, namely, $\nu \ll \nu_{max}$. Similarly for the Wien law $\nu \gg \nu_{max}$.

e—Relation of Radiation Constants to Fundamental Constants. By putting in the explicit form for $B_\nu(T)$ into equation (1.42) we can obtain expressions for a and σ in terms of fundamental constants:

$$\int_0^\infty B_\nu(T)\,d\nu = (2h/c^2)(kT/h)^4 \int_0^\infty \frac{x^3\,dx}{e^x - 1}.$$

The integral can be found in standard integral tables and has a value $\pi^4/15$. Therefore, we have the results

$$\int_0^\infty B_\nu(T)\,d\nu = \frac{2\pi^4 k^4}{15c^2h^3}T^4, \tag{1.58a}$$

$$\sigma = \frac{2\pi^5 k^4}{15c^2h^3}, \qquad a = \frac{8\pi^5 k^4}{15c^3h^3}. \tag{1.58b}$$

Characteristic Temperatures Related to Planck Spectrum

a—Brightness Temperature. One way of characterizing brightness (specific intensity) at a certain frequency is to give the temperature of the blackbody having the same brightness at that frequency. That is, for any value I_ν we define $T_b(\nu)$ by the relation

$$I_\nu = B_\nu(T_b). \tag{1.59}$$

T_b is called the *brightness temperature*. This way of specifying brightness has the advantage of being closely connected with the physical properties of the emitter, and has the simple unit (K) instead of (erg cm^{-2} s^{-1} Hz^{-1} ster^{-1}). This procedure is used especially in radio astronomy, where the Rayleigh–Jeans law is usually applicable, so that

$$I_\nu = \frac{2\nu^2}{c^2}kT_b \tag{1.60a}$$

or

$$T_b = \frac{c^2}{2\nu^2 k} I_\nu \tag{1.60b}$$

for $h\nu \ll kT$.

The transfer equation for thermal emission takes a particularly simple form in terms of brightness temperature in the Rayleigh–Jeans limit [cf. Eq. (1.38)],

$$\frac{dT_b}{d\tau_\nu} = -T_b + T, \tag{1.61}$$

where T is the temperature of the material. When T is constant we have

$$T_b = T_b(0)e^{-\tau_\nu} + T(1 - e^{-\tau_\nu}), \qquad h\nu \ll kT. \tag{1.62}$$

If the optical depth is large, the brightness temperature of the radiation approaches the temperature of the material. We note that the uniqueness of the definition of brightness temperature relies on the monotonicity property of Planck's law. We also note that, in general, the brightness temperature is a function of ν. Only if the source is blackbody is the brightness temperature the same at all frequencies.

In the Wien region of the Planck law the concept of brightness temperature is not so useful because of the rapid decrease of B_ν with ν, and because it is not possible to formulate a transfer equation linear in the brightness temperature.

b—Color Temperature. Often a spectrum is measured to have a shape more or less of blackbody form, but not necessarily of the proper absolute value. For example, by measuring F_ν from an unresolved source we cannot find I_ν unless we know the distance to the source and its physical size. By fitting the data to a blackbody curve without regard to vertical scale, a *color temperature* T_c is obtained. Often the "fitting" procedure is nothing more than estimating the peak of the spectrum and applying Wien's displacement law to find a temperature.

The color temperature T_c will correctly give the temperature of a blackbody source of unknown absolute scale. Also, T_c will give the temperature of a thermal emitter that is optically thin, providing that the optical thickness is fairly constant for frequencies near the peak. In this case the brightness temperature will be less than the temperature of the emitter, since the blackbody spectrum gives the maximum attainable intensity of a thermal emitter at temperature T, by general thermodynamic arguments. (See Problem 1.8).

c—Effective Temperature. The effective temperature of a source T_{eff} is derived from the total amount of flux, integrated over all frequencies, radiated at the source. We obtain T_{eff} by equating the actual flux F to the flux of a blackbody at temperature T_{eff}:

$$F = \int \cos\theta I_\nu \, d\nu \, d\Omega \equiv \sigma T_{\text{eff}}^4. \tag{1.63}$$

Note that both T_{eff} and T_b depend on the magnitude of the source intensity, but T_c depends only on the shape of the observed spectrum.

1.6 THE EINSTEIN COEFFICIENTS

Definition of Coefficients

Kirchhoff's law, $j_\nu = \alpha_\nu B_\nu$, relating emission to absorption for a thermal emitter, clearly must imply some relationship between emission and absorption at a microscopic level. This relationship was first discovered by Einstein in a beautifully simple analysis of the interaction of radiation with an atomic system. He considered the simple case of two discrete energy levels: the first of energy E with statistical weight g_1, the second of energy $E + h\nu_0$ with statistical weight g_2 (see Fig. 1.12). The system makes a transition from 1 to 2 by absorption of a photon of energy $h\nu_0$. Similarly, a transition from 2 to 1 occurs when a photon is emitted. Einstein identified three processes:

1—Spontaneous Emission: This occurs when the system is in level 2 and drops to level 1 by emitting a photon, and it occurs even in the absence of a radiation field. We define the *Einstein A-coefficient* by

$$A_{21} = \text{transition probability per unit time for spontaneous emission (sec}^{-1}\text{)}. \tag{1.64}$$

2—Absorption: This occurs in the presence of photons of energy $h\nu_0$. The system makes a transition from level 1 to level 2 by absorbing a photon. Since there is no self-interaction of the radiation field, we expect

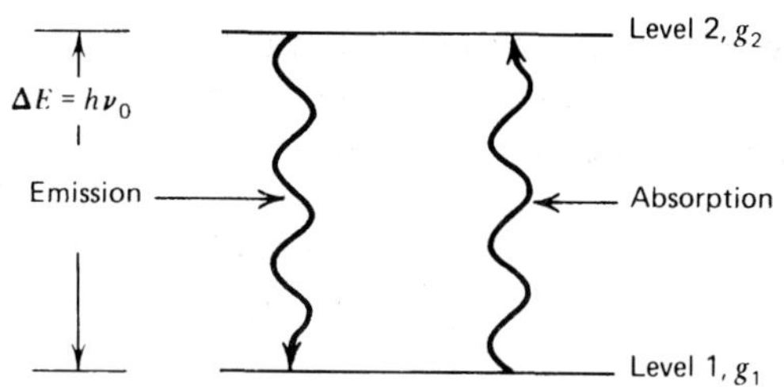

Figure 1.12a Emission and absorption from a two level atom.

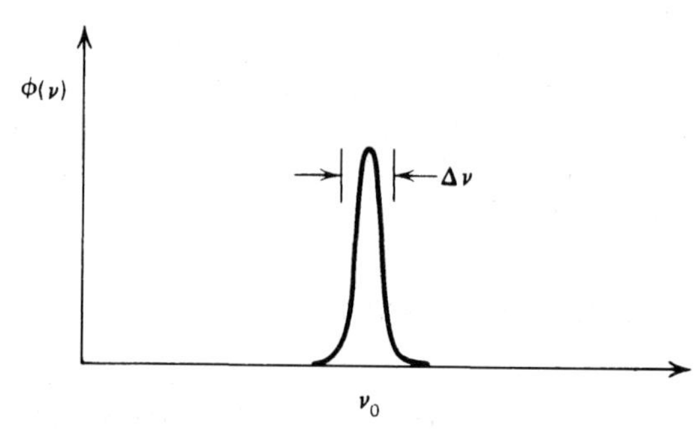

Figure 1.12b Line profile for 12a.

that the probability per unit time for this process will be proportional to the density of photons (or to the mean intensity) at frequency ν_0. To be precise, we must recognize that the energy difference between the two levels is not infinitely sharp but is described by a *line profile function* $\phi(\nu)$, which is sharply peaked at $\nu = \nu_0$ and which is conveniently taken to be normalized:

$$\int_0^\infty \phi(\nu)\,d\nu = 1. \tag{1.65}$$

This line profile function describes the relative effectiveness of frequencies in the neighborhood of ν_0 for causing transitions. The physical mechanisms that determine $\phi(\nu)$ are discussed later in Chapter 10.

These arguments lead us to write

$$B_{12}\bar{J} = \text{transition probability per unit time for absorption,} \tag{1.66}$$

where

$$\bar{J} \equiv \int_0^\infty J_\nu \phi(\nu)\,d\nu. \tag{1.67}$$

The proportionality constant B_{12} is the *Einstein B-coefficient*.

3—Stimulated Emission: Einstein found that to derive Planck's law another process was required that was proportional to $\bar{J}$ and caused *emission* of a photon. As before, we define:

$$B_{21}\bar{J} = \text{transition probability per unit time for stimulated emission.} \tag{1.68}$$

B_{21} is another *Einstein B-coefficient.*

Note that when J_ν changes slowly over the width $\Delta\nu$ of the line, $\phi(\nu)$ behaves like a δ-function, and the probabilities per unit time for absorption and stimulated emission become simply $B_{12}J_{\nu_0}$ and $B_{21}J_{\nu_0}$, respectively. In some discussions of the Einstein coefficients, including Einstein's original one, this assumption is made implicitly. Also be aware that the energy density u_ν is often used instead of J_ν to define the Einstein B-coefficients, which leads to definitions differing by $c/4\pi$, [cf. Eq. (1.7)].

Relations between Einstein Coefficients

In thermodynamic equilibrium we have that the number of transitions per unit time per unit volume out of state 1 = the number of transitions per unit time per unit volume into state 1. If we let n_1 and n_2 be the number densities of atoms in levels 1 and 2, respectively, this reduces to

$$n_1 B_{12}\bar{J} = n_2 A_{21} + n_2 B_{21}\bar{J}. \tag{1.69}$$

Now, solving for $\bar{J}$ from Eq. (1.69):

$$\bar{J} = \frac{A_{21}/B_{21}}{(n_1/n_2)(B_{12}/B_{21}) - 1}.$$

In thermodynamic equilibrium the ratio of n_1 to n_2 is

$$\frac{n_1}{n_2} = \frac{g_1 \exp(-E/kT)}{g_2 \exp[-(E + h\nu_0)/kT]} = \frac{g_1}{g_2}\exp(h\nu_0/kT), \tag{1.70}$$

so that

$$\bar{J} = \frac{A_{21}/B_{21}}{(g_1 B_{12}/g_2 B_{21})\exp(h\nu_0/kT) - 1}. \tag{1.71}$$

But in thermodynamic equilibrium we also know $J_\nu = B_\nu$ [cf. Eq. (1.51)], and the fact that B_ν varies slowly on the scale of $\Delta\nu$ implies that $\bar{J} = B_\nu$.

For the expression in Eq. (1.71) to equal the Planck function for all temperatures we must have the following *Einstein relations*:

$$g_1 B_{12} = g_2 B_{21}, \tag{1.72a}$$

$$A_{21} = \frac{2h\nu^3}{c^2} B_{21}. \tag{1.72b}$$

These connect *atomic properties* A_{21}, B_{21}, and B_{12} and have no reference to the temperature T [unlike Kirchhoff's Law, Eq. (1.37)]. Thus Eq. (1.72) must hold whether or not the atoms are in thermodynamic equilibrium. Equations (1.72) are examples of what are generally known as *detailed balance relations* that connect any microscopic process and its inverse process, here absorption and emission. These Einstein relations are the extensions of Kirchhoff's law to include the nonthermal emission that occurs when the matter is not thermodynamic equilibrium. If we can determine any one of the coefficients A_{21}, B_{21}, or B_{12} these relations allow us to determine the other two; this will be of considerable value to us later on.

Einstein was led to include the process of stimulated emission by the fact that without it he could not get Planck's law, but only Wien's law, which was known to be incorrect. Why does one obtain the Wien law when stimulated emission is neglected? Remember that the Wien law is the expression of the Planck spectrum when $h\nu \gg kT$ [cf. Eq. (1.54)]. But when $h\nu \gg kT$, level 2 is very sparsely populated relative to level 1, $n_2 \ll n_1$. Then, stimulated emission is unimportant compared to absorption, since these are proportional to n_2 and n_1, respectively [cf. Eq. (1.69)]. See Problem 1.7.

A property of stimulated emission that is not clear from the preceding discussion is that it takes place into precisely the same direction and frequency (in fact, into the same photon state). The emitted photon is precisely coherent with the photon that stimulated the emission.

Absorption and Emission Coefficients in Terms of Einstein Coefficients

To obtain the emission coefficient j_ν we must make some assumption about the frequency distribution of the emitted radiation during a spontaneous transition from level 2 to level 1. The simplest assumption is that this emission is distributed in accordance with the same line profile function $\phi(\nu)$ that describes absorption. (This assumption is very often a good one in astrophysics). The amount of energy emitted in volume dV, solid angle $d\Omega$, frequency range $d\nu$, and time dt is, by definition,

$j_\nu dV\,d\Omega\,d\nu\,dt$. Since each atom contributes an energy $h\nu_0$ distributed over 4π solid angle for each transition, this may also be expressed as $(h\nu_0/4\pi)\phi(\nu)n_2A_{21}dV\,d\Omega\,d\nu\,dt$, so that the emission coefficient is

$$j_\nu = \frac{h\nu_0}{4\pi} n_2 A_{21}\phi(\nu). \tag{1.73}$$

To obtain the absorption coefficient we first note from Eqs. (1.66) and (1.67) that the total energy absorbed in time dt and volume dV is

$$dV\,dt\,h\nu_0 n_1 B_{12}(4\pi)^{-1}\int d\Omega \int d\nu\phi(\nu) I_\nu.$$

Therefore, the energy absorbed out of a beam in frequency range $d\nu$ solid angle $d\Omega$ time dt and volume dV is

$$dV\,dt\,d\Omega\,d\nu \frac{h\nu_0}{4\pi} n_1 B_{12}\phi(\nu) I_\nu.$$

Taking the volume element to be that of Fig. 1.4, so that $dV = dA\,ds$, and noting Eqs. (1.2) and (1.20), we have the absorption coefficient (uncorrected for stimulated emission):

$$\alpha_\nu = \frac{h\nu}{4\pi} n_1 B_{12}\phi(\nu). \tag{1.74}$$

What about the stimulated emission? At first sight one might be tempted to add this as a contribution to the emission coefficient; but notice that it is proportional to the intensity and only affects the photons along the given beam, in close analogy to the process of absorption. Thus it is much more convenient to treat stimulated emission as *negative absorption* and include its effect through the absorption coefficient. In operational terms these two processes always occur together and cannot be disentangled by experiments based on Eq. (1.20). By reasoning entirely analogous to that leading to Eq. (1.74) we can find the contribution of stimulated emission to the absorption coefficient. The result for the absorption coefficient, corrected for stimulated emission, is

$$\alpha_\nu = \frac{h\nu}{4\pi}\phi(\nu)(n_1 B_{12} - n_2 B_{21}). \tag{1.75}$$

It is this quantity that will always be meant when we speak simply of the *absorption coefficient*. The form given in Eq. (1.74) will be called the *absorption coefficient uncorrected for stimulated emission.*

It is now possible to write the transfer equation in terms of the Einstein coefficients:

$$\frac{dI_\nu}{ds} = -\frac{h\nu}{4\pi}(n_1 B_{12} - n_2 B_{21})\phi(\nu) I_\nu + \frac{h\nu}{4\pi} n_2 A_{21} \phi(\nu). \tag{1.76}$$

The source function can be obtained by dividing Eq. (1.73) by Eq. (1.75):

$$S_\nu = \frac{n_2 A_{21}}{n_1 B_{12} - n_2 B_{21}}. \tag{1.77}$$

Using the Einstein relations, (1.72), the absorption coefficient and source function can be written

$$\alpha_\nu = \frac{h\nu}{4\pi} n_1 B_{12} (1 - g_1 n_2 / g_2 n_1) \phi(\nu), \tag{1.78}$$

$$S_\nu = \frac{2h\nu^3}{c^2}\left(\frac{g_2 n_1}{g_1 n_2} - 1\right)^{-1}. \tag{1.79}$$

Equation (1.79) is a generalized Kirchhoff's law. Three interesting cases of these equations can be identified.

1—Thermal Emission (LTE): If the matter is in thermal equilibrium with itself (but not necessarily with the radiation) we have

$$\frac{n_1}{n_2} = \frac{g_1}{g_2} \exp\left(\frac{h\nu}{kT}\right).$$

The matter is said to be in *local thermodynamic equilibrium* (*LTE*). In this case,

$$\alpha_\nu = \frac{h\nu}{4\pi} n_1 B_{12}\left[1 - \exp\left(\frac{-h\nu}{kT}\right)\right]\phi(\nu), \tag{1.80}$$

$$S_\nu = B_\nu(T). \tag{1.81}$$

This thermal value for the source function is, of course, just a statement of Kirchhoff's law. A new result is the correction factor $1 - \exp(-h\nu/kT)$ in the absorption coefficient, which is due to stimulated emission.

2—Nonthermal Emission: This term covers all other cases in which

$$\frac{n_1}{n_2} \neq \frac{g_1}{g_2} \exp\left(\frac{h\nu}{kT}\right).$$

For a plasma, for example, this would occur if the radiating particles did not have a Maxwellian velocity distribution or if the atomic populations did not obey the Maxwell–Boltzmann distribution law. The term can also be applied to cases in which scattering is present.

3—Inverted Populations; Masers: For a system in thermal equilibrium we have

$$\frac{n_2 g_1}{n_1 g_2} = \exp\left(\frac{-h\nu}{kT}\right) < 1,$$

so that

$$\frac{n_1}{g_1} > \frac{n_2}{g_2}. \tag{1.82}$$

Even when the material is out of thermal equilibrium, this relation is usually satisfied. In that case we say that there are *normal populations*. However, it is possible to put enough atoms in the upper state so that we have *inverted populations*:

$$\frac{n_1}{g_1} < \frac{n_2}{g_2}. \tag{1.83}$$

In this case the absorption coefficient is *negative*: $\alpha_\nu < 0$, as can be seen from Eq. (1.78). Rather than decrease along a ray, the intensity actually increases. Such a system is said to be a *maser* (*m*icrowave *a*mplification by *s*timulated *e*mission of *r*adiation; also *laser* for *l*ight...).

The amplification involved here can be very large. A negative optical depth of -100, for example, leads to an amplification by a factor of 10^{43}, [cf. equation (1.25)]. The detailed understanding of masers is a specialized field and is not dealt with here. Maser action in molecular lines has been observed in many astrophysical sources.

1.7 SCATTERING EFFECTS; RANDOM WALKS

Pure Scattering

For pure thermal emission the amount of radiation emitted by an element of material is not dependent on the radiation field incident on it—the source function is always $B_\nu(T)$ and depends only on the local temperature. Such an element would emit the same whether it was isolated in free space or imbedded deeply within a star where the ambient radiation field was substantial. This characteristic of thermal radiation makes it particularly easy to treat.

However, another common emission process is *scattering*, which depends completely on the amount of radiation falling on the element. Perhaps the most important mechanism of this type is *electron scattering*, which is treated in detail in Chapter 7. For the present discussion we assume *isotropic scattering*, which means that the scattered radiation is emitted equally into equal solid angles, so that the emission coefficient is independent of direction. We also assume that the total amount of radiation emitted per unit frequency range is just equal to the total amount absorbed in that same frequency range. This is called *coherent scattering*; other terms are *elastic* or *monochromatic scattering*. Scattering from nonrelativistic electrons is very nearly coherent (note, however, that repeated scatterings can build up substantial effects; see Chapter 7).

The emission coefficient for coherent, isotropic scattering can be found simply by equating the power absorbed per unit volume and frequency ranges to the corresponding power emitted. This yields

$$j_\nu = \sigma_\nu J_\nu, \tag{1.84}$$

where σ_ν is the absorption coefficient of the scattering process, also called the *scattering coefficient*. Dividing by the scattering coefficient, we find that the source function for scattering is simply equal to the mean intensity within the emitting material:

$$S_\nu = J_\nu = \frac{1}{4\pi}\int I_\nu \, d\Omega. \tag{1.85}$$

The transfer equation for pure scattering is therefore

$$\frac{dI_\nu}{ds} = -\sigma_\nu (I_\nu - J_\nu). \tag{1.86}$$

This equation cannot simply be solved by the formal solution (1.29), since the source function is not known a priori and depends on the solution I_ν at all directions through a given point. It is now an *integro-differential equation*, which poses a difficult mathematical problem. An approximate method of treating scattering problems, the Eddington approximation, is discussed in §1.8.

A particularly useful way of looking at scattering, which leads to important order-of-magnitude estimates, is by means of *random walks*. It is possible to view the processes of absorption, emission, and propagation in probabilistic terms for a single photon rather than the average behavior of large numbers of photons, as we have been doing so far. For example, the exponential decay of a beam of photons has the interpretation that the probability of a photon traveling an optical depth τ_ν before absorption is just $e^{-\tau_\nu}$. Similarly, when radiation is scattered isotropically we can say that a single photon has equal probabilities of scattering into equal solid angles. In this way we can speak of a typical or sample path of a photon, and the measured intensities can be interpreted as statistical averages over photons moving in such paths.

Now consider a photon emitted in an infinite, homogeneous scattering region. It travels a displacement $\mathbf{r}_1$ before being scattered, then travels in a new direction over a displacement $\mathbf{r}_2$ before being scattered, and so on. The net displacement of the photon after N free paths is

$$\mathbf{R}=\mathbf{r}_1+\mathbf{r}_2+\mathbf{r}_3+\cdots+\mathbf{r}_N. \tag{1.87}$$

We would like to find a rough estimate of the distance $|\mathbf{R}|$ traveled by a typical photon. Simple averaging of Eq. (1.87) over all sample paths will not work, because the average displacement, being a vector, must be zero. Therefore, we first square Eq. (1.87) and then average. This yields the mean square displacement traveled by the photon:

$$\begin{aligned} l_*^2 \equiv \langle \mathbf{R}^2 \rangle = {} & \langle \mathbf{r}_1^2 \rangle + \langle \mathbf{r}_2^2 \rangle + \cdots \langle \mathbf{r}_N^2 \rangle \\ & + 2\langle \mathbf{r}_1 \cdot \mathbf{r}_2 \rangle + 2\langle \mathbf{r}_1 \cdot \mathbf{r}_3 \rangle + \cdots \\ & + \cdots . \end{aligned} \tag{1.88}$$

Each term involving the square of a displacement averages to the mean square of the free path of a photon, which is denoted l^2. To within a factor of order unity, l is simply the mean free path of a photon. The cross terms in Eq. (1.88) involve averaging the cosine of the angle between the directions before and after scattering, and this vanishes for isotropic scattering. (It also vanishes for any scattering with front-back symmetry, as in Thomson or Rayleigh scattering.) Therefore,

$$\begin{aligned} l_*^2 &= N l^2, \\ l_* &= \sqrt{N}\, l. \end{aligned} \tag{1.89}$$

The quantity l_* is the root mean square *net* displacement of the photon, and it increases as the square root of the number of scatterings.

This result can be used to estimate the mean number of scatterings in a finite medium. Suppose a photon is generated somewhere within the medium; then the photon will scatter until it escapes completely. For regions of large optical depth the number of scatterings required to do this is roughly determined by setting $l_*\sim L$, the typical size of the medium. From Eq. (1.89) we find $N\approx L^2/l^2$. Since l is of the order of the mean free path, L/l is approximately the optical thickness of the medium τ. Therefore, we have

$$N\approx\tau^2, \qquad (\tau\gg 1). \tag{1.90a}$$

For regions of small optical thickness the mean number of scatterings is small, of order $1-e^{-\tau}\approx\tau$; that is,

$$N\approx\tau, \qquad (\tau\ll 1). \tag{1.90b}$$

For most order-of-magnitude estimates it is sufficient to use $N\approx\tau^2+\tau$ or $N\approx\max(\tau,\tau^2)$ for any optical thickness.

Combined Scattering and Absorption

The emission and absorption of radiation may be governed by more than one process. As an example, let us treat the case of material with an absorption coefficient α_ν describing thermal emission and a scattering coefficient σ_ν describing coherent isotropic scattering. The transfer equation then has two terms on the right-hand side:

$$\begin{aligned}\frac{dI_\nu}{ds} &= -\alpha_\nu(I_\nu-B_\nu)-\sigma_\nu(I_\nu-J_\nu)\\ &= -(\alpha_\nu+\sigma_\nu)(I_\nu-S_\nu).\end{aligned} \tag{1.91}$$

The source function is [cf. (1.28)],

$$S_\nu=\frac{\alpha_\nu B_\nu+\sigma_\nu J_\nu}{\alpha_\nu+\sigma_\nu} \tag{1.92}$$

and is an average of the two separate source functions, weighted by their respective absorption coefficients.

The net absorption coefficient is $\alpha_\nu+\sigma_\nu$, which can be used to define the optical depth by $d\tau_\nu=(\alpha_\nu+\sigma_\nu)ds$. This net absorption coefficient is often called the *extinction coefficient* to distinguish it from the "true" absorption coefficient α_ν.

If a matter element is deep inside a medium at some constant temperature, we expect that the radiation field will be near to its thermodynamic value $J_\nu=B_\nu(\tau)$. It follows from Eq. (1.92) that $S_\nu=B_\nu(\tau)$ also, as it must in thermal equilibrium. On the other hand, if the element is isolated in free space, where $J_\nu = 0$, then the source function is only a fraction of the Planck function: $S_\nu = \alpha_\nu B_\nu / (\alpha_\nu + \sigma_\nu)$. In general, the source function will not be known a priori but must be calculated as part of a self-consistent solution of the entire radiation field. (See §1.8.)

The random walk arguments can be extended to the case of combined scattering and absorption. The free path of a photon is now determined by the total extinction coefficient $\alpha_\nu + \sigma_\nu$; the mean free path of a photon before scattering or absorption is

$$l_\nu = (\alpha_\nu + \sigma_\nu)^{-1}. \tag{1.93}$$

During the random walk the probability that a free path will end with a true absorption event is

$$\epsilon_\nu = \frac{\alpha_\nu}{\alpha_\nu + \sigma_\nu}, \tag{1.94a}$$

the corresponding probability for scattering being

$$1 - \epsilon_\nu = \frac{\sigma_\nu}{\alpha_\nu + \sigma_\nu}. \tag{1.94b}$$

The quantity $1 - \epsilon_\nu$ is called the *single-scattering albedo*. The source function (1.92) can be written

$$S_\nu = (1 - \epsilon_\nu) J_\nu + \epsilon_\nu B_\nu. \tag{1.95}$$

Let us consider first an infinite homogeneous medium. A random walk starts with the thermal emission of a photon (creation) and ends, possibly after a number of scatterings, with a true absorption (destruction). Since the walk can be terminated with probability ϵ at the end of each free path, the mean number of free paths is $N = \epsilon^{-1}$. From Eq. (1.89) we then have

$$l_*^2 = \frac{l^2}{\epsilon},$$

$$l_* = \frac{l}{\sqrt{\epsilon}}. \tag{1.96}$$

Using Eqs. (1.93) and (1.94a) we have

$$l_* \approx [\alpha_\nu (\alpha_\nu + \sigma_\nu)]^{-1/2}. \tag{1.97}$$

The length l_* represents a measure of the net displacement between the points of creation and destruction of a typical photon; it is variously called the *diffusion length*, *thermalization length*, or *effective mean path*. Note also that l_* is generally frequency dependent.

The behavior of a finite medium also can be discussed in terms of random walks. This behavior depends strongly on whether its size L is larger or smaller than the effective free path l_*. It is convenient to make this distinction in terms of the ratio $\tau_* = L/l_*$, called the *effective optical thickness* of the medium. Using Eq. (1.97) we have the result

$$\tau_* \approx \sqrt{\tau_a(\tau_a + \tau_s)}\ , \tag{1.98}$$

where the absorption and scattering optical thickness are defined by

$$\tau_a = \alpha_\nu L; \qquad \tau_s = \sigma_\nu L. \tag{1.99}$$

When the effective free path is large compared with the size of the medium we have

$$\tau_* \ll 1, \tag{1.100}$$

and the medium is said to be *effectively thin* or *translucent*. Most photons will then escape by random walking out of the medium before being destroyed by a true absorption. The monochromatic luminosity will just be equal to the total radiation created by thermal emission in the medium:

$$\mathcal{L}_\nu = 4\pi\alpha_\nu B_\nu V, \qquad (\tau_* \ll 1) \tag{1.101}$$

where $\mathcal{L}_\nu$ is the emitted power per unit frequency and V is the volume of the medium.

When the effective free path is small compared with the size of the medium we have

$$\tau_* \gg 1, \tag{1.102}$$

and the medium is said to be *effectively thick*. Most photons thermally emitted at depths larger than the effective path length will be destroyed by absorption before they get out. Therefore the physical conditions at large, effective depths approach the conditions for the radiation to come into thermal equilibrium with the matter, and we expect $I_\nu \rightarrow B_\nu$ and $S_\nu \rightarrow B_\nu$. Because of this property the effective path length l_* is sometimes called the *thermalization length*, since it describes the distance over which thermal equilibrium of the radiation is established.

The monochromatic luminosity of an effectively thick medium can be estimated to within factors of order unity by considering the effective emitting volume to be the surface area of the medium times the effective path length. This is because it is only those photons emitted within an effective path length of the boundary that have a reasonable chance of escaping before being absorbed. Thus we have

$$\mathcal{L}_\nu \approx 4\pi\alpha_\nu B_\nu A l_* \approx 4\pi\sqrt{\epsilon_\nu}\, B_\nu A, \qquad (\tau_* \gg 1) \tag{1.103}$$

using Eqs. (1.94a) and (1.97). In the limiting case of no scattering, $\epsilon_\nu \rightarrow 1$, we know that the emission will be that of a blackbody, where $\mathcal{L}_\nu = \pi B_\nu A$, which suggests that the factor 4π in Eq. (1.103) should be replaced by π; however, the form of the exact equation actually depends on ϵ_ν and on geometry in a more complex way, and the equation should be taken only as an estimate. (For a more complete treatment see Problem 1.10).

1.8 RADIATIVE DIFFUSION

The Rosseland Approximation

We have used random walk arguments to show that S_ν approaches B_ν at large effective optical depths in a homogeneous medium. Real media are seldom homogeneous, but often, as in the interiors of stars, there is a high degree of local homogeneity. In such cases it is possible to derive a simple expression for the energy flux, relating it to the local temperature gradient. This result, first derived by Rosseland, is called the *Rosseland approximation*.

First let us assume that the material properties (temperature, absorption coefficient, etc.) depend only on depth in the medium. This is called the *plane-parallel* assumption. Then, by symmetry, the intensity can depend only on a single angle θ, which measures the direction of the ray with respect to the direction normal to the planes of constant properties. (See Fig. 1.13.)

It is convenient to use $\mu = \cos\theta$ as the variable rather than θ itself. We note that

$$ds = \frac{dz}{\cos\theta} = \frac{dz}{\mu}$$

Therefore we have the transfer equation

$$\mu \frac{\partial I_\nu(z,\mu)}{\partial z} = -(\alpha_\nu + \sigma_\nu)(I_\nu - S_\nu). \tag{1.104a}$$

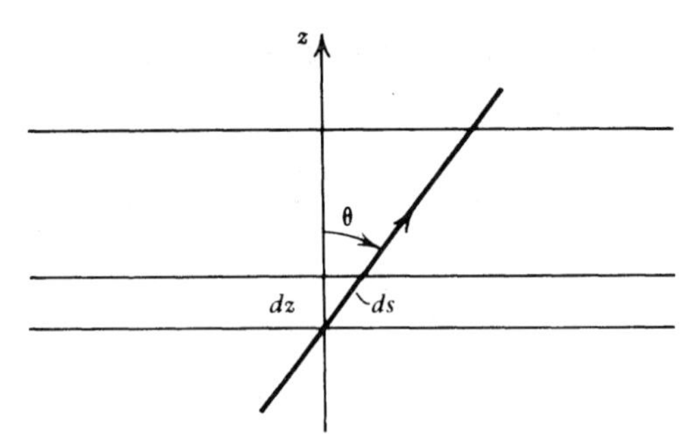

Figure 1.13 Geometry for plane-parallel media.

Let us rewrite this as

$$I_\nu(z,\mu)=S_\nu-\frac{\mu}{\alpha_\nu+\sigma_\nu}\frac{\partial I_\nu}{\partial z}. \tag{1.104b}$$

Now we use the fact that when the point in question is deep in the material the intensity changes rather slowly on the scale of a mean free path. Therefore the derivative term above is small and we write as a "zeroth" approximation,

$$I_\nu^{(0)}(z,\mu)\approx S_\nu^{(0)}(T). \tag{1.105}$$

Since this is independent of the angle μ, the zeroth-order mean intensity is given by $J_\nu^{(0)}=S_\nu^{(0)}$. From Eq. (1.92) this implies $I_\nu^{(0)}=S_\nu^{(0)}=B_\nu$, as we expect from the random walk arguments. We now get a better, "first" approximation by using the value $I_\nu^{(0)}=B_\nu$ in the derivative term:

$$I_\nu^{(1)}(z,\mu)\approx B_\nu(T)-\frac{\mu}{\alpha_\nu+\sigma_\nu}\frac{\partial B_\nu}{\partial z}. \tag{1.106}$$

This is justified, because the derivative term is already small, and any approximation there is not so critical. Note that the angular dependence of the intensity to this order of approximation is linear in $\mu=\cos\theta$.

Let us now compute the flux $F_\nu(z)$ using the above form for the intensity:

$$\begin{aligned} F_\nu(z)&=\int I_\nu^{(1)}(z,\mu)\cos\theta\, d\Omega \\ &=2\pi\int_{-1}^{+1} I_\nu^{(1)}(z,\mu)\mu\, d\mu. \end{aligned} \tag{1.107}$$

The angle-independent part of $I_\nu^{(1)}$ (i.e., B_ν) does not contribute to the flux. Thus we have the result

$$\begin{aligned} F_\nu(z) &= -\frac{2\pi}{\alpha_\nu+\sigma_\nu}\frac{\partial B_\nu}{\partial z}\int_{-1}^{+1}\mu^2\,d\mu \\ &= -\frac{4\pi}{3(\alpha_\nu+\sigma_\nu)}\frac{\partial B_\nu(T)}{\partial z} \\ &= -\frac{4\pi}{3(\alpha_\nu+\sigma_\nu)}\frac{\partial B_\nu(T)}{\partial T}\frac{\partial T}{\partial z}, \end{aligned} \tag{1.108}$$

using the chain rule for differentiation. This is the result for the monochromatic flux.

To obtain the total flux we integrate over all frequencies:

$$\begin{aligned} F(z) &= \int_0^\infty F_\nu(z)\,d\nu \\ &= -\frac{4\pi}{3}\frac{\partial T}{\partial z}\int_0^\infty(\alpha_\nu+\sigma_\nu)^{-1}\frac{\partial B_\nu}{\partial T}\,d\nu. \end{aligned}$$

This can be put into a more convenient form using the result:

$$\int_0^\infty \frac{\partial B_\nu}{\partial T}\,d\nu = \frac{\partial}{\partial T}\int_0^\infty B_\nu\,d\nu = \frac{\partial B(T)}{\partial T} = \frac{4\sigma T^3}{\pi} \tag{1.109}$$

which follows from Eqs. (1.42) and (1.43). Here σ is the Stefan–Boltzmann constant, not to be confused with σ_ν. We then define the *Rosseland mean absorption coefficient* α_R by the relation:

$$\frac{1}{\alpha_R} \equiv \frac{\displaystyle\int_0^\infty(\alpha_\nu+\sigma_\nu)^{-1}\frac{\partial B_\nu}{\partial T}\,d\nu}{\displaystyle\int_0^\infty\frac{\partial B_\nu}{\partial T}\,d\nu}. \tag{1.110}$$

Then we have

$$F(z) = -\frac{16\sigma T^3}{3\alpha_R}\frac{\partial T}{\partial z}. \tag{1.111}$$

This relation is called the *Rosseland approximation* for the energy flux. This equation is often called the *equation of radiative diffusion* [although this term is also used for equations such as (1.119) below]. It shows that radiative energy transport deep in a star is of the same nature as a heat conduction, with an "effective heat conductivity" $= 16\sigma T^3/3\alpha_R$. It also shows that the energy flux depends on only one property of the absorption coefficient, namely, its Rosseland mean. This mean involves a weighted average of $(\alpha_\nu + \sigma_\nu)^{-1}$ so that frequencies at which the extinction coefficient is small (the transparent regions) tend to dominate the averaging process. The weighting function $\partial B_\nu / \partial T$ [see Eq. (1.55)] has a general shape similar to that of the Planck function, but it now peaks at values of $h\nu / kT$ of order 3.8 instead of 2.8.

Although we have assumed a plane-parallel medium to prove the Rosseland formula, the result is quite general: the vector flux is in the direction opposite to the temperature gradient and has the magnitude given above. The only necessary assumption is that all quantities change slowly on the scale of any radiation mean free path.

The Eddington Approximation; Two-Stream Approximation

The basic idea behind the Rosseland approximation was that the intensities approach the Planck function at large effective depths in the medium. In the Eddington approximation, to be considered here, it is only assumed that the intensities approach *isotropy*, and not necessarily their thermal values. Because thermal emission and scattering are isotropic, one expects isotropy of the intensities to occur at depths of order of an ordinary mean free path; thus the region of applicability of the Eddington approximation is potentially much larger than the Rosseland approximation, the latter requiring depths of the order of the effective free path. With the use of appropriate boundary conditions (here introduced through the two-stream approximation) one can obtain solutions to scattering problems of reasonable accuracy at all depths.

The assumption of near isotropy is introduced by considering that the intensity is a power series in μ, with terms only up to linear:

$$I_\nu(\tau, \mu) = a_\nu(\tau) + b_\nu(\tau)\mu. \tag{1.112}$$

We now suppress the frequency variable ν for convenience in the following. Let us take the first three moments of this intensity:

$$J \equiv \tfrac{1}{2}\int_{-1}^{+1} I d\mu = a, \tag{1.113a}$$

$$H \equiv \tfrac{1}{2}\int_{-1}^{+1} \mu I \, d\mu = \frac{b}{3}, \tag{1.113b}$$

$$K \equiv \tfrac{1}{2}\int_{-1}^{+1} \mu^2 I \, d\mu = \frac{a}{3}. \tag{1.113c}$$

J is the mean intensity, and H and K are proportional to the flux and radiation pressure, respectively. Therefore, we have the result, known as the *Eddington approximation*:

$$K = \tfrac{1}{3}J. \tag{1.114}$$

Note the equivalence of this result to Eq. (1.10). The difference is that we have shown Eq. (1.114) to be valid even for slightly nonisotropic fields, containing terms linear in $\cos\theta$. Now defining the normal optical depth

$$d\tau(z) = -(\alpha_\nu + \sigma_\nu)dz, \tag{1.115}$$

we can write Eq. (1.104) as

$$\mu\frac{\partial I}{\partial \tau} = I - S. \tag{1.116}$$

The source function is given by Eq. (1.92) or (1.95) and is isotropic (independent of μ). If we multiply Eq. (1.116) by $\frac{1}{2}$ and integrate over μ from -1 to $+1$ we obtain

$$\frac{\partial H}{\partial \tau} = J - S \tag{1.117}$$

Similarly, multiplying by an extra factor μ before integrating, we obtain

$$\frac{\partial K}{\partial \tau} = H = \frac{1}{3}\frac{\partial J}{\partial \tau}, \tag{1.118}$$

using the Eddington approximation (1.114). These last two equations can be combined to yield

$$\frac{1}{3}\frac{\partial^2 J}{\partial \tau^2} = J - S. \tag{1.119a}$$

Use of Eq. (1.95) then gives a single second-order equation for J:

$$\frac{1}{3}\frac{\partial^2 J}{\partial \tau^2} = \epsilon(J - B). \tag{1.119b}$$

This equation is also sometimes called the *radiative diffusion equation*. Given the temperature structure of the medium, that is, $B(\tau)$, one can solve this equation for J and thus also determine S from Eq. (1.95). Then the problem is essentially solved, because the full intensity field $I(\tau,\mu)$ can be found by formal solution of Eq. (1.116).

An interesting form of Eq. (1.119b) can be derived in the case when ϵ does not depend on depth. Let us define the new optical depth scale

$$\tau_* \equiv \sqrt{3\epsilon}\ \tau = \sqrt{3\tau_a(\tau_a+\tau_s)}\ , \tag{1.120}$$

[cf. Eq. (1.98)]. The transfer equation is then

$$\frac{\partial^2 J}{\partial \tau^2_*} = J - B. \tag{1.121}$$

This equation can be used to demonstrate the properties of τ_* as an effective optical depth (see Problem 1.10).

To solve Eq. (1.119b), boundary conditions must be provided. This can be done in several ways, but here we use the *two-stream approximation*: It is assumed that the entire radiation field can be represented by radiation traveling at just *two* angles, $\mu = \pm 1/\sqrt{3}$. Let us denote the outward and inward intensities by $I^+(\tau) \equiv I(\tau, +1/\sqrt{3}\,)$ and $I^-(\tau) \equiv I(\tau, -1/\sqrt{3}\,)$. In terms of I^+ and I^- the moments J, H, and K have the representations

$$J = \frac{1}{2}(I^+ + I^-), \tag{1.122a}$$

$$H = \frac{1}{2\sqrt{3}}(I^+ - I^-), \tag{1.122b}$$

$$K = \frac{1}{6}(I^+ + I^-) = \frac{1}{3}J. \tag{1.122c}$$

This last equation is simply the Eddington approximation; in fact, the choice of the angles $\mu = \pm 1/\sqrt{3}$ is really motivated by the requirement that this relation be valid.

We now solve Eqs. (1.122a) and (1.122b) for I^+ and I^-, using Eq. (1.118):

$$I^+ = J + \frac{1}{\sqrt{3}}\frac{\partial J}{\partial \tau}, \tag{1.123a}$$

$$I^- = J - \frac{1}{\sqrt{3}}\frac{\partial J}{\partial \tau}. \tag{1.123b}$$

These equations can provide the necessary boundary conditions for the differential Eq. (1.119b). For example, suppose the medium extends from

$\tau=0$ to $\tau=\tau_0$, and there is no incident radiation. Then $I^-(0)=0$ and $I^+(\tau_0)=0$, so that the boundary conditions are

$$\frac{1}{\sqrt{3}}\frac{\partial J}{\partial \tau}=J \quad \text{at} \quad \tau=0, \tag{1.124a}$$

$$\frac{1}{\sqrt{3}}\frac{\partial J}{\partial \tau}=-J \quad \text{at} \quad \tau=\tau_0. \tag{1.124b}$$

These two conditions are sufficient to determine the solution of the second-order differential Eq. (1.119b).

Different methods for obtaining boundary conditions have been proposed; they all give equations of the form (1.124), but with constants slightly different than $1/\sqrt{3}$. For our purposes, it is not worth discussing these alternatives in detail. Examples of the use of the Eddington approximation to solve problems involving scattering are given in Problem 1.10.

PROBLEMS

1.1—A "pinhole camera" consists of a small circular hole of diameter d, a distance L from the "film-plane" (see Fig. 1.14). Show that the flux F_ν at the film plane depends on the brightness field $I_\nu(\theta,\phi)$ by

$$F_\nu=\frac{\pi\cos^4\theta}{4f^2}I_\nu(\theta,\phi),$$

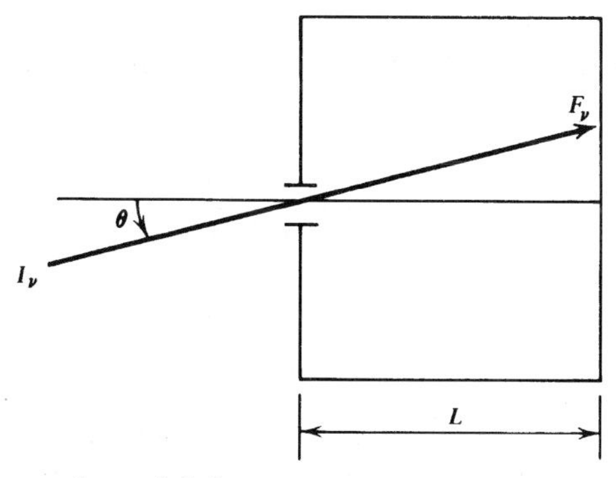

Figure 1.14 Geometry for a pinhole camera.

where the "focal ratio" is $f = L/d$. This is a simple, if crude, method for measuring I_ν.

1.2—Photoionization is a process in which a photon is absorbed by an atom (or molecule) and an electron is ejected. An energy at least equal to the ionization potential is required. Let this energy be $h\nu_0$ and let σ_ν be the cross section for photoionization. Show that the number of photoionizations per unit volume and per unit time is

$$4\pi n_a \int_{\nu_0}^{\infty} \frac{\sigma_\nu J_\nu}{h\nu}\, d\nu = c n_a \int_{\nu_0}^{\infty} \frac{\sigma_\nu u_\nu}{h\nu}\, d\nu,$$

where n_a = number density of atoms.

1.3—X-Ray photons are produced in a cloud of radius R at the uniform rate Γ (photons per unit volume per unit time). The cloud is a distance d away. Neglect absorption of these photons (optically thin medium). A detector at earth has an angular acceptance beam of half-angle $\Delta\theta$ and it has an effective area of ΔA.

a. Assume that the source is completely resolved. What is the observed intensity (photons per unit time per unit area per steradian) toward the center of the cloud.

b. Assume that the source is completely unresolved. What is the observed average intensity when the source is in the beam of the detector?

1.4

a. Show that the condition that an optically thin cloud of material can be ejected by radiation pressure from a nearby luminous object is that the mass to luminosity ratio (M/L) for the object be less than $\kappa/(4\pi Gc)$, where G = gravitational constant, c = speed of light, κ = mass absorption coefficient of the cloud material (assumed independent of frequency).

b. Calculate the terminal velocity v attained by such a cloud under radiation and gravitational forces alone, if it starts from rest a distance R from the object. Show that

$$v^2 = \frac{2GM}{R}\left(\frac{\kappa L}{4\pi GMc} - 1\right).$$

c. A minimum value for κ may be estimated for pure hydrogen as that due to Thomson scattering off free electrons, when the hydrogen is completely ionized. The Thomson cross section is $\sigma_T = 6.65\times10^{-25}$ cm^2. The mass scattering coefficient is therefore $>\sigma_T/m_H$, where m_H = mass of hydrogen atom. Show that the maximum luminosity that a central mass M can have and still not spontaneously eject hydrogen by radiation pressure is

$$L_{EDD} = 4\pi GMcm_H/\sigma_T$$
$$= 1.25\times10^{38}\text{erg s}^{-1}\,(M/M_\odot),$$

where

$$M_\odot \equiv \text{mass of sun} = 2\times10^{33}g.$$

This is called the *Eddington limit.*

1.5—A supernova remnant has an angular diameter $\theta = 4.3$ arc minutes and a flux at 100 MHz of $F_{100} = 1.6\times10^{-19}$ erg cm^{-2} s^{-1} Hz^{-1}. Assume that the emission is thermal.

a. What is the brightness temperature T_b? What energy regime of the blackbody curve does this correspond to?

b. The emitting region is actually more compact than indicated by the observed angular diameter. What effect does this have on the value of T_b?

c. At what frequency will this object's radiation be maximum, if the emission is blackbody?

d. What can you say about the temperature of the material from the above results?

1.6—Prove that the entropy of blackbody radiation S is related to temperature T and volume V by

$$S = \frac{4}{3}aT^3V.$$

1.7

a. Show that if stimulated emission is neglected, leaving only two Einstein coefficients, an appropriate relation between the coefficients will be consistent with thermal equilibrium between the atom and a radiation field of a Wien spectrum, but not of a Planck spectrum.

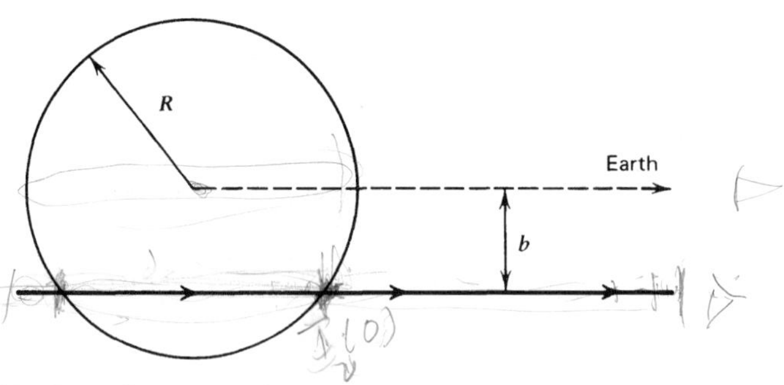

Figure 1.15 Detection of rays from a spherical emitting cloud of radius R.

b. Rederive the relation between the Einstein coefficients by imagining the atom to be in thermal equilibrium with a neutrino field (spin 1/2) rather than a photon field (spin 1).

Hint: Neutrinos are Fermi–Dirac particles and obey the exclusion principle. In addition, their equilibrium intensity is given by

$$I_\nu = \frac{2h\nu^3/c^2}{\exp(h\nu/kT)+1}.$$

1.8—A certain gas emits thermally at the rate $P(\nu)$ (power per unit volume and frequency range). A spherical cloud of this gas has radius R, temperature T and is a distance d from earth ($d \gg R$).

a. Assume that the cloud is optically *thin*. What is the brightness of the cloud as measured on earth? Give your answer as a function of the distance b away from the cloud center, assuming the cloud may be viewed along parallel rays as shown in Fig. 1.15.

b. What is the effective temperature of the cloud?

c. What is the flux F_ν measured at earth coming from the entire cloud?

d. How do the measured brightness temperatures compare with the cloud's temperature?

e. Answer parts (a)–(d) for an optically *thick* cloud.

1.9—A spherical, opaque object emits as a blackbody at temperature T_c. Surrounding this central object is a spherical shell of material, thermally emitting at a temperature T_s ($T_s < T_c$). This shell absorbs in a narrow spectral line; that is, its absorption coefficient becomes large at the frequency ν_0 and is negligibly small at other frequencies, such as ν_1:

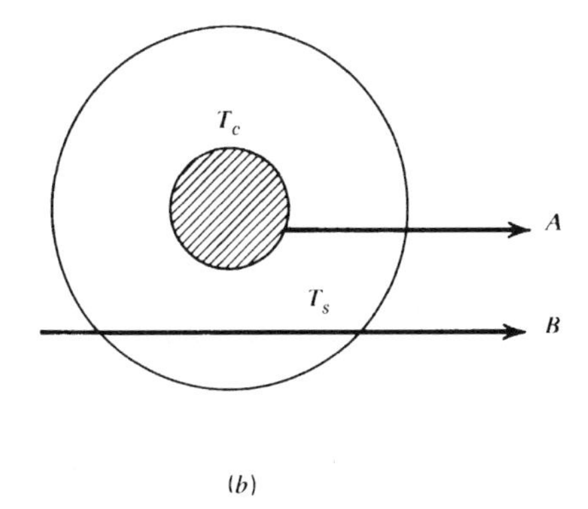

***Figure 1.16a** Blackbody emitter at temperature T_c surrounded by an absorbing shell at temperature T_s, viewed along rays A and B.*

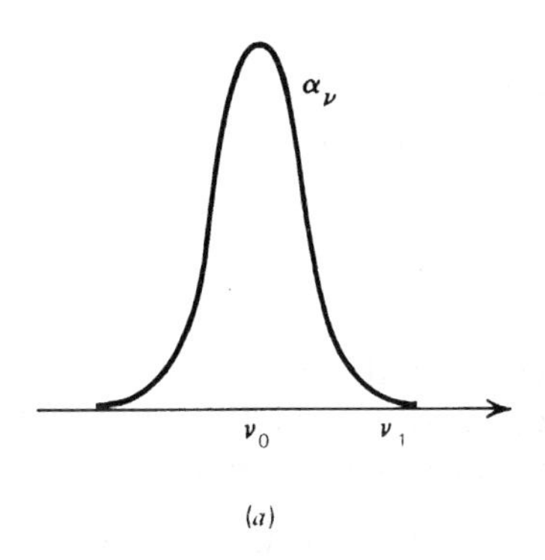

***Figure 1.16b** Absorption coefficient of the material in the shell.*

$\alpha_{\nu_0} \gg \alpha_{\nu_1}$ (see Fig. 1.16). The object is observed at frequencies ν_0 and ν_1 and along two rays A and B shown above. Assume that the Planck function does not vary appreciably from ν_0 to ν_1.

a. At which frequency will the observed brightness be larger when observed along ray A? Along ray B?

b. Answer the preceding questions if $T_s > T_c$.

1.10—Consider a semi-infinite half space in which both scattering (σ) and absorption and emission (α_ν) occur. Idealize the medium as homogeneous and isothermal, so that the coefficients σ and α_ν do not vary with depth. Further assume the scattering is isotropic (which is a good approximation to the forward-backward symmetric Thomson differential cross section).

a. Using the radiative diffusion equation with two-stream boundary conditions, find expressions for the mean intensity $J_\nu(\tau)$ in the medium and the emergent flux $F_\nu(0)$.

b. Show that $J_\nu(\tau)$ approaches the blackbody intensity at an effective optical depth of order $\tau_* = \sqrt{3\tau_a(\tau_a + \tau_s)}$.

REFERENCES

Mihalas, D., 1978, *Stellar Atmospheres*. (Freeman, San Francisco).
Reif, F., 1965, *Fundamentals of Statistical and Thermal Physics* (McGraw-Hill, New York).

2

BASIC THEORY OF RADIATION FIELDS

2.1 REVIEW OF MAXWELL'S EQUATIONS

We open our study of electromagnetic phenomena by a review of the theory applied to nonrelativistic particles. Gaussian units are used throughout.

The operational definitions of the electric field $\mathbf{E}(\mathbf{r},t)$ and the magnetic field $\mathbf{B}(\mathbf{r},t)$ are made through observations on a particle of charge q at point $\mathbf{r}$ with velocity $\mathbf{v}$, and by means of the formula for the *Lorentz force*:

$$\mathbf{F}=q\left(\mathbf{E}+\frac{\mathbf{v}}{c}\times\mathbf{B}\right). \tag{2.1}$$

The rate of work done by the fields on a particle is

$$\mathbf{v}\cdot\mathbf{F}=q\mathbf{v}\cdot\mathbf{E}, \tag{2.2a}$$

because $\mathbf{v}\cdot(\mathbf{v}\times\mathbf{B})=0$. Since $\mathbf{F}=m\,d\mathbf{v}/dt$ for nonrelativistic particles, we have

$$q\mathbf{v}\cdot\mathbf{E}=\frac{d}{dt}\left(\tfrac{1}{2}m\mathbf{v}^2\right). \tag{2.2b}$$

These results may be generalized to total force on a volume element containing many charges. The force per unit volume is

$$\mathbf{f}=\rho\mathbf{E}+\frac{1}{c}\mathbf{j}\times\mathbf{B}, \tag{2.3}$$

where

$$\rho=\lim_{\Delta V\to 0}\frac{1}{\Delta V}\sum_i q_i, \tag{2.4a}$$

$$\mathbf{j}=\lim_{\Delta V\to 0}\frac{1}{\Delta V}\sum_i q_i\mathbf{v}_i, \tag{2.4b}$$

and ΔV is the volume element. ρ and $\mathbf{j}$ are charge and current densities, respectively. In Eqs. (2.3) and (2.4) ΔV must be chosen much smaller than characteristic scales but much larger than the volume containing a single particle.

The rate of work done by the field per unit volume is then

$$\frac{1}{\Delta V}\sum_i q_i\mathbf{v}_i\cdot\mathbf{E}=\mathbf{j}\cdot\mathbf{E}.$$

From Eq. (2.2b) this is also the rate of change of mechanical energy per unit volume due to the fields:

$$\frac{dU_{\text{mech}}}{dt}=\mathbf{j}\cdot\mathbf{E}. \tag{2.5}$$

Maxwell's equations relate $\mathbf{E}$, $\mathbf{B}$, ρ, and $\mathbf{j}$. In Gaussian units, they are

$$\begin{aligned} &\nabla\cdot\mathbf{D}=4\pi\rho && \nabla\cdot\mathbf{B}=0 \\ &\nabla\times\mathbf{E}=-\frac{1}{c}\frac{\partial\mathbf{B}}{\partial t} && \nabla\times\mathbf{H}=\frac{4\pi}{c}\mathbf{j}+\frac{1}{c}\frac{\partial\mathbf{D}}{\partial t}. \end{aligned} \tag{2.6}$$

Here the fields $\mathbf{D}$ and $\mathbf{H}$ can often be related to $\mathbf{E}$ and $\mathbf{B}$ by the linear relations

$$\mathbf{D}=\epsilon\mathbf{E}, \tag{2.7a}$$

$$\mathbf{B}=\mu\mathbf{H}, \tag{2.7b}$$

where ϵ and μ are the dielectric constant and magnetic permeability of the medium, respectively. In the absence of dielectric or permeable media, $\epsilon=\mu=1$.

An immediate consequence of Maxwell's equation is *conservation of charge*: Taking $\nabla\cdot$ of the $\nabla\times\mathbf{H}$ equation gives

$$\nabla\cdot\mathbf{j}+\frac{\partial\rho}{\partial t}=0. \tag{2.8}$$

This expresses conservation of charge for a volume element.

We now give definitions of energy density and energy flux of the electromagnetic field. Consider the work done per unit volume on a particle distribution, [cf. Eq. (2.6)]:

$$\mathbf{j}\cdot\mathbf{E}=\frac{1}{4\pi}\left[c(\nabla\times\mathbf{H})\cdot\mathbf{E}-\mathbf{E}\cdot\frac{\partial\mathbf{D}}{\partial t}\right], \tag{2.9}$$

where we have used Maxwell's equations. Now, use the vector identity:

$$\mathbf{E}\cdot(\nabla\times\mathbf{H})=\mathbf{H}\cdot(\nabla\times\mathbf{E})-\nabla\cdot(\mathbf{E}\times\mathbf{H}),$$

and again use Maxwell's equations to write Eq. (2.9) in the form

$$\mathbf{j}\cdot\mathbf{E}=\frac{1}{4\pi}\left[-\mathbf{H}\cdot\frac{\partial\mathbf{B}}{\partial t}-c\nabla\cdot(\mathbf{E}\times\mathbf{H})-\mathbf{E}\cdot\frac{\partial\mathbf{D}}{\partial t}\right]. \tag{2.10a}$$

Now, if ϵ and μ are independent of time, then the above relation may be written as [cf. Eq. (2.7)]

$$\mathbf{j}\cdot\mathbf{E}+\frac{1}{8\pi}\frac{\partial}{\partial t}\left(\epsilon E^2+\frac{B^2}{\mu}\right)=-\nabla\cdot\left(\frac{c}{4\pi}\mathbf{E}\times\mathbf{H}\right). \tag{2.10b}$$

Equation (2.10b) is *Poynting's theorem* in differential form and can be interpreted as saying that the rate of change of mechanical energy per unit volume plus the rate of change of field energy per unit volume equals minus the divergence of the field energy flux. Accordingly, we set the electromagnetic field energy per unit volume equal to

$$U_{\text{field}}=\frac{1}{8\pi}\left(\epsilon E^2+\frac{B^2}{\mu}\right)=U_E+U_B, \tag{2.11}$$

and the electromagnetic flux vector, or *Poynting vector*, equal to

$$\mathbf{S}=\frac{c}{4\pi}\mathbf{E}\times\mathbf{H}. \tag{2.12}$$

The above can also be understood by integrating over a volume element and using the divergence theorem:

$$\int_V \mathbf{j}\cdot\mathbf{E}dV+\frac{d}{dt}\int_V \frac{\epsilon E^2+B^2/\mu}{8\pi}dV=-\int_\Sigma \mathbf{S}\cdot d\mathbf{A},$$

or

$$\frac{d}{dt}(U_{\text{mech}}+U_{\text{field}})=-\int_\Sigma \mathbf{S}\cdot d\mathbf{A}. \tag{2.13}$$

That is, the rate of change of total (mechanical plus field) energy within the volume V is equal to the net inward flow of energy through the bounding surface Σ.

Although U_{field} is called a field energy, it has contributions from the matter, because ϵ and μ are both macroscopic properties of matter. We are, in effect, putting the energy of the bound charges into the field. If we had treated all charges (free and bound) as part of the mechanical system, then we would use only the *microscopic* fields **E** and **B**. Then **j** would be replaced by the sum of the conduction current and induced molecular currents and $\mathbf{S}\rightarrow(c/4\pi)\mathbf{E}\times\mathbf{B}$. When both matter and fields are present, the allocation of energy into matter and field energies is somewhat arbitrary. What is not arbitrary is that the total energy is conserved.

If we now consider either the microscopic energy flux in the field or the field in vacuum, and use Eq. (1.6) and the fact that $p=E/c$ for photons, then we can write the *momentum per unit volume in the field*, **g** as

$$\mathbf{g}=\frac{1}{4\pi c}\mathbf{E}\times\mathbf{B}. \tag{2.14}$$

The *angular momentum* carried by the field is given by $\mathfrak{L}$, the angular momentum density:

$$\mathfrak{L}=\mathbf{r}\times\mathbf{g}, \tag{2.15}$$

where **r** is the radius vector from the point about which the angular momentum is computed. We do not derive these results in general; however, this identification of momentum and angular momentum for electromagnetic radiation is verified in Problem 2.3.

Returning to the conservation of energy now, we can let the surface Σ approach infinity, and the question arises as to the limit of

$$\int_\Sigma \mathbf{S}\cdot d\mathbf{A}$$

In electrostatics and magnetostatics we recall that both **E** and **B** decrease like r^{-2} as $r\to\infty$. This implies that **S** decreases like r^{-4} in static problems. Thus the above integral goes to zero, since the surface area increases only as r^2. However, for time-varying fields we find that **E** and **B** may decrease only as r^{-1}. Therefore, the integral can contribute a finite amount to the rate of change of energy of the system. This finite energy flowing outward (or inward) at large distances is called *radiation*. Those parts of **E** and **B** that decrease as r^{-1} at large distances are said to constitute the *radiation field*.

2.2 PLANE ELECTROMAGNETIC WAVES

Maxwell's equations in vacuum become [cf. Eqs. (2.6)]

$$\begin{aligned} &\nabla\cdot\mathbf{E}=0 && \nabla\cdot\mathbf{B}=0 \\ &\nabla\times\mathbf{E}=-\frac{1}{c}\frac{\partial\mathbf{B}}{\partial t} && \nabla\times\mathbf{B}=\frac{1}{c}\frac{\partial\mathbf{E}}{\partial t}. \end{aligned} \tag{2.16}$$

A basic feature of these equations is the existence of traveling wave solutions that carry energy. Taking the curl of the third equation and combining it with the fourth, we obtain

$$\nabla\times(\nabla\times\mathbf{E})=-\frac{1}{c^2}\frac{\partial^2\mathbf{E}}{\partial t^2}.$$

If we now use the vector identity

$$\nabla\times(\nabla\times\mathbf{E})=\nabla(\nabla\cdot\mathbf{E})-\nabla^2\mathbf{E}$$

(in Cartesian components) and the first equation, we obtain the *vector wave equation* for **E**:

$$\nabla^2\mathbf{E}-\frac{1}{c^2}\frac{\partial^2\mathbf{E}}{\partial t^2}=0. \tag{2.17}$$

An identical equation holds for **B**, since Eq. (2.16) is invariant under $\mathbf{E}\to\mathbf{B}$, $\mathbf{B}\to-\mathbf{E}$.

Let us now consider solutions of the form

$$\mathbf{E}=\hat{\mathbf{a}}_1 E_0 e^{i(\mathbf{k}\cdot\mathbf{r}-\omega t)}, \tag{2.18a}$$

$$\mathbf{B}=\hat{\mathbf{a}}_2 B_0 e^{i(\mathbf{k}\cdot\mathbf{r}-\omega t)}, \tag{2.18b}$$

where $\hat{\mathbf{a}}_1$ and $\hat{\mathbf{a}}_2$ are unit vectors, E_0 and B_0 are complex constants, and $\mathbf{k}=k\mathbf{n}$ and ω are the "wave vector" and frequency, respectively. Clearly, such solutions represent waves traveling in the $\mathbf{n}$ direction, since surfaces of constant phase advance with time in the $\mathbf{n}$ direction. By superposing such solutions propagating in all directions and with all frequencies, we can construct the most general solution to the source-free Maxwell's equations. Substitution into Maxwell's equations yields:

$$\begin{aligned} &i\mathbf{k}\cdot\hat{\mathbf{a}}_1 E_0=0 && i\mathbf{k}\cdot\hat{\mathbf{a}}_2 B_0=0 \\ &i\mathbf{k}\times\hat{\mathbf{a}}_1 E_0=\frac{i\omega}{c}\hat{\mathbf{a}}_2 B_0 && i\mathbf{k}\times\hat{\mathbf{a}}_2 B_0=-\frac{i\omega}{c}\hat{\mathbf{a}}_1 E_0. \end{aligned} \tag{2.19}$$

The top two equations tell us that both $\hat{\mathbf{a}}_1$ and $\hat{\mathbf{a}}_2$ are *transverse* (perpendicular) to the direction of propagation $\mathbf{k}$. With this information, the cross products in the bottom two equations can be done, and we see that $\hat{\mathbf{a}}_1$ and $\hat{\mathbf{a}}_2$ are perpendicular to each other. The vectors $\hat{\mathbf{a}}_1$, $\hat{\mathbf{a}}_2$, and $\mathbf{k}$ form a right-hand triad of mutually perpendicular vectors. The values of E_0 and B_0 are related by

$$E_0=\frac{\omega}{kc}B_0, \qquad B_0=\frac{\omega}{kc}E_0,$$

so that

$$E_0=\left(\frac{\omega}{kc}\right)^2 E_0$$

and

$$\omega^2=c^2k^2. \tag{2.20a}$$

Taking k and ω positive, as implied by the above discussion, we have

$$\omega=ck. \tag{2.20b}$$

This in turn implies

$$E_0=B_0. \tag{2.21}$$

The waves propagate with a phase velocity that can be found from $v_{\mathrm{ph}}=\omega/k$, so that

$$v_{\mathrm{ph}}=c. \tag{2.22}$$

The waves, as expected, travel at the speed of light. (In a vacuum the group velocity, $v_g \equiv \partial\omega/\partial k$, equals c also.)

We can now compute the energy flux and energy density of these waves. Since **E** and **B** both vary sinusoidally in time, the Poynting vector and the energy density actually fluctuate; however, we take a *time average*, since this is in most cases what is measured.

Now, it can easily be shown (Problem 2.1) that if $A(t)$ and $B(t)$ are two complex quantities with the same sinusoidal time dependence, that is,

$$A(t) = \mathcal{A}e^{i\omega t} \qquad B(t) = \mathcal{B}e^{i\omega t},$$

then the time average of the product of their real parts is

$$\langle \mathrm{Re}\,A(t)\cdot \mathrm{Re}\,B(t)\rangle = \tfrac{1}{2}\,\mathrm{Re}(\mathcal{A}\mathcal{B}^*) = \tfrac{1}{2}\,\mathrm{Re}(\mathcal{A}^*\mathcal{B}). \tag{2.23}$$

We have used * to denote complex conjugation. Thus the time-averaged Poynting vector [cf. Eq. (2.12)] satisfies

$$\langle S\rangle = \frac{c}{8\pi}\mathrm{Re}(E_0B_0^*). \tag{2.24a}$$

Since $E_0 = B_0$,

$$\langle S\rangle = \frac{c}{8\pi}|E_0|^2 = \frac{c}{8\pi}|B_0|^2. \tag{2.24b}$$

Similarly, the time-averaged energy density is [cf. Eq. (2.11)]

$$\langle U\rangle = \frac{1}{16\pi}\mathrm{Re}(E_0E_0^* + B_0B_0^*), \tag{2.25a}$$

or, with $E_0 = B_0$,

$$\langle U\rangle = \frac{1}{8\pi}|E_0|^2 = \frac{1}{8\pi}|B_0|^2. \tag{2.25b}$$

Therefore, the velocity of energy flow is $\langle S\rangle/\langle U\rangle = c$ also.

The above results have all been for propagation in a vacuum. Similar results hold, at least formally, if we use a dielectric constant and permeability that are constants. However, in practice these quantities usually depend on frequency, so a more careful approach is required. Some effects of refraction and dispersion are treated in Chapter 8.

2.3 THE RADIATION SPECTRUM

The spectrum of radiation depends on the *time variation* of the electric field (we can ignore the magnetic field, since it mimics the electric field). A consequence is that one cannot give a meaning to the spectrum of radiation at a precise instant of time, knowing only the electric field at one point. Instead, one must talk about the spectrum of a train of waves, or of the radiation at a point during a sufficiently long time interval Δt. If we have such a time record of the radiation field of length Δt, we still can only define the spectrum to within a frequency resolution $\Delta\omega$ where

$$\Delta\omega\Delta t > 1. \tag{2.26}$$

This uncertainty relation is not necessarily quantum in nature (although it can be proved from the energy-time uncertainty relation), but is a property of any wave theory of light.

Let us assume, for mathematical simplicity, that the radiation is in the form of a finite pulse. (In practice, we only require that $\mathbf{E}(t)$ vanishes sufficiently rapidly for $t \rightarrow \pm\infty$.) Also, let us treat only one of the two independent components of the transverse electric field, say $E(t) \equiv \hat{\mathbf{a}} \cdot \mathbf{E}(t)$. With these assumptions we may express $E(t)$ in terms of a Fourier integral (Fourier transform):

$$\hat{E}(\omega) = \frac{1}{2\pi}\int_{-\infty}^{\infty} E(t) e^{i\omega t} dt. \tag{2.27}$$

The inverse of this is

$$E(t) = \int_{-\infty}^{\infty} \hat{E}(\omega) e^{-i\omega t} d\omega. \tag{2.28}$$

The function $\hat{E}(\omega)$ is complex; however, since $E(t)$ is real we can write

$$\hat{E}(-\omega) = \frac{1}{2\pi}\int_{-\infty}^{\infty} E(t) e^{-i\omega t} dt = \hat{E}^*(\omega),$$

so that the negative frequencies can be eliminated.

Contained in $\hat{E}(\omega)$ is all the information about the frequency behavior of $E(t)$. To convert this into frequency information about the energy we write the energy per unit time per unit area in terms of the Poynting vector:

$$\frac{dW}{dt\,dA} = \frac{c}{4\pi} E^2(t). \tag{2.29}$$

The total energy per unit area in the pulse is

$$\frac{dW}{dA}=\frac{c}{4\pi}\int_{-\infty}^{\infty}E^2(t)\,dt. \tag{2.30}$$

But from Parseval's theorem for Fourier transforms, we know that

$$\int_{-\infty}^{\infty}E^2(t)\,dt=2\pi\int_{-\infty}^{\infty}|\hat{E}(\omega)|^2d\omega. \tag{2.31}$$

By the above symmetry property of $\hat{E}(\omega)$ we have

$$|\hat{E}(\omega)|^2=|\hat{E}(-\omega)|^2,$$

so that

$$\int_{-\infty}^{\infty}E^2(t)\,dt=4\pi\int_{0}^{\infty}|\hat{E}(\omega)|^2d\omega.$$

Thus we have the result

$$\frac{dW}{dA}=c\int_{0}^{\infty}|\hat{E}(\omega)|^2d\omega, \tag{2.32}$$

and we may identify the energy per unit area per unit frequency:

$$\frac{dW}{dA\,d\omega}=c|\hat{E}(\omega)|^2 \tag{2.33}$$

It should be noted that this is the total energy per area per frequency range in the *entire pulse*; we have not written "per unit time." In fact, to write both dt and $d\omega$ would violate the uncertainty relation between ω and t. However, if the pulse repeats on an average time scale T, then we may *formally* write

$$\frac{dW}{dA\,d\omega\,dt}\equiv\frac{1}{T}\frac{dW}{dA\,d\omega}=\frac{c}{T}|\hat{E}(\omega)|^2. \tag{2.34}$$

This formula also can be used to define the spectrum of a portion of length T of a much longer signal. If a very long signal has more or less the same properties over its entire length (property of *time stationarity*) then we expect that the result will be independent of T for large T, and we may write

$$\frac{dW}{dA\,d\omega\,dt}=c\lim_{T\to\infty}\frac{1}{T}|\hat{E}_T(\omega)|^2,$$

where we have written the subscript T on $\hat{E}_T(\omega)$ to emphasize that this is the transform of a portion of the function $E(t)$ of length T. In this way we can generalize our discussion to include infinitely long waves (such as sine waves) using formulas based on finite pulses.

If the properties of $E(t)$ vary with time, then one expects that the spectrum as determined by analyzing a portion of length T will depend on just what portion is analyzed. In that case the whole efficacy of the concept of local spectrum depends on whether the changes of character of $E(t)$ occur on a time scale long enough that one can still define a length T in which a suitable frequency resolution $\Delta\omega \sim 1/T$ can be obtained. If this condition is not met, a local spectrum is not useful, and one must consider the spectrum of the entire pulse as the basic entity.

Let us consider now some typical pulse shapes and their corresponding spectra. (See Figs. 2.1, 2.2, and 2.3.) Study of these should give some

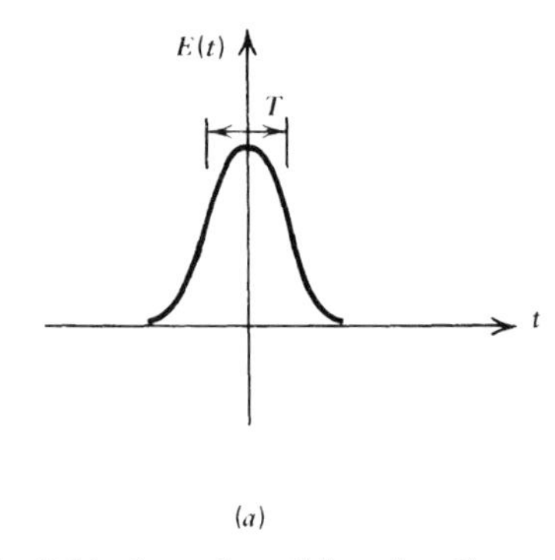

***Figure 2.1a** Electric field of a pulse of duration T.*

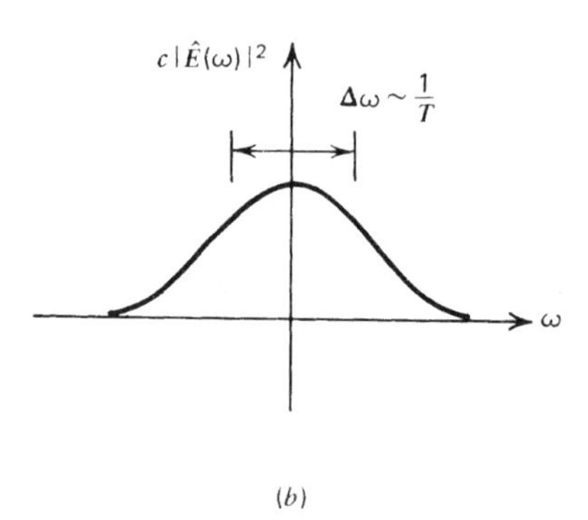

***Figure 2.1b** Power spectrum for a.*

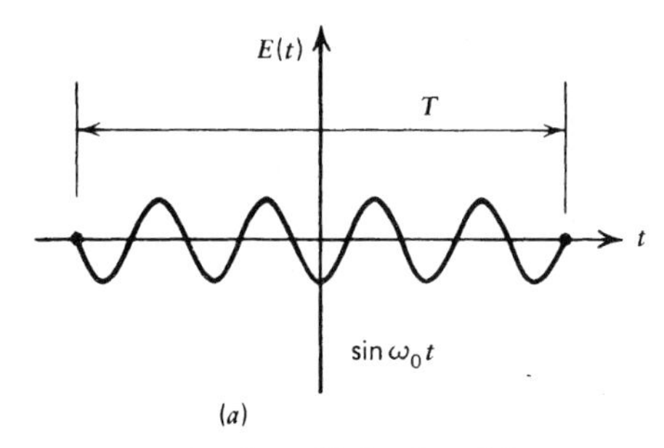

***Figure 2.2a** Electric field of a sinusoidal pulse of frequency ω_0 and duration T.*

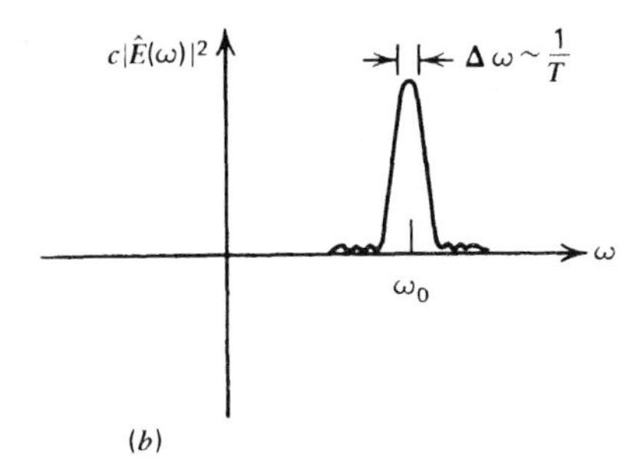

***Figure 2.2b** Power spectrum for a.*

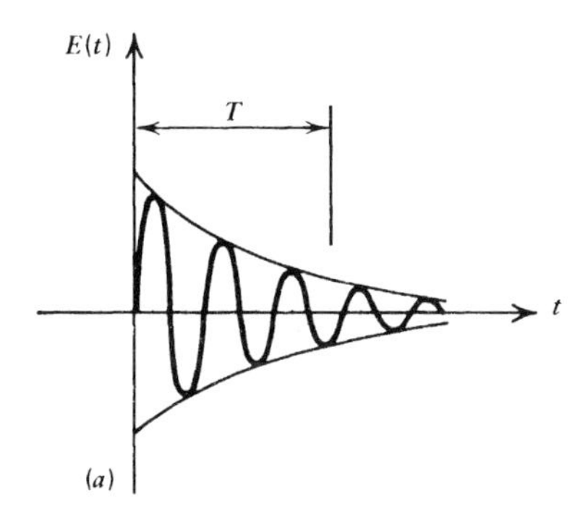

***Figure 2.3a** Electric field of a damped sinusoid of the form $\exp(-t/T)\sin \omega_0 t$.*

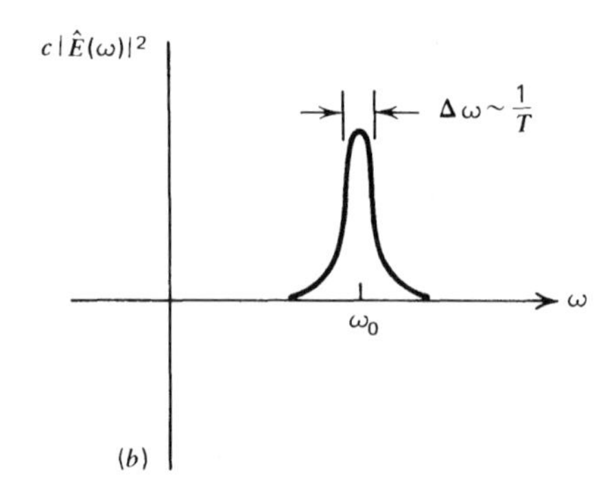

Figure 2.3b Power spectrum for a.

insight into the relationships that are useful in estimating spectra from particular processes. Note that the graphs of $c|\hat{E}(\omega)|^2$ are always symmetric about the origin—sometimes we have drawn the curves for both positive and negative ω for convenience, while in other cases we have only drawn them for positive ω. Only the values for positive ω need concern us.

Some general rules can be seen in these simple examples: First, the time extent of the pulse T determines the width of the finest features in the spectrum by means of $\Delta\omega \sim 1/T$. Second, the existence of a sinusoidal time dependence within the pulse shape causes the spectrum to be concentrated near $\omega \sim \omega_0$.

2.4 POLARIZATION AND STOKES PARAMETERS

Monochromatic Waves

The monochromatic plane waves described in Eq. (2.18) are *linearly polarized*; that is, the electric vector simply oscillates in the direction $\hat{\mathbf{a}}_1$, which, with the propagation direction, defines the *plane of polarization*. By superposing solutions corresponding to two such oscillations in perpendicular directions, we can construct the most general state of polarization for a wave of given $\mathbf{k}$ and ω. We need consider only the electric vector $\mathbf{E}$; the magnetic vector simply stays perpendicular to and has the same magnitude as $\mathbf{E}$. Let us examine the electric vector at an arbitrary point (say, $\mathbf{r}=0$) and choose axes x and y with corresponding unit vectors $\hat{\mathbf{x}}$ and $\hat{\mathbf{y}}$ (see Fig. 2.4). The direction of the wave is out of the page, toward the observer. Then the electric vector is the real part of

$$\mathbf{E} = (\hat{\mathbf{x}}E_1 + \hat{\mathbf{y}}E_2)e^{-i\omega t} \equiv \mathbf{E}_0 e^{-i\omega t}. \tag{2.35}$$

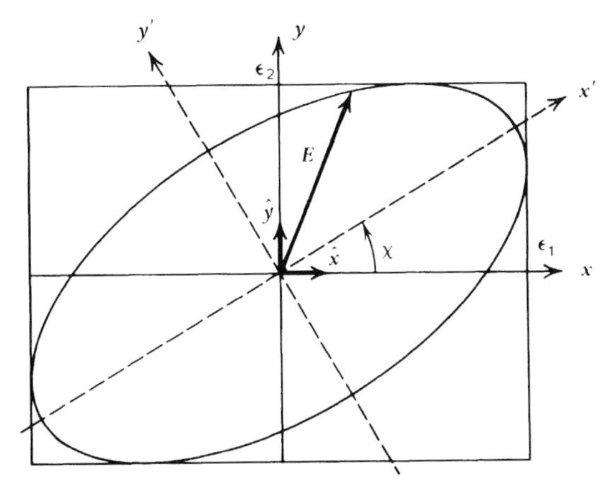

***Figure 2.4** Rotation of x and y electric field components through angle χ to coincide with principal axes of the polarization ellipse.*

This generalization of Eq. (2.18) can be characterized as having replaced $\hat{\mathbf{a}}_1 E_1$ by the general complex vector $\mathbf{E}_0$. The complex amplitudes E_1 and E_2 can be expressed as

$$E_1 = \mathcal{E}_1 e^{i\phi_1}, \qquad E_2 = \mathcal{E}_2 e^{i\phi_2}. \tag{2.36}$$

Taking the real part of $\mathbf{E}$, we find the physical components of the electric field along $\hat{\mathbf{x}}$ and $\hat{\mathbf{y}}$ to be

$$E_x = \mathcal{E}_1 \cos(\omega t - \phi_1), \qquad E_y = \mathcal{E}_2 \cos(\omega t - \phi_2). \tag{2.37}$$

These equations describe the tip of the electric field vector in the x-y plane.

We now show that the figure traced out is an ellipse, and hence the general wave is said to be *elliptically polarized.* First of all, note that the equations for a general ellipse relative to its principal axes x' and y', which are tilted at an angle χ to the x- and y-axes (see Fig. 2.4), can be written

$$E_x' = \mathcal{E}_0 \cos\beta \cos\omega t, \qquad E_y' = -\mathcal{E}_0 \sin\beta \sin\omega t, \tag{2.38}$$

where $-\pi/2 \leqslant \beta \leqslant \pi/2$. The magnitudes of the principal axes are clearly $\mathcal{E}_0 |\cos\beta|$ and $\mathcal{E}_0 |\sin\beta|$, since $(E_x'/\mathcal{E}_0 \cos\beta)^2 + (E_y'/\mathcal{E}_0 \sin\beta)^2 = 1$.The ellipse will be traced out in a clockwise sense for $0 < \beta < \pi/2$ and counter-clockwise sense for $-\pi/2 < \beta < 0$, as viewed by an observer toward whom the wave is propagating. These possibilities are called, respectively, *right-* and *left-handed* elliptical polarization. Other terms are, respectively, *negative* and *positive helicity*.

Two degenerate cases of elliptical polarization can occur: When $\beta = \pm \pi/4$ the ellipse becomes a circle, and the wave is said to be *circularly polarized*. When $\beta = 0$ or $\pm\pi/2$, the ellipse narrows to a straight line, and the wave is said to be *linearly polarized*. In this latter case the wave is neither right-handed nor left-handed.

Let us now make the connections between the quantities that appear in Eq. (2.37) and those defining the principal axes of the ellipse. To do this we transform the electric field components in Eq. (2.38) to the x- and y-axes by rotating through the angle χ (see Fig. 2.4). This yields

$$E_x = \mathcal{E}_0(\cos\beta\cos\chi\cos\omega t + \sin\beta\sin\chi\sin\omega t)$$
$$E_y = \mathcal{E}_0(\cos\beta\sin\chi\cos\omega t - \sin\beta\cos\chi\sin\omega t)$$

These are identical with Eq. (2-37) if we take

$$\mathcal{E}_1\cos\phi_1 = \mathcal{E}_0\cos\beta\cos\chi, \tag{2.39a}$$
$$\mathcal{E}_1\sin\phi_1 = \mathcal{E}_0\sin\beta\sin\chi, \tag{2.39b}$$
$$\mathcal{E}_2\cos\phi_2 = \mathcal{E}_0\cos\beta\sin\chi, \tag{2.39c}$$
$$\mathcal{E}_2\sin\phi_2 = -\mathcal{E}_0\sin\beta\cos\chi. \tag{2.39d}$$

Given $\mathcal{E}_1$, ϕ_1, $\mathcal{E}_2$, ϕ_2 these equations can be solved for $\mathcal{E}_0$, β, and χ. A convenient way of doing this is by means of the *Stokes parameters for monochromatic waves*, which are defined by the equations:

$$I \equiv \mathcal{E}_1^2 + \mathcal{E}_2^2 = \mathcal{E}_0^2 \tag{2.40a}$$
$$Q \equiv \mathcal{E}_1^2 - \mathcal{E}_2^2 = \mathcal{E}_0^2\cos 2\beta\cos 2\chi \tag{2.40b}$$
$$U \equiv 2\mathcal{E}_1\mathcal{E}_2\cos(\phi_1 - \phi_2) = \mathcal{E}_0^2\cos 2\beta\sin 2\chi \tag{2.40c}$$
$$V \equiv 2\mathcal{E}_1\mathcal{E}_2\sin(\phi_1 - \phi_2) = \mathcal{E}_0^2\sin 2\beta. \tag{2.40d}$$

The alternate forms follow from manipulations of Eqs. (2.39). Thus we have

$$\mathcal{E}_0 = \sqrt{I} \tag{2.41a}$$
$$\sin 2\beta = \frac{V}{I} \tag{2.41b}$$
$$\tan 2\chi = \frac{U}{Q}. \tag{2.41c}$$

Pure elliptical polarization is determined solely by three parameters: $\mathcal{E}_0$, β, and χ. Therefore, one expects a relation to exist between the four Stokes parameters in this case; in fact, we have

$$I^2 = Q^2 + U^2 + V^2 \tag{2.42}$$

for a monochromatic wave (pure elliptical polarization.)

The meanings of the Stokes parameters are as follows: I is nonnegative and is proportional to the total energy flux or intensity of the wave. In practice, it is customary to choose a single proportionality factor in all of the definitions of (2.40) so that I is precisely the flux or intensity, but we shall omit it here. V is the *circularity* parameter that measures the ratio of principal axes of the ellipse. The wave has right- or left-handed polarization when V is positive or negative, respectively; $V=0$ is the condition for linear polarization. There is only one remaining independent parameter, Q or U, which measures the orientation of the ellipse relative to the x-axis; $Q=U=0$ is the condition for circular polarization.

Quasi-monochromatic Waves

The monochromatic waves just treated are said to be *completely* or *100%* polarized, since the electric vector displays a simple, nonrandom directional behavior in time. However, in practice we never see a single monochromatic component but rather a superposition of many components, each with its own polarization. An important case of interest occurs when the amplitudes and phases of the wave possess a relatively slow time variation, so that instead of Eq. (2.36) we have

$$E_1(t) = \mathcal{E}_1(t)e^{i\phi_1(t)}, \qquad E_2(t) = \mathcal{E}_2(t)e^{i\phi_2(t)} \tag{2.43}$$

To be precise, we assume that over short times, of order $1/\omega$, the wave looks completely polarized with a definite state of elliptical polarization, but over much longer times, $\Delta t \gg 1/\omega$, characterizing the times over which $\mathcal{E}_1$, $\mathcal{E}_2$, ϕ_1 and ϕ_2 change substantially, this state of polarization can change completely. Such a wave is no longer monochromatic; by the uncertainty relation its frequency spread $\Delta\omega$ about the value ω can be estimated as $\Delta\omega > 1/\Delta t$ so that $\Delta\omega \ll \omega$. For this reason the wave is called *quasi-monochromatic*. The frequency spread $\Delta\omega$ is called the *bandwidth* of the wave, and the time Δt is called the *coherence time*.

The quantitative characterization of quasi-monochromatic waves depends on what kind of measurements can be made. In principle, for strong waves the precise time variations of the quantities $\mathcal{E}_1$, $\mathcal{E}_2$, ϕ_1, and ϕ_2 could be measured; this would be the most detailed characterization possible. On the other hand, most measurements are not so detailed and usually involve some apparatus in which the radiation eventually falls on a detector that measures the time-averaged square of the electric field, for example, the energy flux (2.24b). Before falling on the detector the radiation may pass through a variety of devices that have the effect of forming a linear combination of the two independent electric field components with arbitrary weights and phases. For radio waves such devices include dipole antennas and electric delay lines; the optical equivalents are found in polarizing filters and quarter-wave plates.

If we suppose that any time delays involved are short compared to the coherence time of the wave, then we can show that the outcome of a measurement with such a device depends on simple extensions of the Stokes parameters previously introduced.

We first note that the most general linear transformation of field components by devices of the type described above can be written

$$\begin{aligned} E_1' &= \lambda_{11}E_1 + \lambda_{12}E_2 \\ E_2' &= \lambda_{21}E_1 + \lambda_{22}E_2, \end{aligned} \tag{2.44}$$

where λ_{ij}, $(i,j=1,2)$, are complex constants describing the measuring apparatus. What is measured is the average sum of the squares of the x' and y' components of electric field. The average of the square of the x' component is

$$\begin{aligned} 2\langle[\mathrm{Re}\,E_1'e^{-i\omega t}]^2\rangle = |\lambda_{11}|^2\langle E_1E_1^*\rangle + \lambda_{11}\lambda_{12}^*\langle E_1E_2^*\rangle \\ +\lambda_{12}\lambda_{11}^*\langle E_2E_1^*\rangle \\ +|\lambda_{12}|^2\langle E_2E_2^*\rangle. \end{aligned} \tag{2.45}$$

Eq. (2.23) has been used to average over the "fast" variations in the field described by the $e^{-i\omega t}$ term. The brackets $\langle\ \rangle$ on the right-hand side then refer only to time averaging of the slowly varying combinations of $E_1(t)$ and $E_2(t)$. For example,

$$\langle E_1E_2^*\rangle = \frac{1}{T}\int_0^T E_1(t)E_2^*(t)\,dt, \tag{2.46}$$

where 0 to T is the time interval over which the measurement is made. The average square of the y component yields a result analogous to Eq. (2.45) with λ_{21} and λ_{22} replacing λ_{11} and λ_{12}, respectively.

It is clear from the above that the measurement depends on the radiation field only through the four complex quantities $\langle E_i(t)E_j^*(t)\rangle$, where i, $j=1, 2$. These in turn are equivalent to four real quantities, since $\langle E_1E_1^*\rangle$ and $\langle E_2E_2^*\rangle$ are real and $\langle E_1E_2^*\rangle$ and $\langle E_2E_1^*\rangle$ are complex conjugates. A common and convenient set of four real quantities used to express $\langle E_iE_j^*\rangle$ are the *Stokes parameter for quasi-monochromatic waves*,

$$I \equiv \langle E_1E_1^*\rangle + \langle E_2E_2^*\rangle = \langle \mathcal{E}_1^2 + \mathcal{E}_2^2\rangle \tag{2.47a}$$

$$Q \equiv \langle E_1E_1^*\rangle - \langle E_2E_2^*\rangle = \langle \mathcal{E}_1^2 - \mathcal{E}_2^2\rangle \tag{2.47b}$$

$$U \equiv \langle E_1E_2^*\rangle + \langle E_2E_1^*\rangle = \langle 2\mathcal{E}_1\mathcal{E}_2\cos(\phi_1-\phi_2)\rangle \tag{2.47c}$$

$$V \equiv \frac{1}{i}(\langle E_1E_2^*\rangle - \langle E_2E_1^*\rangle) = \langle 2\mathcal{E}_1\mathcal{E}_2\sin(\phi_1-\phi_2)\rangle, \tag{2.47d}$$

using Eqs. (2.36). We see that these definitions are generalizations of Eqs. (2.40), to which they reduce when $\mathcal{E}_1$, $\mathcal{E}_2$, ϕ_1, and ϕ_2 are time independent. The Stokes parameters are the most complete description of the radiation field, in the sense that two waves having the same parameters cannot be distinguished by any measurements using an apparatus of the type described above.

Equation (2.42) will not hold for arbitrary quasi-monochromatic waves. It is easy to show from the Schwartz inequality, that

$$\langle E_1E_1^*\rangle\langle E_2E_2^*\rangle \geqslant \langle E_1E_2^*\rangle\langle E_2E_1^*\rangle, \tag{2.48}$$

the equality sign holding only when the ratio of $E_1(t)$ to $E_2(t)$ is a complex constant, independent of time. This latter condition implies that the electric vector traces out an ellipse of fixed shape and fixed orientation and only its overall size changes slowly with time. Such a wave is completely equivalent to a pure elliptically polarized (a monochromatic) wave because their Stokes parameters are the same. Summarizing, we have from Eqs. (2.47) and (2.48) that

$$I^2 \geqslant Q^2 + U^2 + V^2, \tag{2.49}$$

the equality holding for a completely elliptically polarized wave.

At the other extreme there is the completely *unpolarized* wave, where the phases between E_1 and E_2 maintain no permanent relation and where there is no preferred orientation in the x-y plane, so that $\langle \mathcal{E}_1^2\rangle = \langle \mathcal{E}_2^2\rangle$. In this case

$$Q = U = V = 0, \tag{2.50a}$$

or

$$Q^2+U^2+V^2=0. \tag{2.50b}$$

An important property of the Stokes parameters is that they are additive for a superposition of independent waves. By independent we mean that there are no permanent phase relations between the various waves, and that over the relevant time scales the relative phases can be assumed to be randomly and uniformly distributed from 0 to 2π. For a superposition of different waves, each having its own $E_1^{(k)}$ and $E_2^{(k)}$, $k=1, 2, 3\cdots$, we have

$$E_1=\sum_k E_1^{(k)}, \qquad E_2=\sum_l E_2^{(l)} \tag{2.51}$$

so that

$$\langle E_i E_j^* \rangle=\sum_k \sum_l \langle E_1^{(k)} E_2^{(l)*} \rangle=\sum_k \langle E_1^{(k)} E_2^{(k)*} \rangle. \tag{2.52}$$

Because of the random phases only terms with $k=l$ survive the averaging, as indicated. It follows that

$$I=\sum I^{(k)} \tag{2.53a}$$

$$Q=\sum Q^{(k)} \tag{2.53b}$$

$$U=\sum U^{(k)} \tag{2.53c}$$

$$V=\sum V^{(k)}, \tag{2.53d}$$

proving the additivity.

By the superposition principle, an arbitrary set of Stokes parameters can be represented as

$$\begin{pmatrix} I \\ Q \\ U \\ V \end{pmatrix} = \begin{pmatrix} I-\sqrt{Q^2+U^2+V^2} \\ 0 \\ 0 \\ 0 \end{pmatrix} + \begin{pmatrix} \sqrt{Q^2+U^2+V^2} \\ Q \\ U \\ V \end{pmatrix}. \tag{2.54}$$

The first term on the right represents the Stokes parameters of a completely unpolarized wave of intensity $I-\sqrt{Q^2+U^2+V^2}$ and the second represents the Stokes parameters of a completely (elliptically) polarized wave of intensity $\sqrt{Q^2+U^2+V^2}$, since it satisfies Eq. (2.42). Therefore an arbitrary wave can be regarded as the independent superposition of a completely polarized and a completely unpolarized wave. With this decomposition the meaning of the Stokes parameters for a quasi-monochromatic wave can be reduced to the meanings previously given for the completely polarized part plus that for the unpolarized part. Such a wave is therefore said to be *partially polarized.* The *degree of polarization* is defined in terms of this representation as the ratio of the intensity of the polarized part to the total intensity:

$$\Pi \equiv \frac{I_{pol}}{I} = \frac{\sqrt{Q^2+U^2+V^2}}{I}. \tag{2.55}$$

This is often given in terms of percentages.

A special case that appears frequently in applications is partial linear polarization, where $V=0$. Such radiation can be analyzed using a single linear polarizing filter (or dipole antenna), which picks out the component of the electric field in one direction. The measurement consists of rotating the filter until the maximum values of intensity are found. The maximum value I_{max} will occur when the filter is aligned with the plane of polarization (the x'-axis), and the minimum value will occur along in the direction perpendicular to it (the y'-axis). The unpolarized intensity only contributes one-half of its intensity to any given measurement, since the total is shared between any two perpendicular directions. Therefore, the maximum and minimum values of intensity are

$$I_{max} = \tfrac{1}{2} I_{unpol} + I_{pol}, \tag{2.56a}$$

$$I_{min} = \tfrac{1}{2} I_{unpol}, \tag{2.56b}$$

where $I_{unpol} = I - \sqrt{Q^2+U^2}$ and $I_{pol} = \sqrt{Q^2+U^2}$. From Eq. (2.55) we have, finally,

$$\Pi = \frac{I_{max} - I_{min}}{I_{max} + I_{min}}. \tag{2.57}$$

One should be cautioned that this formula applies only in cases in which the polarization is known to be of plane type. It will underestimate the true degree of polarization if circular or elliptical polarization is present.

2.5 ELECTROMAGNETIC POTENTIALS

Because of the form of Maxwell's equations, [cf. Eqs. (2.6)], especially the "internal equations," it is found that the **E** and **B** fields may be expressed completely in terms of a *scalar potential* $\phi(r,t)$ and a *vector potential* $\mathbf{A}(r,t)$. There are several reasons for wanting to do this: One scalar plus one vector is simpler than two vectors. Also, the equations determining ϕ and $\mathbf{A}$ are quite a bit simpler than Maxwell's equations for $\mathbf{E}$ and $\mathbf{B}$. Finally, the relativistic formulation of electromagnetic theory is simpler in terms of the potentials than in terms of the electric and magnetic fields.

From Maxwell's equation $\nabla\cdot\mathbf{B}=0$ it follows that $\mathbf{B}$ may be expressed as the curl of some vector field $\mathbf{A}$:

$$\mathbf{B}=\nabla\times\mathbf{A}. \tag{2.58}$$

The $\nabla\times\mathbf{E}$ equation can be written

$$\nabla\times\left(\mathbf{E}+\frac{1}{c}\frac{\partial\mathbf{A}}{\partial t}\right)=0. \tag{2.59}$$

It follows that $\mathbf{E}+\frac{1}{c}\partial\mathbf{A}/\partial t$ may be expressed as the gradient of some scalar field $-\phi$:

$$\mathbf{E}+\frac{1}{c}\frac{\partial\mathbf{A}}{\partial t}=-\nabla\phi,$$

$$\mathbf{E}=-\nabla\phi-\frac{1}{c}\frac{\partial\mathbf{A}}{\partial t}. \tag{2.60}$$

Two of Maxwell's equations have already been satisfied identically by virtue of the definitions of the potentials. The $\nabla\cdot\mathbf{E}$ equation can be written

$$\nabla^2\phi+\frac{1}{c}\frac{\partial}{\partial t}(\nabla\cdot\mathbf{A})=-4\pi\rho, \tag{2.61}$$

where we have used the *microscopic* form of Maxwell's equations ($\rho=\rho_{\text{free}}+\rho_{\text{bound}}$). Equation (2.61) may also be written in the form

$$\nabla^2\phi-\frac{1}{c^2}\frac{\partial^2\phi}{\partial t^2}+\frac{1}{c}\frac{\partial}{\partial t}\left(\nabla\cdot\mathbf{A}+\frac{1}{c}\frac{\partial\phi}{\partial t}\right)=-4\pi\rho. \tag{2.62}$$

The $\nabla\times\mathbf{H}$ equation can be written

$$\nabla\times(\nabla\times\mathbf{A})-\frac{1}{c}\frac{\partial}{\partial t}\left(-\nabla\phi-\frac{1}{c}\frac{\partial\mathbf{A}}{\partial t}\right)=\frac{4\pi}{c}\mathbf{j} \tag{2.63}$$

With the vector identity $\nabla\times(\nabla\times\mathbf{A})=-\nabla^2\mathbf{A}+\nabla(\nabla\cdot\mathbf{A})$ this becomes

$$\nabla^2\mathbf{A}-\frac{1}{c^2}\frac{\partial^2\mathbf{A}}{\partial t^2}-\nabla\left(\nabla\cdot\mathbf{A}+\frac{1}{c}\frac{\partial\phi}{\partial t}\right)=-\frac{4\pi}{c}\mathbf{j}. \tag{2.64}$$

The potentials are not uniquely determined by the conditions imposed above. For example, the addition to **A** of the gradient of an arbitrary scalar function ψ will leave **B** unchanged:

$$\mathbf{A} \rightarrow \mathbf{A} + \nabla\psi, \qquad \mathbf{B} \rightarrow \mathbf{B}.$$

The electric field will also be unchanged if at the same time ϕ is changed by

$$\phi \rightarrow \phi - \frac{1}{c}\frac{\partial\psi}{\partial t}, \qquad \mathbf{E} \rightarrow \mathbf{E}.$$

These alterations of **A** and ϕ are called *Gauge transformations*. Their value for our purposes lies in the possibility of choosing potentials in such a way to simplify the above equations. Note that since we have one free function, we can satisfy one scalar constraint equation. The most important choice made is a gauge for which the *Lorentz condition* is satisfied

$$\nabla \cdot \mathbf{A} + \frac{1}{c}\frac{\partial\phi}{\partial t} = 0. \tag{2.65}$$

The gauge corresponding to Eq. (2.65) is called the *Lorentz gauge*. With this gauge Eqs. (2.62) and (2.64) now become the following inhomogeneous equations:

$$\nabla^2\phi - \frac{1}{c^2}\frac{\partial^2\phi}{\partial t^2} = -4\pi\rho, \tag{2.66a}$$

$$\nabla^2\mathbf{A} - \frac{1}{c^2}\frac{\partial^2\mathbf{A}}{\partial t^2} = -\frac{4\pi}{c}\mathbf{j}. \tag{2.66b}$$

The solutions to Eqs. (2.66) may be written (see, e.g., Jackson 1975) as integrals over the sources:

$$\phi(\mathbf{r},t) = \int \frac{[\rho]\,d^3\mathbf{r}'}{|\mathbf{r}-\mathbf{r}'|}, \tag{2.67a}$$

$$\mathbf{A}(\mathbf{r},t) = \frac{1}{c}\int \frac{[\mathbf{j}]\,d^3\mathbf{r}'}{|\mathbf{r}-\mathbf{r}'|}. \tag{2.67b}$$

Equations (2.67) are the *retarded potentials*. The notation $[Q]$ means that Q is to be evaluated at the *retarded time*

$$[Q] \equiv Q\left(\mathbf{r}', t - \frac{1}{c}|\mathbf{r}-\mathbf{r}'|\right).$$

The retarded time refers to conditions at the point $\mathbf{r}'$ that existed at a time earlier than t by just the time required for light to travel between $\mathbf{r}$ and $\mathbf{r}'$. The interpretation is that information at point $\mathbf{r}'$ propagates at the speed of light, so that the potentials at point $\mathbf{r}$ can only be affected by conditions at point $\mathbf{r}'$ at such a retarded time. (A similar set of solutions with the advanced time $t+c^{-1}|\mathbf{r}-\mathbf{r}'|$ are also possible mathematically, but are ordinarily excluded on the physical grounds of causality.)

We now have a rather straightforward way of finding the electric and magnetic fields due to a given charge and current density: first, find the retarded potentials by means of the above integrals, and then determine $\mathbf{E}$ and $\mathbf{B}$ by their expressions in terms of the potentials. In the next chapter we determine the retarded potentials for a point charge in this way.

2.6 APPLICABILITY OF TRANSFER THEORY AND THE GEOMETRICAL OPTICS LIMIT

Following our discussion of waves, it is now possible to discuss more quantitatively the applicability of geometrical optics. In standard discussions of the propagation, or transfer, of radiation through matter, the specific intensity, with its associated concept of rays, is used as a fundamental variable. However, there are certain limitations imposed on transfer theory by the wave or quantum nature of light. For example, we defined specific intensity by the relation

$$dE = I_\nu \, dA \, d\Omega \, d\nu \, dt,$$

where dA, $d\Omega$, $d\nu$, and dt were presumed to be infinitesimal. However, dA and $d\Omega$ cannot both be made arbitrarily small because of the uncertainty principle for photons:

$$dx\,dp_x\,dy\,dp_y = p^2\,dA\,d\Omega \gtrsim h^2,$$

$$dA\,d\Omega \gtrsim \lambda^2. \tag{2.68}$$

As soon as the size of dA is of order of the square of the wavelength, the direction cannot be defined with any precision and the concept of rays breaks down.

There is another limitation on the sizes of dt and $d\nu$ because of the energy uncertainty principle

$$dE\,dt \gtrsim h,$$

$$d\nu\,dt \gtrsim 1. \tag{2.69}$$

For these reasons, when the wavelength of light is larger than atomic dimensions, as in the optical, we cannot describe the interaction of light on the atomic scale in terms of specific intensity. However, we may still regard transfer theory as a valid macroscopic theory, provided the absorption and emission properties are correctly calculated from electromagnetic theory or quantum theory.

A more precise, classical treatment of the validity of rays is known as the *eikonal approximation*. The essential features of this approach can be seen if we treat a scalar field rather than the vector electromagnetic fields. Rays are curves whose tangents at each point lie along the direction of propagation of the wave. Clearly, these rays are well defined only if the amplitude and direction of the wave is practically constant over a distance of a wavelength λ. This limit is called the *geometrical optics limit*. Let the wave be represented by a function $g(\mathbf{r},t)$ of the form

$$g(\mathbf{r},t)=a(\mathbf{r},t)e^{i\psi(\mathbf{r},t)}, \tag{2.70}$$

where $a(\mathbf{r},t)$ is the slowly varying *amplitude* and $\psi(\mathbf{r},t)$ is the rapidly varying phase. If a were strictly constant, then the local direction of propagation $\mathbf{k}$ of the wave (normal to the surfaces of constant phase ψ), is given by

$$\mathbf{k}=\nabla\psi, \tag{2.71a}$$

and the local frequency, ω, is given by

$$\omega=-\frac{\partial\psi}{\partial t}. \tag{2.71b}$$

The exact behavior of a and ψ is constrained by the *wave* equation for $g(\mathbf{r},t)$,

$$\nabla^2 g(\mathbf{r},t)-\frac{1}{c^2}\frac{\partial^2 g}{\partial t^2}=0,$$

or, substituting in Eq. (2.70) for $g(\mathbf{r},t)$,

$$\nabla^2 a-\frac{1}{c^2}\frac{\partial^2 a}{\partial t^2}+ia\left(\nabla^2\psi-\frac{1}{c^2}\frac{\partial^2\psi}{\partial t^2}\right)+2i\left(\nabla a\cdot\nabla\psi-\frac{1}{c^2}\frac{\partial\psi}{\partial t}\frac{\partial a}{\partial t}\right)$$
$$-a(\nabla\psi)^2+\frac{a}{c^2}\left(\frac{\partial\psi}{\partial t}\right)^2=0. \tag{2.72}$$

The geometrical optics limit can now be made precise. If

$$\frac{1}{a}|\nabla a| \ll |\nabla \psi|, \qquad \frac{1}{a}\left|\frac{\partial a}{\partial t}\right| \ll \left|\frac{\partial \psi}{\partial t}\right|,$$

$$|\nabla^2 \psi| \ll |\nabla \psi|^2, \qquad \left|\frac{\partial^2 \psi}{\partial t^2}\right| \ll \left|\frac{\partial \psi}{\partial t}\right|^2,$$

$$\frac{1}{a}|\nabla^2 a| \ll |\nabla \psi|^2,$$

then the above equation reduces to

$$(\nabla \psi)^2 - \frac{1}{c^2}\left(\frac{\partial \psi}{\partial t}\right)^2 = 0, \tag{2.73}$$

which is the eikonal equation. If Eqs. (2.71) are substituted for the gradients of ψ, we obtain

$$|\mathbf{k}|^2 - \frac{\omega^2}{c^2} = 0,$$

which will be recognized [cf. Eq. (2.20a)] as the relationship between wave number and frequency of a plane wave.

PROBLEMS

2.1—Two oscillating quantities $A(t)$ and $B(t)$ are represented as the real parts of the complex quantities $\mathscr{A}e^{-i\omega t}$ and $\mathscr{B}e^{-i\omega t}$. Show that the time average of AB is given by

$$\langle AB \rangle = \tfrac{1}{2}\,\mathrm{Re}(\mathscr{A}^*\mathscr{B}) = \tfrac{1}{2}\,\mathrm{Re}(\mathscr{A}\mathscr{B}^*).$$

2.2—In certain cases the process of absorption of radiation can be treated by means of the macroscopic Maxwell equations. For example, suppose we have a conducting medium, so that the current density $\mathbf{j}$ is related to the electric field $\mathbf{E}$ by *Ohm's law*:

$$\mathbf{j} = \sigma \mathbf{E},$$

where σ is the conductivity (cgs unit $=$ sec^{-1}). Investigate the propagation of electromagnetic waves in such a medium and show that:

a. The wave vector **k** is complex

$$\mathbf{k}^2 = \frac{\omega^2 m^2}{c^2},$$

where m is the *complex index of refraction*, defined by

$$m^2 = \mu\epsilon\left(1 + \frac{4\pi i\sigma}{\omega\epsilon}\right).$$

b. The waves are attenuated as they propagate, corresponding to an absorption coefficient

$$\alpha_\nu = \frac{2\omega}{c} Im(m).$$

(Note: In some literature, minus signs appear in these formulas. This is because the wave is often taken to be $\exp(-i\mathbf{k}\cdot\mathbf{r} + i\omega t)$ rather than the $\exp(i\mathbf{k}\cdot\mathbf{r} - i\omega t)$ chosen here.)

2.3—This problem is meant to deduce the momentum and angular momentum properties of radiation and does not necessarily represent any real physical system of interest. Consider a charge Q in a viscous medium where the viscous force is proportional to velocity: $\mathbf{F}_{\text{visc}} = -\beta\mathbf{v}$. Suppose a *circularly polarized* wave passes through the medium. The equation of motion of the change is

$$m\frac{d\mathbf{v}}{dt} = \mathbf{F}_{\text{visc}} + \mathbf{F}_{\text{Lorentz}}.$$

We assume that the terms on the right dominate the inertial term on the left, so that approximately

$$0 = \mathbf{F}_{\text{visc}} + \mathbf{F}_{\text{Lorentz}}.$$

Let the frequency of the wave be ω and the strength of the electric field be E.

a. Show that to lowest order (neglecting the magnetic force) the charge moves on a circle in a plane normal to the direction of propagation of the wave with speed QE/β and with radius $QE/\beta\omega$.

b. Show that the power transmitted to the fluid by the wave is Q^2E^2/β.

c. By considering the small magnetic force acting on the particle show that the momentum per unit time (force) given to the fluid by the wave is in the direction of propagation and has the magnitude $Q^2E^2/\beta c$.

d. Show that the angular momentum per unit time (torque) given to the fluid by the wave is in the direction of propagation and has magnitude $\pm Q^2E^2/\beta\omega$, where $(\pm)$ is for $\begin{pmatrix}\text{left}\\\text{right}\end{pmatrix}$ circular polarization.

e. Show that the absorption cross section of the charge is $4\pi Q^2/\beta c$.

f. If we now regard the radiation to be composed of circularly polarized photons of energy $E_\gamma=\hbar\omega$, show that these results imply that the photon has momentum $p=\hbar k=h/\lambda=E_\gamma/c$ and has angular momentum $J=\pm\hbar$ along the direction of propagation.

g. Repeat this problem with appropriate modifications for a linearly polarized wave.

2.4—Show that Maxwell's equations before Maxwell, that is, without the "displacement current" term $c^{-1}\partial\mathbf{D}/\partial t$, unacceptably constrained the sources of the field and also did not permit the existence of waves.

REFERENCES

Born, M. and Wolf, E., 1964, *Principles of Optics* (Macmillan, New York).
Chandrasekhar, S., 1960, *Radiative Transfer*, (Dover, New York).
Jackson, J. D., 1975, *Classical Electrodynamics*, (J. Wiley, New York).

3

RADIATION FROM MOVING CHARGES

3.1 RETARDED POTENTIALS OF SINGLE MOVING CHARGES: THE LIÉNARD-WIECHART POTENTIALS

Consider a particle of charge q that moves along a trajectory $\mathbf{r}=\mathbf{r}_0(t)$. Its velocity at any time is then $\mathbf{u}(t)=\dot{\mathbf{r}}_0(t)$. The charge and current densities are given by

$$\rho(\mathbf{r},t)=q\delta(\mathbf{r}-\mathbf{r}_0(t)), \tag{3.1a}$$

$$\mathbf{j}(\mathbf{r},t)=q\mathbf{u}(t)\delta(\mathbf{r}-\mathbf{r}_0(t)). \tag{3.1b}$$

The δ-function has the property of localizing the charge and current; we also obtain the proper total charge and current by integrating over volume:

$$q=\int\rho(\mathbf{r},t)d^3\mathbf{r},$$

$$q\mathbf{u}=\int\mathbf{j}(\mathbf{r},t)d^3\mathbf{r}.$$

Let us calculate the retarded potentials [Eq. (2.67)] due to these charge and current densities. We use the scalar potential as an example:

$$\phi(\mathbf{r},t)=\int d^3\mathbf{r}'\int dt'\frac{\rho(\mathbf{r}',t')}{|\mathbf{r}-\mathbf{r}'|}\delta(t'-t+|\mathbf{r}-\mathbf{r}'|/c), \tag{3.2}$$

using the property of the δ-function. Substitution of Eq. (3.1a) for the charge density and integration over $\mathbf{r}'$ yields

$$\phi(\mathbf{r},t)=q\int\delta(t'-t+|\mathbf{r}-\mathbf{r}_0(t')|/c)\frac{dt'}{|\mathbf{r}-\mathbf{r}_0(t')|}.$$

This is now an integral over the single variable t'. We now introduce the notations

$$\mathbf{R}(t')=\mathbf{r}-\mathbf{r}_0(t'), \qquad R(t')=|\mathbf{R}(t')|. \tag{3.3}$$

We then have

$$\phi(\mathbf{r},t)=q\int R^{-1}(t')\delta(t'-t+R(t')/c)dt', \tag{3.4a}$$

$$\mathbf{A}(\mathbf{r},t)=\frac{q}{c}\int\mathbf{u}(t')R^{-1}(t')\delta(t'-t+R(t')/c)dt', \tag{3.4b}$$

where we have performed the identical integrations for $\mathbf{A}$. Equations (3.4) are useful forms for the potentials, but they may be simplified still further. Note that the argument of the δ-function vanishes for a value of $t'=t_{\text{ret}}$ given by

$$c(t-t_{\text{ret}})=R(t_{\text{ret}}). \tag{3.5}$$

Let us change variables from t' to $t''=t'-t+[R(t')/c]$, which implies that

$$dt''=dt'+\frac{1}{c}\dot{R}(t')dt'.$$

Since $R^2(t')=\mathbf{R}^2(t')$, it follows that $2R(t')\dot{R}(t')=-2\mathbf{R}(t')\cdot\mathbf{u}(t')$, where $\dot{\mathbf{R}}(t')=-\mathbf{u}(t')$. We also define the unit vector $\mathbf{n}$ by

$$\mathbf{n}=\frac{\mathbf{R}}{R}$$

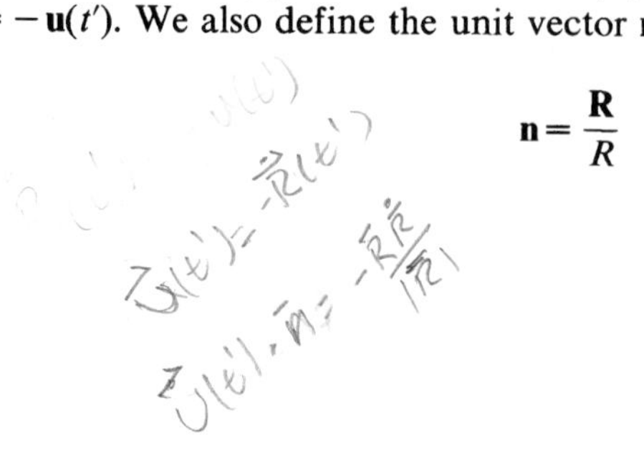

Finally, we obtain

$$dt''=\left[1-\frac{1}{c}\mathbf{n}(t')\cdot\mathbf{u}(t')\right]dt',$$

$$\phi(\mathbf{r},t)=q\int R^{-1}(t')\left[1-\frac{1}{c}\mathbf{n}(t')\cdot\mathbf{u}(t')\right]^{-1}\delta(t'')dt''.$$

Now the integration over the δ-function can be performed by setting $t''=0$, or equivalently by setting $t'=t_{\text{ret}}$. This yields

$$\phi(\mathbf{r},t)=\frac{q}{\kappa(t_{\text{ret}})R(t_{\text{ret}})}$$

where we have used the notation

$$\kappa(t')=1-\frac{1}{c}\mathbf{n}(t')\cdot\mathbf{u}(t'). \tag{3.6}$$

Then, with the brackets denoting retarded times, we have

$$\phi=\left[\frac{q}{\kappa R}\right] \tag{3.7a}$$

$$\mathbf{A}=\left[\frac{q\mathbf{u}}{c\kappa R}\right]. \tag{3.7b}$$

These are called the *Liénard–Wiechart potentials*. These potentials differ from those of static electromagnetic theory in two ways: First, there is the factor $\kappa=1-(\mathbf{n}\cdot\mathbf{u}/c)$. This factor becomes very important at velocities close to that of light, where it tends to concentrate the potentials into a narrow cone about the particle velocity. It is related to the *beaming effect* found in the Lorentz transformation of photon direction of propagation. (See Chapter 4.)

The second difference is that the quantities are all to be evaluated at the retarded time t_{ret}. We have already discussed the meaning of this. The major consequence of retardation is that it makes it possible for a particle to radiate. The potentials roughly fall off as $1/r$ so that differentiation to find the fields would give a $1/r^2$ decrease if this differentiation acted solely on the $1/r$ factor. As we show in the following section retardation allows an implicit dependence on position to occur via the definition of retarded time, and differentiation with respect to this dependence carries the $1/r$ behavior of the potentials into the fields themselves. We have seen that this allows radiation energy to flow to infinite distances.

3.2 THE VELOCITY AND RADIATION FIELDS

The differentiations of the potentials to obtain the fields are straightforward but lengthy, and we omit details (see Jackson, §14.1). The results are as follows: If we want the fields at point r at time t we first must determine the retarded position and time of the particle r_{ret} and t_{ret}. At this time the particle has velocity $\mathbf{u}=\dot{\mathbf{r}}_0(t_{\text{ret}})$ and acceleration $\dot{\mathbf{u}}=\ddot{\mathbf{r}}_0(t_{\text{ret}})$. We introduce the notation

$$\boldsymbol{\beta}\equiv\frac{\mathbf{u}}{c}, \qquad \kappa\equiv 1-\mathbf{n}\cdot\boldsymbol{\beta}. \tag{3.8}$$

Then the fields are

$$\mathbf{E}(\mathbf{r},t)=q\left[\frac{(\mathbf{n}-\boldsymbol{\beta})(1-\beta^2)}{\kappa^3R^2}\right]+\frac{q}{c}\left[\frac{\mathbf{n}}{\kappa^3R}\times\{(\mathbf{n}-\boldsymbol{\beta})\times\dot{\boldsymbol{\beta}}\}\right], \tag{3.9a}$$

$$\mathbf{B}(\mathbf{r},t)=\left[\mathbf{n}\times\mathbf{E}(\mathbf{r},t)\right]. \tag{3.9b}$$

Note from Figure 3.1 that at time t the particle is at some point further along its path, but only the conditions at the retarded time determine the fields at point r at time t. The magnetic field is always perpendicular to both $\mathbf{E}$ and $\mathbf{n}$.

The electric field appears above as composed of two terms: the first, the *velocity field*, falls off as $1/R^2$ and is just the generalization of the Coulomb law to moving particles: for $u \ll c$ this becomes precisely Coulomb's law. When the particle moves with constant velocity it is only this term that contributes to the fields. A remarkable fact in this case is that the electric field always points along the line toward the *current* position of the particle. This follows from the fact that the displacement to

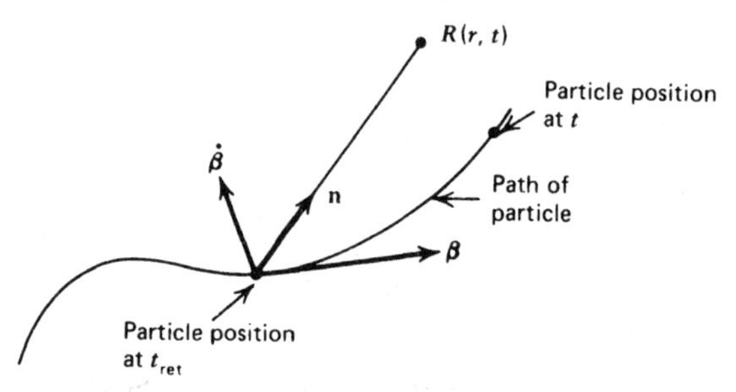

***Figure 3.1** Geometry for calculation of the radiation field at R from the position of the radiating particle at the retarded time.*

the field point from the retarded point is $\mathbf{n}c\bar{t}$, where $\bar{t}=t-t_{\text{ret}}$ is the light travel time. In the same time the particle undergoes a displacement $\boldsymbol{\beta}c\bar{t}$. The displacement between the field point and the current position is thus $(\mathbf{n}-\boldsymbol{\beta})c\bar{t}$, which is seen to be the direction of the velocity field in Eq. (3.9a).

The second term, the *acceleration field*, falls off as $1/R$, is proportional to the particle's acceleration and is perpendicular to $\mathbf{n}$. This electric field, along with the corresponding magnetic field, constitutes the *radiation field*:

$$\mathbf{E}_{\text{rad}}(\mathbf{r},t)=\frac{q}{c}\left[\frac{\mathbf{n}}{\kappa^3 R}\times\{(\mathbf{n}-\boldsymbol{\beta})\times\dot{\boldsymbol{\beta}}\}\right], \tag{3.10a}$$

$$\mathbf{B}_{\text{rad}}(\mathbf{r},t)=[\mathbf{n}\times\mathbf{E}_{\text{rad}}]. \tag{3.10b}$$

Note that $\mathbf{E}$, $\mathbf{B}$ and $\mathbf{n}$ form a right-hand triad of mutually perpendicular vectors, and that $|\mathbf{E}_{\text{rad}}|=|\mathbf{B}_{\text{rad}}|$. These properties are consistent with the radiation solutions of the source-free Maxwell equations.

Figure 3.2 demonstrates geometrically how an acceleration can give rise to a transverse field that decreases as $1/R$, rather than the $1/R^2$ decrease of a nonaccelerated charge. The particle originally moved with constant velocity along the x-axis and stopped at $x=0$ at time $t=0$. At $t=1$ the field outside of a radius c is radial and points to the position where the particle would have been had there been no deceleration, since no information of the latter has yet propagated to this distance. On the other hand, the field inside radius c is "informed" and is radially directed to the true position of the particle. There is only one way these two fields can be

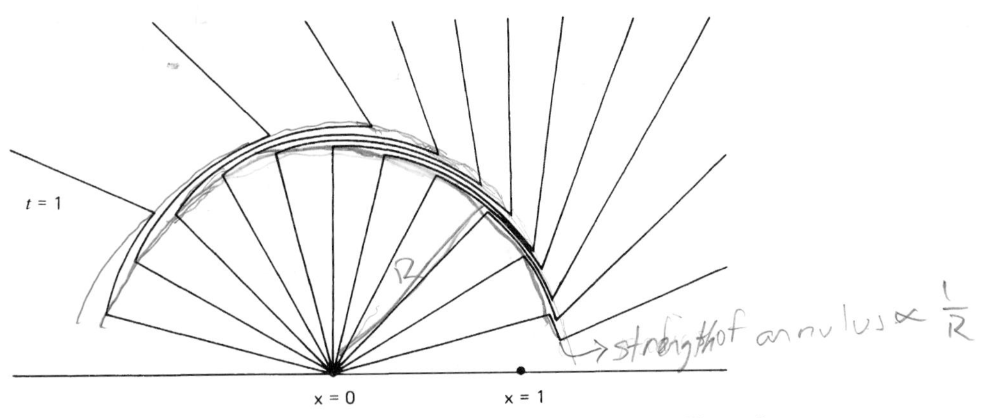

***Figure 3.2** Graphical demonstration of the $1/R$ acceleration field. Charged particle moving at uniform velocity in positive x direction is stopped at $x=0$ and $t=0$.*

connected that is consistent with Gauss's law and flux conservation: it is graphically illustrated in the figure. It can be seen that a transition zone (whose radial thickness is the time interval over which the deceleration occurs) propagates outward. In this zone the field is almost transverse and is much stronger (closely packed flux lines) than the radial fields outside the zone. Further geometrical arguments can be used to show that the field intensity in this zone is proportional to $1/ct=1/R$. If one looks at an annular ring centered on and perpendicular to the line of travel, containing all the flux lines in the passing wavefront, then the thickness of the ring is constant (light travel distance during acceleration time), and the radius of the ring varies as R. Since the total number of flux lines is conserved, the strength of the field varies as $1/R$.

A useful result is obtained by considering the energy per unit frequency per unit solid angle corresponding to the radiation field of a single particle [cf. Eqs. (3.10a) and (2.33)]:

$$\frac{dW}{d\omega\, d\Omega}=\frac{c}{4\pi^2}\left|\int [R\mathbf{E}(t)]e^{i\omega t}dt\right|^2 \tag{3.11a}$$

$$=\frac{q^2}{4\pi^2 c}\left|\int \left[\mathbf{n}\times\{(\mathbf{n}-\boldsymbol{\beta})\times\dot{\boldsymbol{\beta}}\}\kappa^{-3}\right]e^{i\omega t}dt\right|^2 \tag{3.11b}$$

where the expression in the brackets is evaluated at the retarded time $t'=t-R(t')/c$. Now, changing variables from t to t' in the integral, $dt=\kappa\, dt'$, and using the expansion $R(t')\approx|\mathbf{r}|-\mathbf{n}\cdot\mathbf{r}_0$, valid for $|\mathbf{r}_0|\ll|\mathbf{r}|$, we have

$$\frac{dW}{d\omega\, d\Omega}=\frac{q^2}{4\pi^2 c}\left|\int \mathbf{n}\times\{(\mathbf{n}-\boldsymbol{\beta})\times\dot{\boldsymbol{\beta}}\}\kappa^{-2}\exp\left[i\omega(t'-\mathbf{n}\cdot\mathbf{r}_0(t')/c)\right]dt'\right|^2. \tag{3.12}$$

Finally, we may integrate Eq. (3.12) by parts to obtain an expression involving only $\boldsymbol{\beta}$. Using the identity $\mathbf{n}\times\{(\mathbf{n}-\boldsymbol{\beta})\times\dot{\boldsymbol{\beta}}\}\kappa^{-2}=d/dt'\kappa^{-1}\mathbf{n}\times(\mathbf{n}\times\boldsymbol{\beta})$, Eq. (3.12) becomes

$$\frac{dW}{d\omega\, d\Omega}=(q^2\omega^2/4\pi^2 c)\left|\int \mathbf{n}\times(\mathbf{n}\times\boldsymbol{\beta})\exp\left[i\omega(t'-\mathbf{n}\cdot\mathbf{r}_0(t')/c)\right]dt'\right|^2. \tag{3.13}$$

3.3 RADIATION FROM NONRELATIVISTIC SYSTEMS OF PARTICLES

Using the above formulas we could discuss many radiation processes involving moving charges, including particles moving relativistically. However, the interpretation of many of these results would be made easier after the section on special relativity. Therefore, for the moment, we shall specialize the discussion to nonrelativistic particles, that is, the case

$$|\beta| = \frac{u}{c} \ll 1.$$

Let us compare the order of magnitude of the two fields $E_{\rm rad}$ and $E_{\rm vel}$: taking the leading terms we obtain

$$\frac{E_{\rm rad}}{E_{\rm vel}} \sim \frac{R\dot{u}}{c^2}. \tag{3.14a}$$

Now, if we focus on the particular Fourier component of frequency ν, or if the particle has a characteristic frequency of oscillation ν, then $\dot{u} \sim u\nu$, and Eq. (3.14a) becomes

$$\frac{E_{\rm rad}}{E_{\rm vel}} \sim \frac{Ru\nu}{c^2} = \frac{u}{c}\frac{R}{\lambda}. \tag{3.14b}$$

Thus for field points inside the "near zone", $R \lesssim \lambda$, the velocity field is stronger than the radiation field by a factor $\gtrsim c/u$; whereas for field points sufficiently far in the "far zone," $R \gg \lambda(c/u)$, the radiation field dominates and increases its domination linearly with R.

Larmor's Formula

When $\beta \ll 1$ we can simplify equations (3.10) to

$$\mathbf{E}_{\rm rad} = \left[(q/Rc^2)\mathbf{n} \times (\mathbf{n} \times \dot{\mathbf{u}})\right] \tag{3.15a}$$

$$\mathbf{B}_{\rm rad} = \left[\mathbf{n} \times \mathbf{E}_{\rm rad}\right]. \tag{3.15b}$$

This is illustrated in Fig. 3.3, which has been drawn in the plane of $\mathbf{n}$ and $\dot{\mathbf{u}}$. We note that $\mathbf{E}_{\rm rad}$ is also in this plane in the orientation indicated, and $\mathbf{B}_{\rm rad}$ is into the plane of the diagram. The magnitudes of $\mathbf{E}_{\rm rad}$ and $\mathbf{B}_{\rm rad}$ are

$$|\mathbf{E}_{\rm rad}| = |\mathbf{B}_{\rm rad}| = \frac{q\dot{u}}{Rc^2}\sin\Theta. \tag{3.16}$$

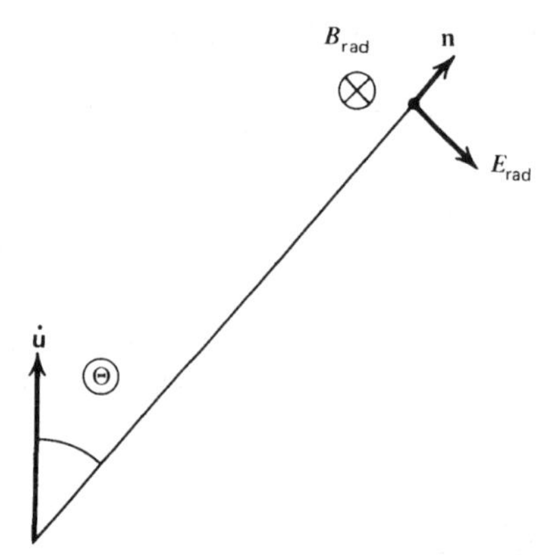

***Figure 3.3** Electric and magnetic radiation field configurations for a slowly moving particle. The direction of $\mathbf{B}_{rad}$ is into the page.*

The Poynting vector is in the direction of **n** and has the magnitude

$$S = \frac{c}{4\pi} E_{\text{rad}}^2 = \frac{c}{4\pi} \frac{q^2 \dot{u}^2}{R^2 c^4} \sin^2 \Theta. \tag{3.17}$$

This corresponds to an outward flow of energy, along the direction **n**. We can put this into the form of an emission coefficient. The energy dW emitted per unit time into solid angle $d\Omega$ about **n** can be evaluated by multiplying the Poynting vector (erg s^{-1} cm^{-2}) by the area $dA = R^2 d\Omega$ represented by Ω at the field point:

$$\frac{dW}{dt\, d\Omega} = \frac{q^2 \dot{u}^2}{4\pi c^3} \sin^2 \Theta. \tag{3.18}$$

We may obtain the total power emitted into all angles by integrating this over solid angles:

$$P = \frac{dW}{dt} = \frac{q^2 \dot{u}^2}{4\pi c^3} \int \sin^2 \Theta \, d\Omega$$

$$= \frac{q^2 \dot{u}^2}{2c^3} \int_{-1}^{1} (1 - \mu^2)\, d\mu.$$

Thus we have *Larmor's formula* for emission from a single accelerated charge q:

$$P = \frac{2q^2 \dot{u}^2}{3c^3}. \tag{3.19}$$

There are several points to notice about Eqs. (3.18) and (3.19):

1. The power emitted is proportional to the square of the charge and the square of the acceleration.
2. We have the characteristic dipole pattern $\propto \sin^2\Theta$: no radiation is emitted along the direction of acceleration, and the maximum is emitted perpendicular to acceleration.
3. The instantaneous direction of $\mathbf{E}_{\text{rad}}$ is determined by $\dot{\mathbf{u}}$ and $\mathbf{n}$. If the particle accelerates along a line, the radiation will be 100% linearly polarized in the plane of $\dot{\mathbf{u}}$ and $\mathbf{n}$.

The dipole approximation

When there are many particles with positions $\mathbf{r}_i$, velocities $\mathbf{u}_i$, and charges q_i, $i=1,2\ldots N$, we can find the radiation field at large distances by simply adding together the $\mathbf{E}_{\text{rad}}$ from each particle. However, there is a complication here, because the above expressions for the radiation fields refer to conditions at *retarded times*, and these retarded times will differ for each particle. Another way of stating the complication is that we must keep track of the phase relations between the various pieces of the radiating system introduced by retardation.

There are situations, however, in which it is possible to ignore this difficulty. Let the typical size of the system be L, and let the typical time scale for changes within the system be τ. If τ is much longer than the time it takes light to travel a distance L, $\tau \gg L/c$, then the differences in retarded time across the source are negligible. We may also characterize τ as the time scale over which significant changes in the radiation field $\mathbf{E}_{\text{rad}}$ occur, and this in turn determines the typical characteristic frequency of the emitted radiation. Calling this frequency ν, we write

$$\nu \approx \frac{1}{\tau}.$$

Combining this with the above we obtain

$$\frac{c}{\nu} \gg L,$$

or

$$\lambda \gg L, \tag{3.20}$$

that is, the differences in retarded times can be ignored when the size of the system is small compared to a wavelength.

We may also characterize τ as the time a particle takes to change its motion substantially. Letting l be a characteristic scale of the particle's orbit and u be a typical velocity, then $\tau \sim l/u$. The condition $\tau \gg L/c$ then implies $u/c \ll l/L$. But since $l < L$, this is simply equivalent to the nonrelativistic condition

$$u \ll c.$$

We may therefore consistently use the nonrelativistic form of the radiation fields for these problems. With the above conditions met we can write

$$\mathbf{E}_{\text{rad}} = \sum_i \frac{q_i}{c^2} \frac{\mathbf{n} \times (\mathbf{n} \times \dot{\mathbf{u}}_i)}{R_i}. \tag{3.21}$$

Let R_0 be the distance from some point in the system to the field point (see Fig. 3.4). Since the differences in the actual R_i are negligible as $R_0 \to \infty$, we have

$$\mathbf{E}_{\text{rad}} = \frac{\mathbf{n} \times (\mathbf{n} \times \ddot{\mathbf{d}})}{c^2 R_0}, \tag{3.22a}$$

where the *dipole moment* is

$$\mathbf{d} = \sum_i q_i \mathbf{r}_i. \tag{3.22b}$$

The right-hand side of Eqs. (3.22) must still be evaluated at a retarded time, but this time can be evaluated using any point within the region, say, the point used to define R_0.

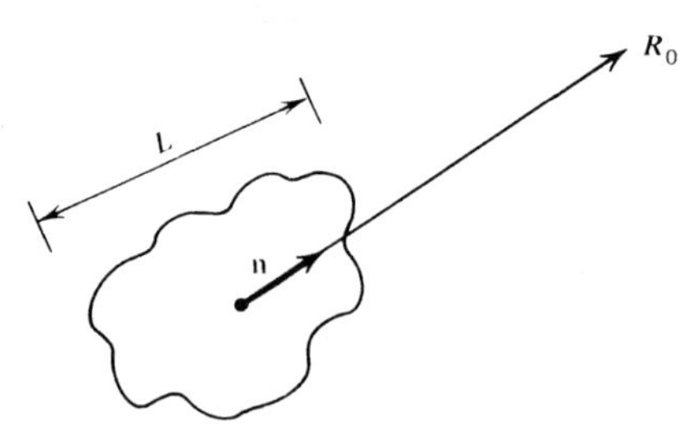

***Figure 3.4** Radiation from a medium of size L.*

As before, we find

$$\frac{dP}{d\Omega}=\frac{\ddot{\mathbf{d}}^2}{4\pi c^3}\sin^2\Theta, \tag{3.23a}$$

$$P=\frac{2\ddot{\mathbf{d}}^2}{3c^3}. \tag{3.23b}$$

This is called the *dipole approximation* and is a generalization of the formulas [Eqs. (3.18) and (3.19)] for a single nonrelativistic particle. The instantaneous polarization of **E** lies in the plane of $\ddot{\mathbf{d}}$ and **n** (see Fig. 3.5).

As an application of the preceding analysis, let us consider the spectrum of radiation in the dipole approximation. For simplicity we assume that **d** always lies in a single direction. Then from Eq. (3.22a), we have

$$E(t)=\ddot{d}(t)\frac{\sin\Theta}{c^2R_0}, \tag{3.24}$$

where $E(t)$ and $d(t)$ are the magnitudes of $\mathbf{E}(t)$ and $\mathbf{d}(t)$, respectively. The

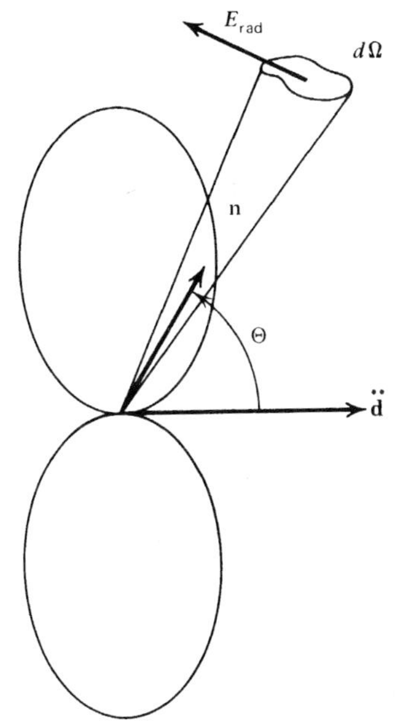

Figure 3.5 ***Geometry and emission pattern for dipole radiation.***

Fourier transform of $d(t)$ can be defined so that

$$d(t)=\int_{-\infty}^{\infty} e^{-i\omega t}\hat{d}(\omega)\,d\omega.$$

Then we have the relations

$$\ddot{d}(t)=-\int_{-\infty}^{\infty}\omega^2\hat{d}(\omega)e^{-i\omega t}\,d\omega, \tag{3.25a}$$

$$\hat{E}(\omega)=-\frac{1}{c^2R_0}\omega^2\hat{d}(\omega)\sin\Theta. \tag{3.25b}$$

For the energy per unit solid angle per frequency range and for the total energy per frequency range we have, using Eqs. (2.33), (3.25), and $dA=R_0^2\,d\Omega$,

$$\frac{dW}{d\omega\,d\Omega}=\frac{1}{c^3}\omega^4|\hat{d}(\omega)|^2\sin^2\Theta, \tag{3.26a}$$

$$\frac{dW}{d\omega}=\frac{8\pi\omega^4}{3c^3}|\hat{d}(\omega)|^2. \tag{3.26b}$$

These formulas describe an interesting property of dipole radiation, namely, that the spectrum of the emitted radiation is related directly to the frequencies of oscillation of the dipole moment. This property is not true for particles with relativistic velocities.

The general multipole expansion

In the above treatment of the dipole approximation we have argued only qualitatively. We would like to be slightly more explicit and indicate the features of the general case. Since **E** and **B** are simply related well outside of the source, we may consider the vector potential **A** to contain all of the necessary information. Consider a Fourier analysis of the sources and fields [cf. Eq. (2.3)]:

$$\mathbf{j}_\omega(\mathbf{r})=\int \mathbf{j}(\mathbf{r},t)e^{i\omega t}\,dt, \tag{3.27a}$$

$$\mathbf{A}_\omega(\mathbf{r})=\int \mathbf{A}(\mathbf{r},t)e^{i\omega t}\,dt. \tag{3.27b}$$

Then, using the equation analogous to Eq. (3.2) for the vector potential

$$\mathbf{A}(\mathbf{r},t)=\frac{1}{c}\int d^3\mathbf{r}'\int dt'\frac{\mathbf{j}(\mathbf{r}',t')}{|\mathbf{r}-\mathbf{r}'|}\delta(t'-t+|\mathbf{r}-\mathbf{r}'|/c),$$

and taking the Fourier transform of this equation, using Eqs. (3.27), we obtain

$$\mathbf{A}_{\omega}(\mathbf{r})=\frac{1}{c}\int\frac{\mathbf{j}_{\omega}(\mathbf{r}')}{|\mathbf{r}-\mathbf{r}'|}e^{ik|\mathbf{r}-\mathbf{r}'|}d^3\mathbf{r}', \tag{3.28}$$

where $k\equiv\omega/c$. Note that our equations now relate single Fourier components of $\mathbf{j}$ and $\mathbf{A}$.

Let us choose an origin of coordinates inside the source of size L. Then, at field points such that $r\gg L$, we have the approximation

$$|\mathbf{r}-\mathbf{r}'|\approx r-\mathbf{n}\cdot\mathbf{r}', \tag{3.29}$$

where $\mathbf{n}$ points toward the field point $\mathbf{r}$ and where $r\equiv|\mathbf{r}|$. Substituting Eq. (3.29) into (3.28), we obtain

$$\mathbf{A}_{\omega}(\mathbf{r})\approx(e^{ikr}/cr)\int\mathbf{j}_{\omega}(\mathbf{r}')e^{-ik\mathbf{n}\cdot\mathbf{r}'}d^3\mathbf{r}'. \tag{3.30}$$

The factor $\exp(ikr)$ outside the integral expresses the effect of retardation from the source as a whole. The factor $\exp(-ik\mathbf{n}\cdot\mathbf{r}')$ inside the integral expresses the *relative* retardation of each element of the source. In our slow-motion approximation, $kL\ll 1$. Thus, expanding the exponential in the integral:

$$\mathbf{A}_{\omega}(\mathbf{r})=\frac{e^{ikr}}{cr}\sum_{n=0}^{\infty}\frac{1}{n!}\int\mathbf{j}_{\omega}(\mathbf{r}')(-ik\mathbf{n}\cdot\mathbf{r}')^n d^3\mathbf{r}'. \tag{3.31}$$

Equation (3.31) is clearly an expansion in the small dimensionless parameter $kL=2\pi L/\lambda$. The *dipole approximation* results from taking just the first term in the expansion ($n=0$):

$$\mathbf{A}_{\omega}(\mathbf{r})\big|_{\text{dipole}}=\frac{e^{ikr}}{cr}\int\mathbf{j}_{\omega}(\mathbf{r}')d^3\mathbf{r}'. \tag{3.32}$$

The *quadrupole* term is the second term in the expansion ($n=1$):

$$\mathbf{A}_{\omega}(\mathbf{r})\big|_{\text{quad}}=\frac{-ike^{ikr}}{cr}\int\mathbf{j}_{\omega}(\mathbf{r}')(\mathbf{n}\cdot\mathbf{r}')d^3\mathbf{r}'. \tag{3.33}$$

Although it is true that the frequencies present in the vector potential (and hence in the radiation) are identical to those in the current density, it should be pointed out that these frequencies may differ from the frequencies of particle orbits in the source. For example, in the case of a particle orbiting in a circle with angular frequency ω_0, the function $\mathbf{j}_\omega(\mathbf{r})$ actually contains frequencies not only at ω_0 but also at all harmonics $2\omega_0$, $3\omega_0 \ldots$. In the dipole approximation only ω_0 contributes, in the quadrupole approximation only $2\omega_0$ contributes, and so on (see problem 3.7).

3.4 THOMSON SCATTERING (ELECTRON SCATTERING)

An important application of the dipole formula is to the process in which a free charge radiates in response to an incident electromagnetic wave. If the charge oscillates at nonrelativistic velocities, $v \ll c$, then we may neglect magnetic forces, since $E = B$ for an electromagnetic wave. Thus the force due to a *linearly polarized wave* is

$$\mathbf{F} = e\boldsymbol{\epsilon} E_0 \sin \omega_0 t, \tag{3.34}$$

where e is the charge and $\boldsymbol{\epsilon}$ is the $\mathbf{E}$-field direction. (See Fig. 3.6.) From Eq. (3.34), we have

$$m\ddot{\mathbf{r}} = e\boldsymbol{\epsilon} E_0 \sin \omega_0 t.$$

In terms of the dipole moment, $\mathbf{d} = e\mathbf{r}$, we have

$$\ddot{\mathbf{d}} = \frac{e^2 E_0}{m} \boldsymbol{\epsilon} \sin \omega_0 t,$$

$$\mathbf{d} = -\left(\frac{e^2 E_0}{m\omega_0^2}\right) \boldsymbol{\epsilon} \sin \omega_0 t,$$

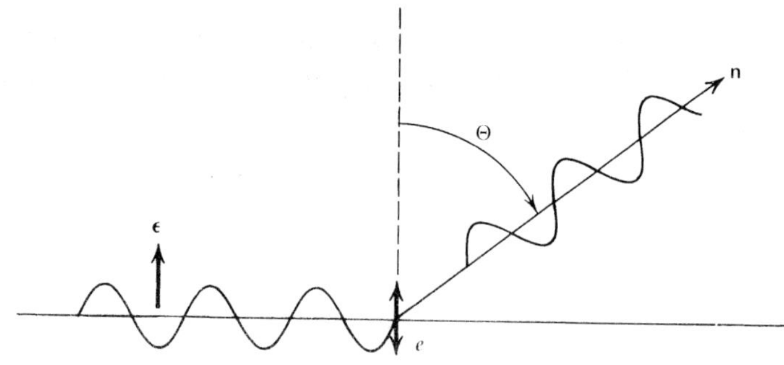

Figure 3.6 Scattering of polarized radiation by a charged particle.

which describes an oscillating dipole of amplitude

$$\mathbf{d}_0 = \frac{e^2 E_0}{m\omega_0^2}\boldsymbol{\epsilon}.$$

From our previous results of Eqs. (3.23), we can write the time-averaged power as

$$\frac{dP}{d\Omega} = \frac{e^4 E_0^2}{8\pi m^2 c^3}\sin^2\Theta, \tag{3.35a}$$

$$P = \frac{e^4 E_0^2}{3m^2 c^3}, \tag{3.35b}$$

where the time average of $\sin^2\omega_0 t$ gives a factor $\frac{1}{2}$. Note that the incident flux is $\langle S\rangle = (c/8\pi)E_0^2$. Defining the *differential cross section* $d\sigma$ for scattering into $d\Omega$ we have

$$\frac{dP}{d\Omega} = \langle S\rangle\frac{d\sigma}{d\Omega} = \frac{cE_0^2}{8\pi}\frac{d\sigma}{d\Omega}. \tag{3.36}$$

Therefore, we have the relation

$$\left(\frac{d\sigma}{d\Omega}\right)_{\text{polarized}} = \frac{e^4}{m^2c^4}\sin^2\Theta = r_0^2\sin^2\Theta, \tag{3.37}$$

where

$$r_0 \equiv \frac{e^2}{mc^2}. \tag{3.38}$$

The quantity r_0 gives a measure of the "size" of the point charge, assuming its rest energy mc^2 is purely electromagnetic in origin. For an electron r_0 is called the *classical electron radius* and has a value $r_0 = 2.82\times 10^{-13}$ cm. The total cross section can be found by integrating over solid angle, using $\mu \equiv \cos\Theta$,

$$\sigma = \int \frac{d\sigma}{d\Omega}d\Omega = 2\pi r_0^2\int_{-1}^{1}(1-\mu^2)\,d\mu.$$

This gives the result

$$\sigma = \frac{8\pi}{3}r_0^2. \tag{3.39}$$

(Alternatively, one can obtain σ from $P = \langle S\rangle\sigma$.)

For an electron $\sigma=\sigma_T$ = *Thomson cross section* $=0.665\times10^{-24}$ cm^2. The above scattering process is then called *Thomson scattering* or *electron scattering.*

Note that the total and differential cross sections above are frequency independent, so that the scattering is equally effective at all frequencies. However, this is really only valid for sufficiently low frequencies, so that a classical description is valid. At high frequencies, where the energy of emitted photons $h\nu$ becomes comparable to or larger than mc^2, then the quantum mechanical cross sections must be used; this occurs for X-rays of energies $h\nu \gtrsim 0.511$ MeV for electron scattering (see Chapter 7). Also, for sufficiently intense radiation fields the electron moves relativistically; then the dipole approximation ceases to be valid.

We note that the scattered radiation is linearly polarized in the plane of the incident polarization vector $\boldsymbol{\epsilon}$ and the direction of scattering $\mathbf{n}$.

It is easy to get the differential cross section for scattering of *unpolarized radiation* by recognizing that an unpolarized beam can be regarded as the independent superposition of two linear-polarized beams with perpendicular axes. Let us choose one such beam along $\boldsymbol{\epsilon}_1$, which is in the plane of the incident and scattered directions, and the second along $\boldsymbol{\epsilon}_2$, perpendicular to this plane. (See Fig. 3.7.) Let Θ be the angle between $\boldsymbol{\epsilon}_1$ and $\mathbf{n}$. Note that the angle between $\boldsymbol{\epsilon}_2$ and $\mathbf{n}$ is $\pi/2$. We also have introduced the angle $\theta=\pi/2-\Theta$, which is the angle between the scattered wave and incident wave. Now the differential cross section for unpolarized radiation is the average of the cross sections for scattering of linear-polarized radiation

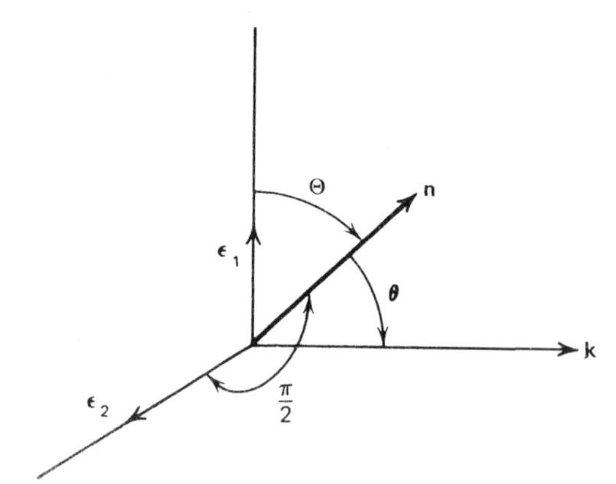

Figure 3.7 Geometry for scattering unpolarized radiation.

through angles Θ and $\pi/2$. Thus we have the result

$$\left(\frac{d\sigma}{d\Omega}\right)_{\text{unpol}}=\frac{1}{2}\left[\left(\frac{d\sigma(\Theta)}{d\Omega}\right)_{\text{pol}}+\left(\frac{d\sigma(\pi/2)}{d\Omega}\right)_{\text{pol}}\right]$$

$$=\tfrac{1}{2}r_0^2(1+\sin^2\Theta)$$

$$=\tfrac{1}{2}r_0^2(1+\cos^2\theta), \tag{3.40}$$

which depends only on the angle between the incident and scattered directions, as it should for unpolarized radiation.

There are several features of electron scattering of unpolarized radiation which we now point out:

1. Forward-backward symmetry: The scattering cross section, Eq. (3.40), is symmetric under the reflection $\theta\rightarrow-\theta$.
2. Total cross section: The total scattering cross section of unpolarized incident radiation is the same as that for polarized incident radiation $\sigma_{\text{unpol}}=\sigma_{\text{pol}}=(8\pi/3)r_0^2$. This is because the electron at rest has no net direction intrinsically defined.
3. Polarization of scattered radiation: The two terms in Eq. (3.40) clearly refer to intensities in two perpendicular directions in the plane normal to $\mathbf{n}$, since they arise from the two perpendicular components of the incident wave. Since the polarized intensities in the plane and perpendicular to the plane of scattering are in the ratio $\cos^2\theta$: 1, the degree of polarization of the scattered wave is [cf. Eq. (2.57)]

$$\Pi=\frac{1-\cos^2\theta}{1+\cos^2\theta}. \tag{3.41}$$

Since $\Pi\geqslant 0$, we have the interesting result that electron scattering of a completely unpolarized incident wave produces a scattered wave with some degree of polarization, the degree depending on the viewing angle with respect to the incident direction. If we look along the incident direction ($\theta=0$) we see no net polarization, since, by symmetry, all directions in the plane are equivalent. If we look perpendicular to the incident wave ($\theta=\pi/2$) we see 100% polarization, since the electron's motion is confined to a plane normal to the incident direction.

3.5 RADIATION REACTION

The energy radiated away by an accelerating charge must come from the particle's own energy or from the agency maintaining the particle's energy.

We conclude that there must be a force acting on a particle by virtue of the radiation it produces. This is called the *radiation reaction force*. The full treatment of this effect from first principles involves the calculation of the force on one part of the charge by the fields of another part, including retardation within the particle itself. Throughout the calculation the size of the particle is kept as nonzero. Afterwards the size can be set to zero, or at least to some small value such as r_0. Here we derive the main result using energy considerations alone.

We first delineate those regimes in which radiation reaction may be considered as a *perturbation* on the particle's motion. Let T be the time interval over which the kinetic energy of the particle is changed substantially by the emission of radiation. Then from Eq. (3.19), with $a=\dot{u}$,

$$T \sim \frac{mv^2}{P_{rad}} \sim \frac{3mc^3}{2e^2}\left(\frac{v}{a}\right)^2,$$

where m is the mass of the particle, and v its velocity. We estimate $v/a \sim t_p$ as the typical orbital time scale for the particle. Then the condition $T/t_p \gg 1$ requires that $t_p \gg \tau$, where, for an electron,

$$\tau \equiv \frac{2e^2}{3mc^3} \sim 10^{-23} s. \tag{3.42}$$

Thus as long as we are considering processes that occur on a time scale much longer than τ, we can treat radiation reaction as a perturbation. It should be noticed that τ is the time for radiation to cross a distance comparable to the classical electron radius, the "size" of the electron: [cf. Eq. (3.38)]

$$\tau \sim \frac{r_0}{c}.$$

We can infer the formula for the radiation reaction force from elementary considerations of energy balance. When the radiation reaction force is relatively small, we may sensibly define the force as a term added onto the existing external force, such that the energy radiated must be compensated for by the work done against the radiation reaction force. Thus we are tempted to set

$$-\mathbf{F}_{\mathrm{rad}} \cdot \mathbf{u} = \frac{2e^2\dot{u}^2}{3c^3}. \tag{3.43}$$

However, one can see that there is no $\mathbf{F}_{\text{rad}}$ that can instantaneously satisfy this equation: $\mathbf{F}_{\text{rad}}$ cannot depend on $\mathbf{u}$, because this would imply a preferred frame relative to which $\mathbf{u}$ is measured. But then one side of Eq. (3.43) explicitly depends on $\mathbf{u}$ whereas the other does not, a contradiction. The best we can do is to satisfy this equation in some average sense, the remaining energy fluctuations being taken up in the nonradiation fields. Integrate the above equation over a time interval t_1 to t_2, with $(t_2-t_1)\gg\tau$. Integrating by parts, we obtain:

$$-\int_{t_1}^{t_2}\mathbf{F}_{\text{rad}}\cdot\mathbf{u}\,dt=\frac{2e^2}{3c^3}\int_{t_1}^{t_2}\dot{\mathbf{u}}\cdot\dot{\mathbf{u}}\,dt$$

$$=\frac{2e^2}{3c^3}\left[\dot{\mathbf{u}}\cdot\mathbf{u}\Big|_{t_1}^{t_2}-\int_{t_1}^{t_2}\ddot{\mathbf{u}}\cdot\mathbf{u}\,dt\right]. \tag{3.44}$$

If we assume that the initial and final states are the same (so that the nonradiation fields are the same and do not contribute to the energy difference) or that $\dot{\mathbf{u}}\cdot\mathbf{u}(t_1)=\dot{\mathbf{u}}\cdot\mathbf{u}(t_2)$, the first term on the right-hand side of Eq. (3.44) vanishes, leaving

$$-\int_{t_1}^{t_2}\left(\mathbf{F}_{\text{rad}}-\frac{2e^2\ddot{\mathbf{u}}}{3c^3}\right)\cdot\mathbf{u}\,dt=0.$$

Thus we take

$$\mathbf{F}_{\text{rad}}=\frac{2e^2\ddot{\mathbf{u}}}{3c^3}=m\tau\ddot{\mathbf{u}}, \tag{3.45}$$

where Eq. (3.45) now represents the radiation force in some *time-averaged*, approximate sense.

This formula for the radiation reaction force depends on the derivative of acceleration, that is, the third derivative of position. This increases the degree of the equation of motion of a particle and can lead to some nonphysical behavior if not used properly and consistently.

For example, the equation of motion for a particle with applied force $\mathbf{F}$ is

$$m(\dot{\mathbf{u}}-\tau\ddot{\mathbf{u}})=\mathbf{F}.$$

Suppose $\mathbf{F}=0$; then a solution is the obvious

$$\mathbf{u}=\text{constant},$$

which is also the physically correct solution. However, there is also another solution

$$\mathbf{u}=\mathbf{u}_0 e^{t/\tau},$$

which rapidly becomes exceedingly large ("runaway" solution). We must exclude such solutions from consideration. We note that they violate the restriction on the motion that it not change on a time scale short compared to τ. Furthermore, $\dot{\mathbf{u}}\cdot\mathbf{u}(t_2)\neq\dot{\mathbf{u}}\cdot\mathbf{u}(t_1)$. We can thus argue that such solutions are spurious, on the mathematical grounds that they violate the assumptions on which the equations were based.

3.6 RADIATION FROM HARMONICALLY BOUND PARTICLES

Undriven Harmonically Bound Particles

A particle that is harmonically bound to a center of force (i.e., $\mathbf{F}=-k\mathbf{r}=-m\omega_0^2\mathbf{r}$) will oscillate sinusoidally with frequency ω_0. Such a system, although rarely found in nature, is interesting because it gives the only possible *classical* model of a spectral line. Many of the quantum results are stated against the framework of this model ("oscillator strengths," "classical damping widths"). Since there is always a small damping of the oscillations by the radiation reaction force, the oscillation will not be purely harmonic. We assume that $\omega_0\tau\ll 1$, so that the radiation reaction formula is valid. If the oscillations are along the x axis, [cf. Eq. (3.45)]

$$-\tau\dddot{x}+\ddot{x}+\omega_0^2 x=0. \tag{3.46}$$

This is a third-order differential equation with constant coefficients. Since the term involving the third derivative is small, a convenient approximation is that the motion will be harmonic to first order, with $x(t)\propto\cos(\omega_0 t+\phi_0)$. Therefore, we approximate the damping implied by the third derivative by a damping in the first derivative, through

$$\dddot{x}\approx-\omega_0^2\dot{x}. \tag{3.47}$$

This approximation preserves an important feature of damping: it is expressed as an odd number of time derivatives and is therefore not time reversible. Therefore, our equation becomes

$$\ddot{x}+\omega_0^2\tau\dot{x}+\omega_0^2 x=0. \tag{3.48}$$

This may be solved by assuming $x(t)$ has the form $e^{\alpha t}$, where α is found from

$$\alpha^2+\omega_0^2\tau\alpha+\omega_0^2=0,$$

which has the solution

$$\alpha=\pm i\omega_0-\tfrac{1}{2}\omega_0^2\tau+0\left(\omega_0^3\tau^2\right)$$

when expanded in powers of $\omega_0\tau$. Taking as initial conditions for $t=0$, $x(0)=x_0$, $\dot{x}(0)\approx 0$, we have

$$\begin{aligned} x(t)&=x_0e^{-\Gamma t/2}\cos\omega_0 t\\ &=\tfrac{1}{2}x_0\left(e^{-\frac{1}{2}\Gamma t+i\omega_0 t}+e^{-\frac{1}{2}\Gamma t-i\omega_0 t}\right), \end{aligned}\tag{3.49}$$

where

$$\Gamma\equiv\omega_0^2\tau=\frac{2e^2\omega_0^2}{3mc^3}.\tag{3.50}$$

The Fourier transform of $x(t)$ is, [cf. Eq. (2.27)],

$$\hat{x}(\omega)=\frac{1}{2\pi}\int_0^\infty x(t)e^{i\omega t}dt=\frac{x_0}{4\pi}\left[\frac{1}{\Gamma/2-i(\omega+\omega_0)}+\frac{1}{\Gamma/2-i(\omega-\omega_0)}\right].\tag{3.51}$$

This becomes large in the vicinity of $\omega=\omega_0$ and $\omega=-\omega_0$. Since we are ultimately interested only in positive frequencies, and only in regions in which the values become large, let us make the approximations

$$\begin{aligned} \hat{x}(\omega)&\approx\frac{x_0}{4\pi}\frac{1}{\Gamma/2-i(\omega-\omega_0)},\\ |\hat{x}(\omega)|^2&=\left(\frac{x_0}{4\pi}\right)^2\frac{1}{(\omega-\omega_0)^2+(\Gamma/2)^2}. \end{aligned}\tag{3.52}$$

The energy radiated per unit frequency is then [cf. Eq. (3.26b)]

$$\frac{dW}{d\omega}=\frac{8\pi\omega^4}{3c^3}\frac{e^2x_0^2}{(4\pi)^2}\frac{1}{(\omega-\omega_0)^2+(\Gamma/2)^2}.\tag{3.53}$$

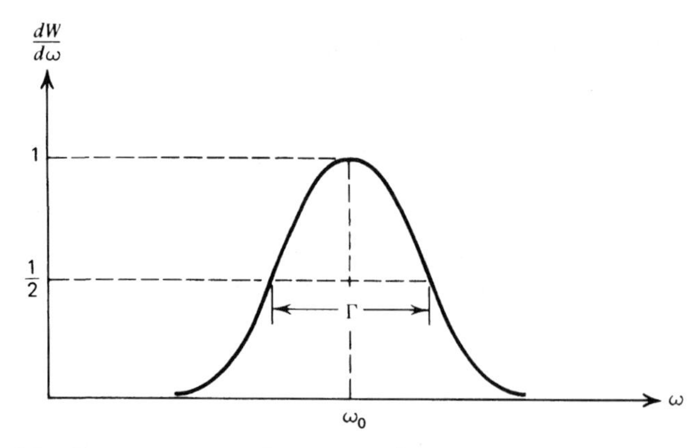

***Figure 3.8** Power spectrum for an undriven, harmonically bound particle damped by radiation reaction.*

Equation (3.53) gives the frequency spectrum typical of a "decaying oscillator." Note that this has a sharp maximum in the neighborhood of $\omega=\omega_0$, since $\Gamma/\omega_0 \ll 1$. This is illustrated in Fig. 3.8, where it is seen that Γ is the full width at half maximum (FWHM).

Using the definition of Γ and $k \equiv m\omega_0^2$ = spring constant, we can write Eq. (3.53) in the form

$$\frac{dW}{d\omega}=\left(\tfrac{1}{2}kx_0^2\right)\frac{\Gamma/2\pi}{(\omega-\omega_0)^2+(\Gamma/2)^2}. \tag{3.54}$$

The first factor gives the initial potential energy of the particle (energy stored in spring). The second factor gives the distribution of the radiated energy over frequency. The integral over ω can be performed easily, if we note that the range of integration can be taken as infinite, since the function is confined essentially to a small region about ω_0:

$$\int_{-\infty}^{\infty}\frac{\Gamma/2\pi}{(\omega-\omega_0)^2+(\Gamma/2)^2}\,d\omega=\frac{1}{\pi}\tan^{-1}\left[\frac{2(\omega-\omega_0)}{\Gamma}\right]_{-\infty}^{\infty}=1.$$

Thus we find that

$$W=\int_0^{\infty}\frac{dW}{d\omega}\,d\omega=\tfrac{1}{2}kx_0^2 \tag{3.55}$$

is the total emitted energy, as it should by conservation of energy.

The profile of the emitted spectrum,

$$\frac{\Gamma/2\pi}{(\omega-\omega_0)^2+(\Gamma/2)^2}, \tag{3.56}$$

is known as a *Lorentz profile*.

The classical line breadth $\Delta\omega=\Gamma$ for electronic oscillators is a universal constant when expressed in terms of wavelength:

$$\begin{aligned}\Delta\lambda &= 2\pi c\frac{\Delta\omega}{\omega^2}\\ &= 2\pi c\tau = 1.2\times 10^{-4}\text{Å},\end{aligned} \tag{3.57}$$

where

$$1\text{ Å} \equiv 10^{-8}\text{ cm}.$$

Driven Harmonically Bound Particles

We have just computed the radiation from the *free* oscillations of a harmonic oscillator. Now, we wish to consider *forced* oscillations, when the forcing is due to an incident beam of radiation. This will give the *scattered* radiation from the incident beam. Let us now write

$$m\ddot{x} = -m\omega_0^2 x + m\tau\dddot{x} + eE_0\cos\omega t, \tag{3.58}$$

where the last term is the force due to a sinusoidally varying incident field. Here we have left the radiation reaction term as a third derivative. With the usual trick of representing x by a complex variable, we have

$$\ddot{x} - \tau\dddot{x} + \omega_0^2 x = \frac{eE_0}{m}e^{i\omega t}, \tag{3.59}$$

where we take the real part of x. The steady-state solution of this equation is

$$x = x_0 e^{i\omega t} \equiv |x_0| e^{i(\omega t+\delta)}, \tag{3.60a}$$

where

$$x_0 = -\left(\frac{eE_0}{m}\right)\left(\omega^2-\omega_0^2-i\omega_0^3\tau\right)^{-1} \tag{3.60b}$$

$$\delta = \tan^{-1}\left(\frac{\omega^3\tau}{\omega^2-\omega_0^2}\right) \tag{3.60c}$$

Note that there is a *phase shift* in the response of the particle displacement to the driving force, caused by the odd time derivative damping term. For $\omega > \omega_0$ the particle "leads" the driving force and for $\omega < \omega_0$ it "lags." Taking the real part of x we see that we have an oscillating dipole of charge e and amplitude $|x_0|$ with frequency ω. The time-averaged total power radiated is therefore

$$P = \frac{e^2|x_0|^2\omega^4}{3c^3} = \frac{e^4E_0^2}{3m^2c^3}\frac{\omega^4}{(\omega^2-\omega_0^2)^2+(\omega_0^3\tau)^2}. \tag{3.61}$$

Dividing Eq. (3.61) by the time-average Poynting vector $\langle S\rangle = (c/8\pi)E_0^2$, we obtain the cross section for scattering as a function of frequency:

$$\sigma(\omega) = \sigma_T\frac{\omega^4}{(\omega^2-\omega_0^2)^2+(\omega_0^3\tau)^2}. \tag{3.62}$$

Here σ_T is the Thomson cross section. Three interesting regimes for ω can be identified (see Fig. 3.9):

1—$\omega \gg \omega_0$. In this case $\sigma(\omega) \to \sigma_T$, the value for free electrons. This is to be expected, since at high incident energies the binding becomes negligible.

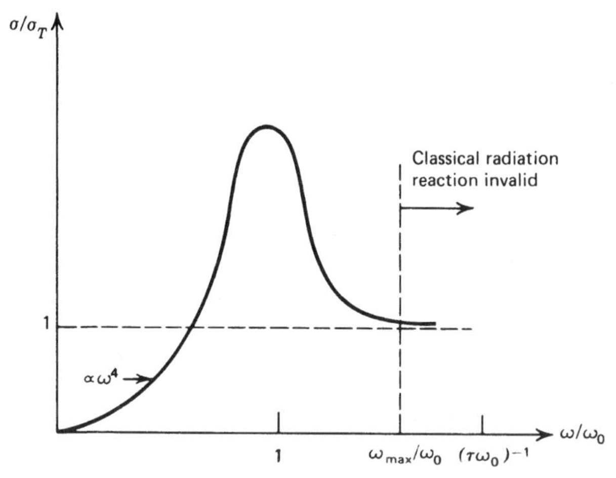

***Figure 3.9** Scattering cross section for a driven, harmonically bound particle as a function of the driving frequency. Here ω_0 and σ_T are the natural frequency and Thomson cross section, respectively.*

2—$\omega \ll \omega_0$. Here we have

$$\sigma(\omega) \rightarrow \sigma_T \left(\frac{\omega}{\omega_0} \right)^4. \tag{3.63}$$

This case corresponds to the electron responding directly to the incident field with no inertial effects, so that $kx \approx eE$. (Since $\omega \ll \omega_0$, the electric field appears nearly static and produces a nearly static force.) The dipole moment is then directly proportional to the incident field and therefore is describable in terms of a static *polarizability*. In such cases the scattered radiation will always go as ω^4, and the scattering is called *Rayleigh scattering*. It is responsible for the blue color of the sky and the red color of the sun at sunrise and sunset, because it favors the scattering of higher frequency (bluer) light.

3—$\omega \approx \omega_0$. This case is dominated by the closeness of $\omega^2 - \omega_0^2$ to zero. Thus we write

$$\omega^2 - \omega_0^2 = (\omega - \omega_0)(\omega + \omega_0)$$

and leave the factor $(\omega - \omega_0)$, but in every other appearance of ω we set $\omega = \omega_0$. This leads to the approximation

$$\sigma(\omega) \approx \frac{\pi \sigma_T}{2\tau} \frac{\Gamma / 2\pi}{(\omega - \omega_0)^2 + (\Gamma/2)^2},$$

using $\Gamma = \omega_0^2 \tau$. With the definitions of σ_T and τ, this can be written

$$\sigma(\omega) = \frac{2\pi^2 e^2}{mc} \frac{\Gamma / 2\pi}{(\omega - \omega_0)^2 + (\Gamma/2)^2}. \tag{3.64}$$

In the neighborhood of the resonance the shape of the scattering cross section is the same as the emission from the free oscillations of the oscillator [cf. Eq. (3.56)]. This can easily be explained, since the free oscillations can be excited by a pulse of radiation, $E(t) \propto \delta(t)$. The spectrum of this pulse is independent of ω (white spectrum), so that the free oscillations may be regarded as the scattering of a white spectrum, yielding emission proportional to the scattering cross section.

An interesting result obtains from integrating $\sigma(\omega)$ over ω:

$$\int_0^\infty \sigma(\omega)\, d\omega = \frac{2\pi^2 e^2}{mc} \tag{3.65a}$$

or in terms of frequency ν,

$$\int_0^\infty \sigma(\nu)\,d\nu = \frac{\pi e^2}{mc}. \tag{3.65b}$$

In evaluating this integral we have apparently neglected a divergence, since the cross section actually approaches σ_T for large ω. This may be justified as follows: the radiation reaction formula is only valid for $\omega\tau \ll 1$, so that we must cut off the integral at a $\omega_{\max}$ such that $\omega_{\max} \ll 1/\tau$. The contribution to the integral from the Thomson limit is less than

$$\int_0^{\omega_{\max}} \sigma_T\,d\omega = \sigma_T \omega_{\max}.$$

This is negligible in comparison to the value of the integral in Eq. (3.65a), since $\sigma_T \omega_{\max} \ll \sigma_T/\tau = 4\pi e^2/mc$.

In the quantum theory of spectral lines we obtain similar formulas, which are conveniently stated in terms of the above classical results as

$$\int_0^\infty \sigma(\nu)\,d\nu = \frac{\pi e^2}{mc} f_{nn'}, \tag{3.66}$$

where $f_{nn'}$ is called the *oscillator strength* or *f-value* for the transition between states n and n' (see Chapter 10).

PROBLEMS

3.1—A pulsar is conventionally believed to be a rotating neutron star. Such a star is likely to have a strong magnetic field, B_0, since it traps lines of force during its collapse. If the magnetic axis of the neutron star does not line up with the rotation axis, there will be magnetic dipole radiation from the time-changing magnetic dipole, $\mathbf{m}(t)$. Assume that the mass and radius of the neutron star are M and R, respectively; that the angle between the magnetic and rotation axes is α; and that the rotational angular velocity is ω.

a. Find an expression for the radiated power P in terms of ω, R, B_0 and α.

b. Assuming that the rotational energy of the pulsar is the ultimate source of the radiated power, find an expression for the slow-down time scale, τ, of the pulsar.

c. For $M=1M_{\odot}\equiv 2\times 10^{33}$ g, R$=10^6$ cm, $B_0=10^{12}$ gauss, $\alpha=90°$, find P and τ for $\omega=10^4\ \mathrm{s}^{-1}$, $10^3\ \mathrm{s}^{-1}$; $10^2\ \mathrm{s}^{-1}$. (The highest rate, $\omega=10^4\ \mathrm{s}^{-1}$, is believed to be typical of newly formed pulsars.)

3.2—A particle of mass m and charge e moves at constant, nonrelativistic speed v_1 in a circle of radius a.

a. What is the power emitted per unit solid angle in a direction at angle θ to the axis of the circle?

b. Describe qualitatively and quantitatively the polarization of the radiation as a function of the angle θ.

c. What is the spectrum of the emitted radiation?

d. Suppose a particle is moving nonrelativistically in a constant magnetic field B. Show that the frequency of circular motion is $\omega_B = eB/mc$, and that the total emitted power is

$$P=\tfrac{2}{3}r_0^2 c(v_\perp/c)^2 B^2,$$

and is emitted solely at the frequency ω_B. (This nonrelativistic form of synchrotron radiation is called *cyclotron* or *gyro* radiation).

e. Find the differential and total cross sections for Thomson scattering of circularly polarized radiation. Use these results to find the cross sections for unpolarized radiation.

3.3—Two oscillating dipole moments (radio antennas) $\mathbf{d}_1$ and $\mathbf{d}_2$ are oriented in the vertical direction and are a horizontal distance L apart. They oscillate in phase at the same frequency ω. Consider radiation at angle θ with respect to the vertical and in the vertical plane containing the two dipoles.

a. Show that

$$\frac{dP}{d\Omega}=\frac{\omega^4\sin^2\theta}{8\pi c^3}\left(d_1^2+2d_1d_2\cos\delta+d_2^2\right),$$

where

$$\delta\equiv\frac{\omega L\sin\theta}{c}.$$

b. Thus show directly that when $L \ll \lambda$, the radiation is the same as from a single oscillating dipole of amplitude $d_1 + d_2$.

3.4—An optically thin cloud surrounding a luminous object is estimated to be 1 pc in radius and to consist of ionized plasma. Assume that electron scattering is the only important extinction mechanism and that the luminous object emits unpolarized radiation.

a. If the cloud is unresolved (angular size smaller than angular resolution of detector), what is the net polarization observed?

b. If the cloud is resolved, what is the polarization direction of the observed radiation as a function of position on the sky? Assume only single scattering occurs.

c. If the central object is clearly seen, what is an upper bound for the electron density of the cloud, assuming that the cloud is homogeneous?

3.5—A plane-polarized wave is incident on a sphere of radius a, composed of a solid material. We assume that the wavelength λ is large compared with a. In that case it is known that the electric field at any instant of time is constant throughout the sphere and has the value $E' = E/(1 + 4\pi\alpha/3)$, where E is the external (applied) field and α is the polarizability of the material. The dipole moment per unit volume is simply proportional to the internal electric field $P = \alpha E'$. Show that the total cross section for scattering the radiation is

$$\sigma = \pi a^2 Q_{\text{scatt}}$$

where

$$Q_{\text{scatt}} = \frac{8(ka)^4}{3(1+3/4\pi\alpha)^2}.$$

3.6—Consider a medium containing a large number of radiating particles. (For definiteness you may wish to imagine electrons emitting bremsstrahlung.) Each particle emits a pulse of radiation with an electric field $E_0(t)$ as a function of time. An observer will detect a series of such pulses, all with the same shape but with random arrival times $t_1, t_2, t_3, \ldots, t_N$. The measured electric field will be

$$E(t) = \sum_{i=1}^{N} E_0(t - t_i).$$

a. Show that the Fourier transform of $E(t)$ is

$$\hat{E}(\omega) = \hat{E}_0(\omega) \sum_{i=1}^{N} e^{i\omega t_i},$$

where $\hat{E}_0(\omega)$ is the Fourier transform of $E_0(t)$.

b. Argue that

$$\left| \sum_{i=1}^{N} e^{i\omega t_i} \right|^2 = N$$

when averaged over the random arrival times.

c. Thus show that the measured spectrum is simply N times the spectrum of an individual pulse. (Note that this result still holds if the pulses overlap.)

d. By contrast, show that if all the particles are in a region much smaller than a wavelength and they emit their pulses simultaneously, then the measured spectrum will be N^2 times the spectrum of an individual pulse.

3.7—Consider a charge e moving around a circle of radius r_0 at frequency ω_0. By consideration of the current density and its Fourier transform, show that the Fourier transform of the vector potential, $\mathbf{A}_\omega(\mathbf{x})$, is nonzero only at $\omega = \omega_0$ in the dipole approximation, nonzero only at $\omega = 2\omega_0$ in the quadrupole approximation and so on.

REFERENCE

Jackson, J. D., 1975, *Classical Electrodynamics*, (Wiley, New York).

4

RELATIVISTIC COVARIANCE AND KINEMATICS

4.1 REVIEW OF LORENTZ TRANSFORMATIONS

The special theory of relativity is based on two postulates:

1. The laws of nature are the same in two frames of reference in uniform relative motion with no rotation.
2. The speed of light is c in all such frames.

Let us consider two frames K and K', as shown in Fig. 4.1, with a relative uniform velocity v along the x axis. The origins are assumed to coincide at $t=0$. If a pulse of light is emitted at the origin at $t=0$, each observer will see an expanding sphere centered on his own origin. This is a consequence of postulate 2 and is inconsistent with classical concepts, which would have the sphere always centered on a point at rest in the "ether." The reconciliation of this result requires us to view both space and time as quantities peculiar to each observer and not universal. Therefore, we have the equations of the expanding sphere in each frame

$$x^2+y^2+z^2-c^2t^2=0, \qquad x'^2+y'^2+z'^2-c^2t'^2=0, \tag{4.1}$$

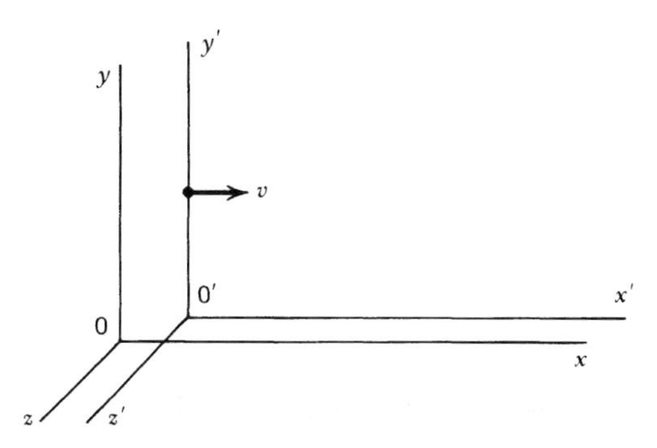

Figure 4.1 ***Two inertial frames with a relative velocity v along the x axis.***

where t' does not equal t, as in Newtonian physics. The actual relations between $x\, y\, z\, t$ and $x'\, y'\, z'\, t'$ can be deduced by fairly elementary means if some further postulates (homogeneity and isotropy of space) are introduced. The result is called the *Lorentz transformation*:

$$x' = \gamma(x - vt) \tag{4.2a}$$

$$y' = y \tag{4.2b}$$

$$z' = z \tag{4.2c}$$

$$t' = \gamma\left(t - \frac{v}{c^2}x\right), \tag{4.2d}$$

where

$$\gamma \equiv \left(1 - \frac{v^2}{c^2}\right)^{-1/2}. \tag{4.2e}$$

The inverse of this transformation is easily found:

$$x = \gamma(x' + vt'), \qquad y = y'$$

$$z = z', \qquad t = \gamma\left(t' + \frac{v}{c^2}x'\right).$$

It should be noted that this inverse has the same form as the original except that the primed and unprimed variables are interchanged, and v is replaced by $-v$.

Since space and time are both subject to transformation, the basic unit is now an *event*, specified by a location in space and by its time of occurrence. Lorentz transformations always refer to events.

We now consider some elementary consequences of Lorentz transformations.

1. Length Contraction (Lorentz–Fitzgerald Contraction)

Suppose a rigid rod of length $L_0 = x_2' - x_1'$ is carried at rest in the frame K'. What is the length as measured in K? This length is equal to $L = x_2 - x_1$, where x_2 and x_1 are the positions of the ends of the rod *at the same time t* in the frame K. Thus we have the result

$$L_0 = x_2' - x_1' = \gamma(x_2 - x_1) = \gamma L,$$

$$L = \left(1 - \frac{v^2}{c^2}\right)^{1/2} L_0. \tag{4.3}$$

The rod appears shorter by a factor $\gamma^{-1} = (1 - v^2/c^2)^{1/2}$. The effect is really symmetric between the two observers. If the rod were carried by K, then K' would see its length contracted. How then can both take place together? If both carry rods (of the same length when compared at rest—say, meter sticks) each thinks the other's rod has shrunk! The point here is that each observer would object to the manner in which the other has carried out the measurement, since it would appear to each that the two ends of the moving stick were not marked *at the same time* by the other observer. This accounts for the apparent lack of symmetry implied by the contraction. (Since the Lorentz transformation of time depends on position, temporal simultaneity is not Lorentz invariant.)

2. Time Dilation

Suppose a device (clock) at rest at the origin of K' measures off an interval of time $T_0 = t_2' - t_1'$. What is the interval of time measured in K? Note that in K', the device moves so that $x' = 0$. Thus we obtain

$$T = t_2 - t_1 = \gamma(t_2' - t_1') = \gamma T_0. \tag{4.4}$$

The interval measured has increased by a factor $\gamma = (1 - v^2/c^2)^{-1/2}$, so that the moving clock appears to have slowed down. Again, the effect is symmetrical between the two observers: K' thinks clocks in K have slowed down, too. The resolution of this apparent contradiction is again a result of looking at the manner of measuring an interval of time between two events separated in space. K measures t_1 as the moving clock passes x_1, then measures t_2 as it passes x_2; he simply subtracts $t_2 - t_1$ on the assumption that his own two clocks at x_1 and x_2 are *synchronized.* K' will object to this, since according to his observations the two clocks in K are not synchronized at all.

In both the time-dilation and length-contraction effects we can see the powerful role played by the questions of synchronization of clocks and of the whole concept of simultaneity. Many of the apparent contradictions of special relativity are simply a result of the *relativity of simultaneity* between two events separated in space.

3. Transformation of Velocities

If a point has velocity $\mathbf{u}'$ in frame K', what is its velocity $\mathbf{u}$ in frame K (Fig. 4.2)? Writing Lorentz transformations for differentials [cf. Eqs. (4.2)]

$$dx = \gamma(dx' + v\,dt'), \qquad dy = dy'$$

$$dz = dz', \qquad dt = \gamma\left(dt' + \frac{v}{c^2}dx'\right).$$

We then have the relations

$$u_x = \frac{dx}{dt} = \frac{\gamma(dx' + v\,dt')}{\gamma(dt' + v\,dx'/c^2)} = \frac{u_x' + v}{1 + vu_x'/c^2}, \tag{4.5a}$$

$$u_y = \frac{u_y'}{\gamma(1 + vu_x'/c^2)}, \tag{4.5b}$$

$$u_z = \frac{u_z'}{\gamma(1 + vu_x'/c^2)}. \tag{4.5c}$$

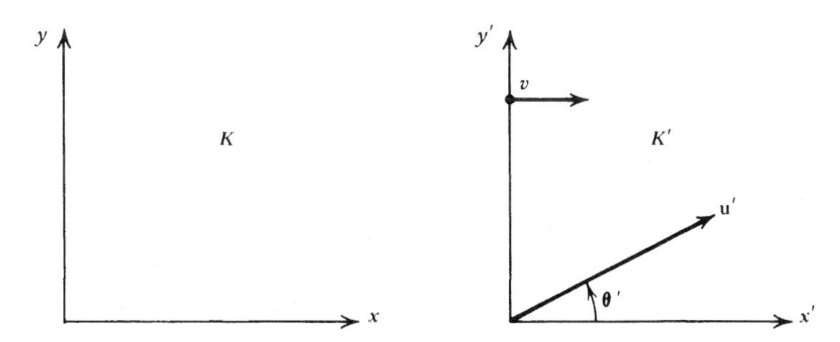

***Figure 4.2** Lorentz transformation of velocities.*

The generalization of these equations to an arbitrary velocity **v**, not necessarily along the x axis, can be stated in terms of the components of **u** perpendicular to and parallel to **v**:

$$u_{\parallel}=\frac{u'_{\parallel}+v}{\left(1+vu'_{\parallel}/c^2\right)}, \qquad u_{\perp}=\frac{u'_{\perp}}{\gamma\left(1+vu'_{\parallel}/c^2\right)}. \tag{4.6}$$

The directions of the velocities in the two frames are related by the *aberration formula,*

$$\tan\theta=\frac{u_{\perp}}{u_{\parallel}}=\frac{u'\sin\theta'}{\gamma(u'\cos\theta'+v)}, \tag{4.7}$$

where $u'\equiv|\mathbf{u}'|$. The azimuthal angle ϕ remains unchanged. An interesting application is for the case $u'=c$, where

$$\tan\theta=\frac{\sin\theta'}{\gamma(\cos\theta'+v/c)}, \tag{4.8a}$$

$$\cos\theta=\frac{\cos\theta'+v/c}{1+(v/c)\cos\theta'}. \tag{4.8b}$$

Equations (4.8) represent the *aberration of light.*

It is instructive to set $\theta'=\pi/2$, that is, a photon is emitted at right angles to v in K'. Then we have

$$\tan\theta=\frac{c}{\gamma v}, \tag{4.9a}$$

$$\sin\theta=\frac{1}{\gamma}. \tag{4.9b}$$

Now for highly relativistic speeds, $\gamma\gg 1$, θ becomes small:

$$\theta\sim\frac{1}{\gamma}. \tag{4.10}$$

If photons are emitted isotropically in K', then half will have $\theta'<\pi/2$ and half $\theta'>\pi/2$ (see Fig. 4.3). Equation (4.10) shows that in frame K photons are concentrated in the forward direction, with half of them lying within a cone of half-angle $1/\gamma$. Very few photons will be emitted having $\theta\gg 1/\gamma$. This is called the *beaming effect*.

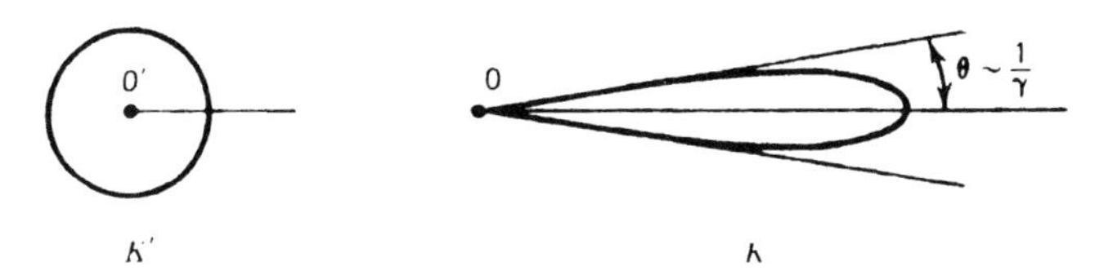

***Figure 4.3** Relativistic beaming of radiation emitted isotropically in the rest frame K′.*

4. Doppler Effect

We have seen that any periodic phenomenon in the moving frame K' will appear to have a longer period by a factor γ when viewed by local observers in frame K. If, on the other hand, we measure the *arrival times* of pulses or other indications of the periodic phenomenon that propagate with the velocity of light, then there will be an additional effect on the observed period due to the delay times for light propagation. The joint effect is called the *Doppler effect*.

In the rest frame of the observer K imagine that the moving source emits one period of radiation as it moves from point 1 to point 2 at velocity v. If the frequency of the radiation in the rest frame of the source is ω' then the time taken to move from point 1 to point 2 in the observer's frame is given by the time-dilation effect:

$$\Delta t = \frac{2\pi\gamma}{\omega'}$$

Now consider Fig. 4.4 and note $l = v\,\Delta t$ and $d = v\,\Delta t\cos\theta$. The difference in arrival times Δt_A of the radiation emitted at 1 and 2 is equal to Δt minus the time taken for radiation to propagate a distance d. Thus we have

$$\Delta t_A = \Delta t - \frac{d}{c} = \Delta t\left(1 - \frac{v}{c}\cos\theta\right).$$

Therefore, the observed frequency ω will be

$$\omega = \frac{2\pi}{\Delta t_A} = \frac{\omega'}{\gamma\left(1 - \frac{v}{c}\cos\theta\right)}. \tag{4.11}$$

This is the relativistic Doppler formula. The factor γ^{-1} is purely a relativistic effect, whereas the $1-(v/c)\cos\theta$ factor appears even classically. One distinction between the classical and relativistic points of view should be mentioned, however. The classical Doppler effect (say, for sound

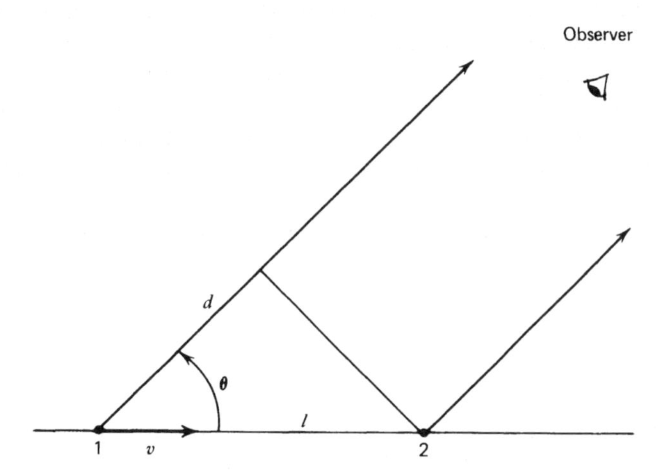

Figure 4.4 Geometry for the Doppler effect.

waves) requires knowledge not only of the relative velocity between source and observer but also the velocities of source and observer relative to the medium (say, air) carrying the waves. The relativistic formula has no reference to an underlying medium for the propagation of light, and only the relative velocity of source and observer appears.

We can also write the Doppler formula as

$$\omega' = \omega\gamma\left(1 - \frac{v}{c}\cos\theta\right). \tag{4.12a}$$

It is easy to show that the inverse of this is

$$\omega = \omega'\gamma\left(1 + \frac{v}{c}\cos\theta'\right). \tag{4.12b}$$

5. Proper Time

Although intervals of space and time are separately subject to Lorentz transformation and thus have differing values in differing frames of reference, there are some quantities that are the same in all Lorentz frames. An important such *Lorentz invariant* is the quantity $d\tau$ defined by

$$c^2 d\tau^2 = c^2 dt^2 - (dx^2 + dy^2 + dz^2). \tag{4.13}$$

This is called the *proper time* element between the events differing by dx, dy, dz in space and dt in time. It is easily shown from Eqs. (4.2) that $d\tau$ is left unchanged under Lorentz transformations, $d\tau = d\tau'$.

The quantity $d\tau$ is called a proper time interval, because it measures time intervals between events occurring at the same spatial location ($dx = dy = dz = 0$), that is, ticks of clocks carried by an observer, which measure his own time.

If the coordinate differentials refer to the position of the origin of another reference frame traveling with velocity v, then

$$d\tau = dt\left(1 - \frac{v^2}{c^2}\right)^{1/2}. \tag{4.14}$$

Equation (4.14) is just the time dilation formula (4.4) in which $d\tau$ is the time interval measured by the observer in motion.

4.2 FOUR-VECTORS

We could continue to find Lorentz transformation properties of physical quantities using ad hoc methods, as in the preceding sections. However, a great deal of order can be brought to this task by introducing the concept of *four-vectors*. A four-vector has transformation properties that are identical to the transformation of coordinates of events [Eq. (4.2)]. Once it is established that a certain quantity is a four-vector, its transformation properties are fully defined. Most physical quantities can be related to four-vectors or to their generalizations—the tensors. It is easy to construct invariants from vectors and tensors, and in this way a physical result can often be obtained without using the Lorentz transformation at all.

The squared length of the three-dimensional vector $\mathbf{x}$, namely, $x^2+y^2+z^2$, is an invariant with respect to three-dimensional rotations. By analogy, the invariance of the quantity $s^2 = -c^2\tau^2 = -c^2t^2 + x^2 + y^2 + z^2$ suggests that the quantities x, y, z and t can be formed into a vector in a four-dimensional space, and that Lorentz transformations correspond to rotations in this space. Let us define

$$\begin{aligned} x^0 &= ct \\ x^1 &= x \\ x^2 &= y \\ x^3 &= z. \end{aligned} \tag{4.15}$$

The quantities x^μ for $\mu=0,1,2,3$ define coordinates of an event in *space-time*. Just as x, y, and z form the components of a three-dimensional spatial vector $\mathbf{x}$, we shall say that x^μ are the components of a four-dimensional space-time vector $\vec{x}$, or simply a *four-vector*.

The fact that the expression for s^2 contains a minus sign in front of c^2t^2 means that space-time is not a Euclidean space; it is a special space called *Minkowski space*. Such a space can be handled in two ways, either by including $\sqrt{-1}$ in the definition of the time component or by the introduction of a *metric*. Although the former method has some simplifying features, the latter method lends itself to the transition to general relativity, and so we adopt it here. Once the notational difficulties of the metric approach are mastered, it is not much more complicated than the $\sqrt{-1}$ approach.

Let us define the *Minkowski metric*. In Cartesian coordinates, the components of $\eta_{\mu\nu}$ are:

$$\eta_{\mu\nu}=\eta^{\mu\nu}=\begin{cases}-1, & \text{if } \mu=\nu=0\\ +1, & \text{if } \mu=\nu=1,2,3\\ 0 & \text{if } \mu\neq\nu.\end{cases} \tag{4.16a}$$

The distinction between superscripted and subscripted indices is explained shortly. The metric $\eta_{\mu\nu}$ can be presented as the 4×4 array (matrix):

$$\eta_{\mu\nu}=\eta^{\mu\nu}=\begin{bmatrix}-1 & 0 & 0 & 0\\ 0 & 1 & 0 & 0\\ 0 & 0 & 1 & 0\\ 0 & 0 & 0 & 1\end{bmatrix}. \tag{4.16b}$$

Note that this metric is symmetric: $\eta_{\mu\nu}=\eta_{\nu\mu}$. The invariant $s^2=-c^2t^2+x^2+y^2+z^2$ can now be written in terms of the metric:

$$s^2=\sum_{\mu=0}^{3}\sum_{\nu=0}^{3}\eta_{\mu\nu}x^\mu x^\nu. \tag{4.17a}$$

An important and beautiful notational advance (originated by Einstein) is the *summation convention*: In any single term containing a repeated Greek index, a summation is implied over that index with values 0, 1, 2, and 3. Therefore, we can write Eq. (4.17a) without the summation signs, since both μ and ν are repeated, once in $\eta_{\mu\nu}$ and then in x^μ or x^ν:

$$s^2=\eta_{\mu\nu}x^\mu x^\nu. \tag{4.17b}$$

We shall henceforth use the summation convention unless otherwise stated.

A few remarks should be made about the summation convention. Since a repeated index is summed over, its exact designation is irrelevant. Therefore, it is often called a *dummy index*, and any Greek letter can be used for it. Equation (4.17b) can also be written $s^2 = \eta_{\sigma\tau} x^\sigma x^\tau$, for example. Another point is that an index cannot be repeated more than twice in a single term; for example, the combination $\eta_{\mu\mu} x^\mu$ is regarded as meaningless.

An equivalent way to use the Minkowski metric is to define another set of components of the vector $\vec{x}$, denoted by x_μ, where

$$\begin{aligned} x_0 &= -ct, \\ x_1 &= x, \\ x_2 &= y, \\ x_3 &= z. \end{aligned} \tag{4.18}$$

These differ from the superscripted components x^μ only in the sign of the time component. The superscripted components are called the *contravariant components*, and the subscripted components are called the *covariant components*. The relation between the two can be written

$$x_\mu = \eta_{\mu\nu} x^\nu, \tag{4.19a}$$

$$x^\mu = \eta^{\mu\nu} x_\nu. \tag{4.19b}$$

Thus the metric can be used to *raise* or *lower* indices. Now the invariant s^2 can be written simply

$$s^2 = x^\mu x_\mu.$$

(Summation on indices occurs only between contravariant and covariant indices. As we show later, this ensures Lorentz invariance.)

The Lorentz transformation (4.2) (corresponding to a *boost* along the x axis) can be written simply in terms of a set of coefficients defined by the array (with $\beta \equiv v/c$)

$$\Lambda^\mu{}_\nu = \begin{bmatrix} \gamma & -\beta\gamma & 0 & 0 \\ -\beta\gamma & \gamma & 0 & 0 \\ 0 & 0 & 1 & 0 \\ 0 & 0 & 0 & 1 \end{bmatrix}. \tag{4.20}$$

Then Eq. (4.2) can be written

$$x'^{\mu} = \Lambda^{\mu}{}_{\nu} x^{\nu}. \tag{4.21}$$

In fact, any arbitrary Lorentz transformation in Cartesian coordinates can be written in the form (4.21), since the spatial three-dimensional rotations necessary to align the x axes before and after the boost are also of linear form. The coefficients $\Lambda^{\mu}{}_{\nu}$ of such an arbitrary Lorentz transformation will, in general, not be given by Eq. (4.20), but will be more complicated.

The transformation law (4.21) defines the transformation of the contravariant components of the vector $\vec{x}$. Since the transformation must leave the quantity s^2 invariant, we must have

$$\eta_{\mu\nu} x^{\mu} x^{\nu} = \eta_{\sigma\tau} x'^{\sigma} x'^{\tau} = \eta_{\sigma\tau} \Lambda^{\sigma}{}_{\mu} \Lambda^{\tau}{}_{\nu} x^{\mu} x^{\nu}.$$

This can be true for arbitrary x^{μ} only if

$$\eta_{\mu\nu} = \Lambda^{\sigma}{}_{\mu} \Lambda^{\tau}{}_{\nu} \eta_{\sigma\tau}. \tag{4.22}$$

This equation can be regarded as the condition on the coefficients $\Lambda^{\mu}{}_{\nu}$ that yields the most general kind of Lorentz transformation. The transformations of interest to us are of a more restrictive nature, however. Note that Eq. (4.22) can be written in matrix form as $\eta = \Lambda^T \eta \Lambda$, where Λ^T is the transpose matrix of Λ. Taking determinants of this yields the result that $\det \Lambda = \pm 1$. We restrict ourselves to *proper* Lorentz transformations, for which

$$\det \Lambda = +1. \tag{4.23a}$$

This rules out reflections, such as $x \to -x$, that would change a right-handed coordinate system into a left-handed one. We also assume *isochronous* Lorentz transformations, for which

$$\Lambda^{0}{}_{0} \geqslant 1, \tag{4.23b}$$

so that the sense of flow of time is the same in K and K'. Note that the boost (4.20) satisfies both (4.23a) and (4.23b).

The transformation of the covariant components x_{μ} of the vector can be deduced from Eq. (4.19b) and (4.21):

$$x'_{\mu} = \tilde{\Lambda}_{\mu}{}^{\nu} x_{\nu} \tag{4.24}$$

where the coefficients $\tilde{\Lambda}_{\mu}{}^{\nu}$ are simply related to the $\Lambda^{\mu}{}_{\nu}$ by

$$\tilde{\Lambda}_{\mu}{}^{\nu}=\eta_{\mu\tau}\Lambda^{\tau}{}_{\sigma}\eta^{\sigma\nu}. \tag{4.25}$$

From the invariance of $s^2=x^{\mu}x_{\mu}$ we easily deduce

$$\Lambda^{\sigma}{}_{\nu}\tilde{\Lambda}_{\sigma}{}^{\mu}=\delta^{\mu}{}_{\nu}, \tag{4.26}$$

where we have introduced the *Kronecker-δ*:

$$\delta^{\mu}{}_{\nu}=\begin{cases}1, & \mu=\nu\\ 0, & \mu\neq\nu.\end{cases} \tag{4.27}$$

These are the components of the 4×4 unit matrix, which accounts for the substitution property of $\delta^{\mu}{}_{\nu}$: For any arbitrary quantity Q^{μ} we have

$$Q^{\mu}=\delta^{\mu}{}_{\nu}Q^{\nu}.$$

Note the useful result

$$\eta^{\mu\sigma}\eta_{\sigma\nu}=\delta^{\mu}{}_{\nu}. \tag{4.28}$$

Multiplying Eq. (4.21) by $\tilde{\Lambda}_{\mu}{}^{\alpha}$ and using Eq. (4.26) yields the inverse transformation:

$$x^{\alpha}=\tilde{\Lambda}_{\mu}{}^{\alpha}x'^{\mu}. \tag{4.29}$$

Everything so far has referred to the vector $\vec{x}$ alone. We now wish to define a general *four-vector* $\vec{A}$ as having four contravariant components A^{μ} in each Lorentz frame, such that the transformation of components between any two frames is given by the same transformation law as applies to x^{μ}, namely, Eq. (4.21):

$$A'^{\mu}=\Lambda^{\mu}{}_{\nu}A^{\nu}. \tag{4.30}$$

The covariant components of $\vec{A}$ are found from the equation analogous to Eq. (4.19a),

$$A_{\mu}=\eta_{\mu\nu}A^{\nu}. \tag{4.31}$$

These transform according to

$$A'_{\mu}=\tilde{\Lambda}_{\mu}{}^{\nu}A_{\nu}. \tag{4.32}$$

Let us consider another four-vector $\vec{B}$ having covariant components B_μ, which transform like $B'_\mu = \tilde{\Lambda}_\mu{}^\sigma B_\sigma$. Multiplying this equation by Eq. (4.30) yields, with the use of Eq. (4.26),

$$A'^\mu B'_\mu = \Lambda^\mu{}_\nu \tilde{\Lambda}_\mu{}^\sigma A^\nu B_\sigma = \delta^\sigma{}_\nu A^\nu B_\sigma = A^\nu B_\nu.$$

Thus the *scalar product* of $\vec{A}$ and $\vec{B}$,

$$\vec{A}\cdot\vec{B} = A^\nu B_\nu = A'^\nu B'_\nu \tag{4.33}$$

is a *Lorentz invariant* or *scalar*. In particular, the "square" of a vector $\vec{A}^2 = A^\mu A_\mu$ is an invariant. Thus our starting point, the invariance of $\vec{x}^2$, is seen to be a general property of four-vectors. We should point out that in Minkowski space, where the metric is not wholly positive, it is possible for the "square" of a four-vector to be positive, zero, or even negative; these possibilities are associated with what are called, respectively, a *spacelike*, *null*, or *timelike* four-vector.

The zeroth component of any four-vector $\vec{A}$ is called the *time-component* A^0, while the first, second, and third form an ordinary three-vector $\mathbf{A}$, called the *space-components*. Often it is convenient to use latin indices to describe the space part, so that these always range over the values 1, 2, and 3. For example, we write

$$\vec{A}\cdot\vec{B} = -A^0B^0 + \mathbf{A}\cdot\mathbf{B} = -A^0B^0 + A^iB_i. \tag{4.34}$$

Three-vectors are always denoted by a boldfaced symbol, whereas four-vectors are denoted by an arrow over the symbol. It should be understood, however, that the division of a four-vector into spatial and time components is dependent on the coordinate system. It is clear that a boost will mix these parts, although spatial rotations will not; for this reason the division will only depend on the velocity of the frame of reference but not on its orientation.

Let us introduce some physically interesting four-vectors other than the prototype $\vec{x}$. First of all we see that the difference between the coordinates of two different events $x_2^\mu - x_1^\mu$ is also a vector, since each term transforms by the same linear transformation. In particular, the difference between two infinitesimally neighboring events dx^μ constitutes a four-vector. Dividing now by the invariant $d\tau$ clearly also yields a four-vector, the *four-velocity* $\vec{U}$, for which

$$U^\mu \equiv \frac{dx^\mu}{d\tau}. \tag{4.35}$$

The zeroth component of this is

$$U^0 = \frac{dx^0}{d\tau} = \frac{c\,dt}{d\tau} = c\gamma_u, \tag{4.36a}$$

where $\gamma_u = (1 - u^2/c^2)^{-1/2}$, and u is the magnitude of the ordinary velocity $\mathbf{u} = d\mathbf{x}/dt$. The spatial components are

$$U^i = \frac{dx^i}{d\tau} = \gamma_u u^i. \tag{4.36b}$$

We may write

$$\vec{U} = \gamma_u \binom{c}{\mathbf{u}}. \tag{4.37}$$

Thus the spatial part of $\vec{U}$ is γ_u times the *ordinary* velocity, whereas the time component is γ_u times c. In this way we have promoted the ordinary velocity into a four-vector. The transformation of U^μ under the boost (4.20) is

$$\begin{aligned} U'^0 &= \gamma(U^0 - \beta U^1), \\ U'^1 &= \gamma(-\beta U^0 + U^1), \\ U'^2 &= U^2, \\ U'^3 &= U^3. \end{aligned}$$

With the above definitions we have

$$\begin{aligned} \gamma_{u'} c &= \gamma(c\gamma_u - \beta\gamma_u u^1), \\ \gamma_{u'} u'^1 &= \gamma(-\beta c\gamma_u + \gamma_u u^1), \\ \gamma_{u'} u'^2 &= \gamma_u u^2, \\ \gamma_{u'} u'^3 &= \gamma_u u^3. \end{aligned}$$

The first two of these are

$$\gamma_{u'} = \gamma\gamma_u(1 - vu^1/c^2), \tag{4.38a}$$

$$\gamma_{u'} u'^1 = \gamma\gamma_u(u^1 - v). \tag{4.38b}$$

Since $u^1 = u\cos\theta$, we obtain the transformation for speed in terms of the γ's:

$$\gamma_{u'} = \gamma\gamma_u\left(1 - \frac{uv}{c^2}\cos\theta\right). \tag{4.39}$$

Dividing (4.38b) by (4.38a) yields the previously derived formula (4.5a):

$$u'^1 = \frac{u^1 - v}{1 - vu'/c^2}.$$

The "length" of $\vec{U}$ is found from

$$\vec{U}\cdot\vec{U} = U^\mu U_\mu = -(\gamma_u c)^2 + (\gamma_u \mathbf{u})^2 = -c^2, \tag{4.40}$$

which is clearly Lorentz invariant.

The four-velocity takes a particularly simple form in a frame in which the ordinary velocity $\mathbf{u}$ vanishes (the rest frame). In that case, we have

$$U'^\mu = c\begin{pmatrix}1\\0\\0\\0\end{pmatrix}. \tag{4.41}$$

Only the time component is nonzero. This property makes $\vec{U}$ a useful tool in picking out the time component of an arbitrary vector as measured by an observer with four-velocity $\vec{U}$:

$$A'^0 = -\frac{1}{c}U'_\mu A'^\mu.$$

But since $U'_\mu A'^\mu = \vec{U}\cdot\vec{A}$ is an invariant, we can write generally

$$A'^0 = -\frac{1}{c}\vec{U}\cdot\vec{A}, \tag{4.42}$$

where $\vec{U}\cdot\vec{A}$ can be evaluated in *any* convenient frame, not necessarily the rest frame. Two examples of this formula can be checked immediately: First, set $\vec{A} = \vec{U}$, and we obtain the trivial result $U'^0 = c$. Set $\vec{A} = \vec{x}$, and we find

$$x'^0 = -\frac{1}{c}x_\mu\frac{dx^\mu}{d\tau} = -\frac{1}{2c}\frac{d}{d\tau}(x_\mu x^\mu)$$
$$= -\frac{1}{2c}\frac{d}{d\tau}(-c^2\tau^2) = c\tau,$$

which is correct, since the proper time is physically equal to the time of a clock in the rest frame.

Another four-vector can be introduced by the following indirect arguments: An electromagnetic wave of plane type has space and time dependence proportional to $\exp(i\mathbf{k}\cdot\mathbf{x}-i\omega t)$. The phase of this wave must be an invariant to all observers, since the vanishing of the electric and magnetic fields in one frame implies their vanishing in all frames. (A charged particle moving on an unaccelerated straight-line trajectory in one frame must have such a trajectory in all frames, by the relativity principle.) Notice that we may write

$$\mathbf{k}\cdot\mathbf{x}-\omega t = k_\mu x^\mu = \vec{k}\cdot\vec{x},$$

where

$$k^\mu = \begin{pmatrix} \omega/c \\ \mathbf{k} \end{pmatrix}. \tag{4.43}$$

It can be shown easily that since the product $\vec{k}\cdot\vec{x}$ is an invariant and $\vec{x}$ is an arbitrary four-vector, then $\vec{k}$ must be a four-vector also. Therefore, we can write the transformation for $\vec{k}$ immediately

$$k'^0 = \gamma(k^0 - \beta k^1), \tag{4.44a}$$

$$k'^1 = \gamma(-\beta k^0 + k^1), \tag{4.44b}$$

$$k'^2 = k^2, \tag{4.44c}$$

$$k'^3 = k^3. \tag{4.44d}$$

Since $|\mathbf{k}| = \omega/c$ for electromagnetic waves, we have $k^1 = (\omega/c)\cos\theta$, so that the zeroth component of the transformation reduces to the Doppler formula

$$\omega' = \omega\gamma\left(1 - \frac{v}{c}\cos\theta\right). \tag{4.45}$$

Another way of deriving (4.45) is to apply (4.42) with $A^\mu = k^\mu$.

Note that $\vec{k}$ is a null vector, since

$$\vec{k}\cdot\vec{k} = |\mathbf{k}|^2 - \frac{\omega^2}{c^2} = 0 \tag{4.46}$$

where the last quantity vanishes by Eq. (2.20a).

The construction of four-vectors is by no means an automatic procedure, as our experience so far has shown. In two cases (x^μ and k^μ) we have simply used a known three-vector for the spatial part and added an appropriate time component. In one case (U^μ) we had to multiply by an appropriate factor γ_u to make the resultant a four-vector. In some cases to be treated presently (electric and magnetic fields) there is *no* four-vector that corresponds to a given three-vector. The systematic construction of four-vectors is best accomplished by means of *tensor analysis*, which we now consider.

4.3 TENSOR ANALYSIS

We are already familiar with some kinds of tensors: A *zeroth-rank tensor* is precisely what we have been calling a Lorentz invariant or Lorentz scalar. A *first-rank tensor* is precisely what we have been calling a four-vector.

Let us now define a *second-rank tensor*. The contravariant components of such a tensor, say T, are given by the sixteen numbers $T^{\mu\nu}$, where, as usual, μ and ν take on the values 0, 1, 2, and 3. The defining transformation properties of T are given by

$$T'^{\mu\nu} = \Lambda^\mu{}_\sigma \Lambda^\nu{}_\tau T^{\sigma\tau}. \tag{4.47}$$

We can define an associated set of covariant components $T_{\mu\nu}$ by lowering indices with the Minkowski metric

$$T_{\mu\nu} = \eta_{\mu\sigma}\eta_{\nu\tau} T^{\sigma\tau}. \tag{4.48}$$

It is easy to show that these components transform as

$$T'_{\mu\nu} = \tilde{\Lambda}_\mu{}^\sigma \tilde{\Lambda}_\nu{}^\tau T_{\sigma\tau}. \tag{4.49}$$

It is also possible to define *mixed components* such as

$$T^\mu{}_\nu = \eta_{\nu\tau} T^{\mu\tau}, \qquad T_\mu{}^\nu = \eta_{\mu\sigma} T^{\sigma\nu}. \tag{4.50}$$

These have the transformation properties

$$T'^\mu{}_\nu = \Lambda^\mu{}_\sigma \tilde{\Lambda}_\nu{}^\tau T^\sigma{}_\tau, \tag{4.51a}$$

$$T'_\mu{}^\nu = \tilde{\Lambda}_\mu{}^\sigma \Lambda^\nu{}_\tau T_\sigma{}^\tau. \tag{4.51b}$$

The position of the tensor index, as a superscript or subscript, determines whether it is contravariant or covariant in its transformations.

Since second-rank tensors are perhaps less familiar than vectors, let us give several examples:

1. The sixteen quantities $A^\mu B^\nu$, formed from the components A^μ and B^ν of two vectors. This can be proved by multiplying the transformation laws for the vector components:

$$A'^\mu B'^\nu = \Lambda^\mu{}_\sigma \Lambda^\nu{}_\tau A^\sigma B^\tau$$

   This is precisely of the form (4.47).

2. The Minkowski metric $\eta^{\mu\nu}$. The transformation of components of the second rank $\eta^{\mu\nu}$ transform by

$$\eta'^{\alpha\beta} = \Lambda^\alpha{}_\mu \Lambda^\beta{}_\nu \eta^{\mu\nu}.$$

   By comparison with Eq. (4.26), $\eta'^{\alpha\beta} = \eta^{\alpha\beta}$. Thus $\eta^{\alpha\beta}$ has the same components in all frames, as we have assumed.

3. The Kronecker–delta δ^μ_ν. A proof similar to the preceding one for the metric can be given starting with Eq. (4.26). This shows that δ^μ_ν forms the components of a mixed second-rank tensor.

Higher-rank tensors can be defined in a similar fashion. The transformation law involves a factor Λ for each contravariant index and a factor $\tilde{\Lambda}$ for each covariant index.

There are a number of simple and useful rules of *tensor analysis* that can be used to form tensors from other tensors:

1. *Addition*. Two tensors of the same type, having the same free indices, can be added to form another tensor of that same type. Examples: $A^\mu + B^\mu$; $F^\mu{}_\nu + G^\mu{}_\nu$. The proof follows from the linearity of the transformations.

2. *Multiplication*. Given two tensors having distinct free indices, multiplication will yield a tensor of rank equal to the sum of the ranks of the two tensors. Examples: $A^\mu B^\nu$ is a second-rank tensor; also $F^{\mu\nu}G_{\sigma\tau}$ is a fourth-rank tensor. The general proof follows the lines outlined above for $A^\mu B^\nu$.

3. *Raising and Lowering Indices*. The Minkowski metric can be used to change contravariant indices into covariant ones, and vice versa, by the processes of *raising* and *lowering*. For example, see Eqs. (4.19),

(4.31), and (4.48). The proof of this result depends on the results

$$\eta_{\mu\nu}\Lambda^{\mu}{}_{\sigma}=\tilde{\Lambda}_{\nu}{}^{\tau}\eta_{\tau\sigma}, \tag{4.52a}$$

$$\eta^{\mu\nu}\tilde{\Lambda}_{\mu}{}^{\sigma}=\Lambda^{\nu}{}_{\tau}\eta^{\tau\sigma}, \tag{4.52b}$$

which follow from Eqs. (4.25) and (4.28). This means the lowering operator $\eta_{\mu\nu}$, in commuting with the Lorentz transformation coefficients Λ, changes them to $\tilde{\Lambda}$, and this changes a contravariant index into a covariant one. A similar statement holds for the raising operator $\eta^{\mu\nu}$.

4. *Contraction*. Consider a tensor having at least two indices, one of which is contravariant and the other covariant. If these two indices are set equal, implying a summation over that index, then the result is a tensor of rank two less. Examples: The scalar product of two vectors $A^{\mu}B_{\mu}$ can be regarded as the contraction of the second-rank tensor $A^{\mu}B_{\nu}$. If $T^{\mu\nu}{}_{\sigma}$ is a third-order tensor, then $T^{\mu\nu}{}_{\nu}$ is a vector. Note that contraction can be used more than once in a single term. Thus starting with the fourth-rank tensor $F^{\mu\nu}G_{\sigma\tau}$ we can form the invariant $F^{\mu\nu}G_{\mu\nu}$. Let us prove this property of contraction for the above example of $T^{\mu\nu}{}_{\nu}$. From the transformation law for $T^{\mu\nu}{}_{\sigma}$ we obtain

$$T'^{\mu\nu}{}_{\nu}=\Lambda^{\mu}{}_{\alpha}\Lambda^{\nu}{}_{\beta}\tilde{\Lambda}_{\nu}{}^{\tau}T^{\alpha\beta}{}_{\tau}.$$

But $\Lambda^{\nu}{}_{\beta}\tilde{\Lambda}_{\nu}{}^{\tau}=\delta^{\tau}_{\beta}$ [cf Eq. (4.26)], so that

$$T'^{\mu\nu}{}_{\nu}=\Lambda^{\mu}{}_{\alpha}T^{\alpha\beta}{}_{\beta},$$

showing that $T^{\mu\nu}{}_{\nu}$ is indeed a vector. The general proof of this property follows along similar lines.

5. *Gradients of Tensor Fields*. A *tensor field* is defined as a tensor that is a function of the spacetime coordinates x^0,x^1,x^2,x^3. Then the gradient operation $\partial/\partial x^{\mu}$ acting on such a field produces a tensor field of one higher rank with μ as a new *covariant* index. A convenient notation for the gradient operation is a comma followed by the index μ. Thus, for example, if λ is a scalar, then $\lambda_{,\mu}\equiv\partial\lambda/\partial x^{\mu}$ is a covariant vector. Similarly $\mathrm{T}^{\mu\nu}{}_{,\sigma}\equiv\partial\tau^{\mu\nu}/\partial x^{\sigma}$ is a third-rank tensor. We shall prove this rule for the special case of the vector field A^{μ}. Differentiating the transformation

$$A'^{\mu}=\Lambda^{\mu}{}_{\sigma}A^{\sigma},$$

gives

$$\frac{\partial A'^{\mu}}{\partial x'^{\nu}}=\Lambda^{\mu}{}_{\sigma}\frac{\partial A^{\sigma}}{\partial x'^{\nu}}=\Lambda^{\mu}{}_{\sigma}\frac{\partial x^{\alpha}}{\partial x'^{\nu}}\frac{\partial A^{\sigma}}{\partial x^{\alpha}}=\Lambda^{\mu}{}_{\sigma}\tilde{\Lambda}_{\nu}{}^{\alpha}\frac{\partial A^{\sigma}}{\partial x^{\alpha}},$$

where we have used Eq. (4.29) to evaluate $\partial x^\alpha/\partial x'^\nu$. This is recognized as the transformation for a second-rank tensor with contravariant index μ and covariant index ν. Note that we have assumed that the components of Λ are constant, a result in Cartesian coordinate systems but not in general (e.g., spherical) coordinate systems. In non-Cartesian systems, partial derivatives do not form the components of a tensor.

The above rules of tensor analysis are extremely useful in practice. Once they have been mastered they become almost automatic; the notation itself almost provides sufficient guidance as to the correct forms. In this regard we note that although the summation convention allows summation over any two indices, only when it involves a subscript-superscript pair is the result assured as a tensor. (See Problem 4.5.) Thus we have always been careful to define quantities with superscripts and subscripts in such a way as to satisfy this requirement.

Some further definitions concerning tensors follows: Tensors of second rank $T^{\mu\nu}$ are *symmetric* or *antisymmetric* if $T^{\mu\nu}=T^{\nu\mu}$ or if $T^{\mu\nu}=-T^{\nu\mu}$, respectively. The *divergence* of a tensor field is a gradient followed by a contraction of the gradient index with one of the other contravariant indices; For example, $A^{\mu}{}_{,\mu}\equiv$divergence of the vector A^μ; $T^{\mu\nu}{}_{,\nu}\equiv$ divergence of the tensor $T^{\mu\nu}$.

A *tensor equation* is a statement that two tensors of the same rank and type are equal. A fundamental property of a tensor equation is that *if it is true in one Lorentz frame, then it is true in all Lorentz frames*. This is clearly true, since each side transforms in the same way. For this reason tensor equations automatically obey the postulate of relativity, which makes them an ideal way to state the laws of nature. This property of the equations of physics under Lorentz transformation is called *invariance of form* or *Lorentz covariance* or simply *covariance*. (This use of the word "covariance" has nothing to do with covariant components of tensors.) Covariance plays a powerful role in helping decide what the proper equations of physics are; in the next section we see this role clearly.

4.4 COVARIANCE OF ELECTROMAGNETIC PHENOMENA

It is empirically found that Maxwell's equations are valid in all Lorentz frames. The two parameters that enter Maxwell's equations and the Lorentz force equation are c and e, the velocity of light and charge, respectively. If Maxwell's equations are to be Lorentz invariant in form, then c and e must be Lorentz scalars; c is invariant by one of the postulates of special relativity. Also, it is an empirical fact that e is invariant. If ρ is a charge density, then $de = \rho\, dx_1 dx_2 dx_3$ is a Lorentz invariant. But the four-volume element $dx_0 dx_1 dx_2 dx_3$ is an invariant, since the Jacobian of the transformation from x_μ to x'_μ is simply the determinant of Λ, which has been shown [Eq. (4.23a)] to be unity. Thus ρ must transform in the same manner as the zeroth component of a four-vector.

To find the other three components, note that the equation of charge conservation

$$\frac{\partial \rho}{\partial t} + \nabla \cdot \mathbf{j} = 0$$

can be written as a tensor equation,

$$j^{\mu}{}_{,\mu} = 0, \tag{4.53}$$

where $\vec{j}$ has components

$$j^{\mu} = \begin{pmatrix} \rho c \\ \mathbf{j} \end{pmatrix}. \tag{4.54}$$

This four-vector is called the *four-current*.

We next look at the set of vector and scalar wave equations in the Lorentz gauge, Eqs. (2.66):

$$\nabla^2 \mathbf{A} - \frac{1}{c^2}\frac{\partial^2 \mathbf{A}}{\partial t^2} = -\frac{4\pi}{c}\mathbf{j},$$

$$\nabla^2 \phi - \frac{1}{c^2}\frac{\partial^2 \phi}{\partial t^2} = -4\pi\rho.$$

If we define the *four-potential*

$$A^{\mu} = \begin{pmatrix} \phi \\ \mathbf{A} \end{pmatrix}, \tag{4.55}$$

then the wave equations may be written as the tensor equations

$$A^{\beta,\alpha}_{\ \ ,\alpha} = -\frac{4\pi}{c} j^{\beta}. \tag{4.56}$$

The Lorentz gauge

$$\nabla \cdot \mathbf{A} + \frac{1}{c}\frac{\partial \phi}{\partial t} = 0,$$

should be preserved under Lorentz transformations, since it was used to obtain the tensor equations (4.56). Indeed, it can be written as a scalar equation,

$$A^{\alpha}{}_{,\alpha}=0. \tag{4.57}$$

What is the tensor representing the fields themselves, **E** and **B**? Since these fields are obtained from derivatives of **A** and ϕ, they should be expressible in terms of derivatives of the four-potential $A_{\mu,\nu}$. Since **E** and **B** have six components all together, we consider the *antisymmetric* tensor

$$F_{\mu\nu}\equiv A_{\nu,\mu}-A_{\mu,\nu}, \tag{4.58}$$

because a rank two antisymmetric tensor has exactly six independent components. From the relationship between the fields and potentials, (2.58) and (2.60), we may write the components as

$$F_{\mu\nu}=\begin{bmatrix} 0 & -E_x & -E_y & -E_z \\ E_x & 0 & B_z & -B_y \\ E_y & -B_z & 0 & B_x \\ E_z & B_y & -B_x & 0 \end{bmatrix}. \tag{4.59}$$

To check that $F_{\mu\nu}$ is the object we want, let us see that it can be used to write Maxwell's equations in tensor form: The two Maxwell equations containing sources,

$$\nabla\cdot\mathbf{E}=4\pi\rho, \qquad \nabla\times\mathbf{B}-\frac{1}{c}\frac{\partial\mathbf{E}}{\partial t}=\frac{4\pi}{c}\mathbf{j}$$

can be written as

$$F_{\mu\nu}{}^{,\nu}=\frac{4\pi}{c}j_{\mu}, \tag{4.60}$$

as can easily be checked. Note that Eq. (4.53), (4.56), (4.57), and (4.60) all involve tensor divergences. The conservation of charge, Eq. (4.53), easily follows from Eq. (4.60):

$$\frac{4\pi}{c}j^{\mu}{}_{,\mu}=F^{\mu\nu}{}_{,\mu\nu}=0,$$

where the last relation follows from the fact that

$$F^{\mu\nu}{}_{,\mu\nu}=-F^{\nu\mu}{}_{,\mu\nu}=-F^{\nu\mu}{}_{,\nu\mu}.$$

The "internal" Maxwell equations,

$$\nabla \cdot \mathbf{B} = 0, \qquad \nabla \times \mathbf{E} + \frac{1}{c}\frac{\partial \mathbf{B}}{\partial t} = 0,$$

can be written as

$$F_{\mu\nu,\sigma} + F_{\sigma\mu,\nu} + F_{\nu\sigma,\mu} = 0. \tag{4.61a}$$

This equation can be written concisely as

$$F_{[\mu\nu,\sigma]} = 0, \tag{4.61b}$$

where [] around indices denote all permutations of indices, with even permutations contributing with a positive sign and odd permutations with a negative sign, for example,

$$A_{[\alpha\beta]} = A_{\alpha\beta} - A_{\beta\alpha}. \tag{4.62}$$

Using the same notation, we can write

$$F_{\mu\nu} = A_{[\nu,\mu]}. \tag{4.63}$$

Since $F_{\mu\nu}$ is a second-rank tensor, its components transform in the usual way, that is,

$$F'_{\mu\nu} = \tilde{\Lambda}_{\mu}{}^{\alpha}\tilde{\Lambda}_{\nu}{}^{\beta}F_{\alpha\beta}. \tag{4.64}$$

Using this transformation law and the definition of $F_{\mu\nu}$ we obtain the transformation law for the fields **E** and **B**. For a pure boost with velocity $\mathbf{v} = c\boldsymbol{\beta}$, these equations can be written in the form:

$$\mathbf{E}'_{\parallel} = \mathbf{E}_{\parallel} \qquad \mathbf{B}'_{\parallel} = \mathbf{B}_{\parallel} \tag{4.65a}$$

$$\mathbf{E}'_{\perp} = \gamma(\mathbf{E}_{\perp} + \boldsymbol{\beta} \times \mathbf{B}) \qquad \mathbf{B}'_{\perp} = \gamma(\mathbf{B}_{\perp} - \boldsymbol{\beta} \times \mathbf{E}). \tag{4.65b}$$

One immediate consequence of these equations is that the concept of a pure electric or pure magnetic field is not Lorentz invariant. If the field is purely electric ($\mathbf{B} = 0$) in one frame, in another frame it will be, in general, a mixed electric and magnetic field. Thus the general term *electromagnetic* field.

Any scalar formed from $F_{\mu\nu}$ represents a function of **E** and **B** which is a Lorentz invariant. One such scalar is just the dot product of F with itself, or "square" of F

$$F_{\mu\nu}F^{\mu\nu} = 2(\mathbf{B}^2 - \mathbf{E}^2). \tag{4.66}$$

Thus $\mathbf{B}^2 - \mathbf{E}^2 = \mathbf{B}'^2 - \mathbf{E}'^2$ is invariant under Lorentz transformations. Another scalar which can be obtained from F is just the determinant of F:

$$\det F = (\mathbf{E}\cdot\mathbf{B})^2. \tag{4.67}$$

Thus $\mathbf{E}\cdot\mathbf{B} = \mathbf{E}'\cdot\mathbf{B}'$ is also an invariant. It is easy to show that the determinant of any second-rank tensor is a scalar, since

$$\begin{aligned}\det A_{\alpha\beta} &= \det A'_{\mu\nu}\tilde{\Lambda}_{\alpha}{}^{\mu}\tilde{\Lambda}_{\beta}{}^{\nu} = (\det\tilde{\Lambda})^2 \det A'_{\mu\nu} \\ &= \det A'_{\mu\nu}.\end{aligned}$$

4.5 A PHYSICAL UNDERSTANDING OF FIELD TRANSFORMATIONS

It is sometimes useful to understand Lorentz transformations of quantities in terms of a piecemeal intuitive approach, as well as in terms of the elegant language of tensor transformations. For example, by means of a simple physical model we can derive the transformation of the electromagnetic fields $\mathbf{E}$ and $\mathbf{B}$ represented in Eqs. (4.65) for the case of an initially pure electric field ($\mathbf{B}=0$). Consider a charged capacitor with plates perpendicular to the x axis in its rest frame K'. Let σ be the surface charge density (esu/cm^2). Then it is known that the electric field inside is $E = 4\pi\sigma$, independent of the separation of the plates d and has a direction normal to the plates.

In frame K' the capacitor is moving with velocity v and the plates are separated by d/γ. The surface charge density is unchanged $\sigma' = \sigma$, because the net charge on a surface element is invariant, and the surface area of the element is also invariant, because the y and z components are unchanged. Since the field depends only on surface charge density and not on plate separation we have $E' = E$, so that in general we have

$$\mathbf{E}'_{\parallel} = \mathbf{E}_{\parallel}$$

as we had previously found.

Now consider the capacitor turned so that the plates are perpendicular to the y axis. The charge density σ is now increased by a factor γ because of length contraction, and we also have a surface current density of magnitude $\mu' = -\sigma' v$, which gives rise to a magnetic field in the z direction of magnitude $B'_z = -(4\pi/c)\mu'$. Thus for this case we have

$$\mathbf{E}'_\perp = \gamma \mathbf{E}_\perp, \qquad \mathbf{B}'_\perp = -\gamma \boldsymbol{\beta} \times \mathbf{E}_\perp.$$

It is also possible to treat the case of an initially pure magnetic field by a similar model, and thus to derive Eqs. (4.65) by superposition. However, we omit the details here.

4.6 FIELDS OF A UNIFORMLY MOVING CHARGE

Let us apply Eqs. (4.65) to find the fields of a charge moving with constant velocity v along the x axis. In the rest frame of the particle the fields are

$$E'_x = \frac{qx'}{r'^3} \qquad B'_x = 0$$

$$E'_y = \frac{qy'}{r'^3} \qquad B'_y = 0$$

$$E'_z = \frac{qz'}{r'^3} \qquad B'_z = 0$$

where

$$r'^3 = \left(x'^2 + y'^2 + z'^2\right)^{3/2}.$$

The inverse of the transformation of the fields Eq. (4.65) is simply found by intercharging primed and unprimed quantities and reversing the sign of v. Then it follows that

$$E_x = \frac{qx'}{r'^3} \qquad B_x = 0$$

$$E_y = \frac{q\gamma y'}{r'^3} \qquad B_y = -\frac{q\gamma\beta z'}{r'^3}$$

$$E_z = \frac{q\gamma z'}{r'^3} \qquad B_z = \frac{q\gamma\beta y'}{r'^3}.$$

These are given in terms of the primed coordinates. We can Lorentz transform the coordinates to give

$$E_x = \frac{q\gamma(x - vt)}{r^3} \qquad B_x = 0$$

$$E_y = \frac{q\gamma y}{r^3} \qquad B_y = -\frac{q\gamma\beta z}{r^3} \tag{4.68}$$

$$E_z = \frac{q\gamma z}{r^3} \qquad B_z = \frac{q\gamma\beta y}{r^3}$$

where

$$r^3 = \left[\gamma^2(x - vt)^2 + y^2 + z^2\right]^{3/2}.$$

Now, we may show that Eqs. (4.68) are precisely what one obtains from the fields given by the Liénard–Wiechert potentials Eqs. (3.7a) and (3.7b). To do this, let us first find where the retarded position of the particle is. For simplicity, assume $z = 0$. Then we have (Fig. 4.5)

$$t_{\text{ret}} = t - \frac{R}{c}$$

$$R^2 = y^2 + (x - vt_{\text{ret}})^2$$

$$= y^2 + \left(x - vt + \frac{vR}{c}\right)^2.$$

Solving for R, we obtain

$$R = \gamma^2\beta\bar{x} + \gamma(y^2 + \gamma^2 x^2)^{1/2},$$

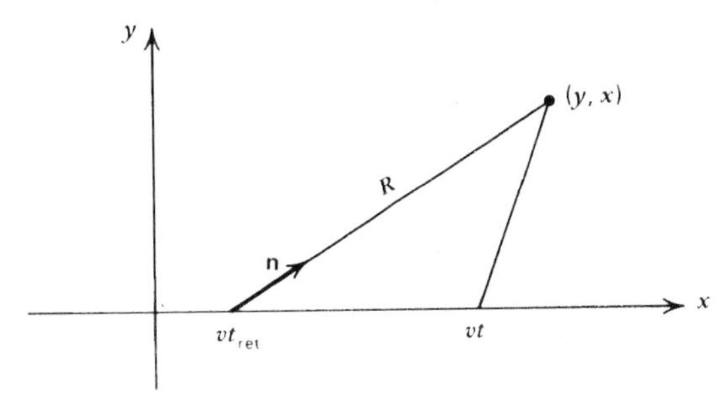

Figure 4.5 ***Evaluation of the radiation field from the retarded position of the particle.***

where

$$\bar{x} \equiv x - vt.$$

We can write the unit vector $\mathbf{n}$ as

$$\mathbf{n} = \frac{y\hat{\mathbf{y}} + (x - vt + vR/c)\hat{\mathbf{x}}}{R} \tag{4.69a}$$

and κ as:

$$\begin{aligned}\kappa &= 1 - \mathbf{n} \cdot \boldsymbol{\beta} \\ &= \frac{(y^2 + \gamma^2 \bar{x}^2)^{1/2}}{\gamma R}.\end{aligned}$$

Thus we have the result

$$\frac{q}{\gamma^2 R^2 \kappa^3} = \frac{\gamma R q}{(y^2 + \gamma^2 \bar{x}^2)^{3/2}}. \tag{4.69b}$$

Using Eqs. (4.69a) and (4.69b), and Eq. (4.68), we find that

$$\mathbf{E} = q\left[\frac{(\mathbf{n} - \boldsymbol{\beta})(1 - \beta^2)}{\kappa^3 R^2}\right],$$

which is identical to the field components of Eq. (3.10).

An important application of these results is the case of a highly relativistic charge, $\gamma \gg 1$. For simplicity, let us choose the field point to be a distance b from the origin along the y axis; this involves no loss in generality (see Fig. 4.6). Then we have the results

$$E_x = -\frac{qv\gamma t}{(\gamma^2 v^2 t^2 + b^2)^{3/2}} \qquad B_x = 0 \tag{4.70a}$$

$$E_y = \frac{q\gamma b}{(\gamma^2 v^2 t^2 + b^2)^{3/2}} \qquad B_y = 0 \tag{4.70b}$$

$$E_z = 0 \qquad B_z = \beta E_y. \tag{4.70c}$$

For large γ we have $\beta \approx 1$ and $E_y \approx B_z$. In Fig. 4.7 E_x and E_y are plotted as functions of time. We see that the fields are strong only when t is of the same order as $b/\gamma v$. This means that the fields of the moving charge are

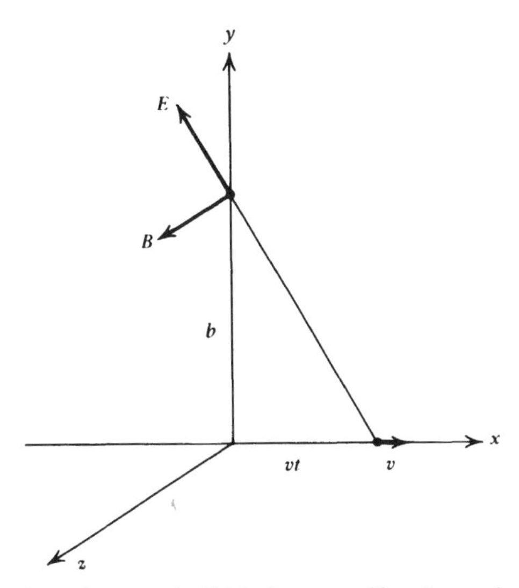

Figure 4.6 Electric and magnetic fields from a uniformly moving particle.

concentrated in the plane transverse to its motion, in fact, into an angle of order $1/\gamma$. The fields are also mostly transverse, since E_x is at maximum only of order q/b^2. Therefore, the field of a highly relativistic charge appears to be a pulse of radiation traveling in the same direction as the charge and confined to the transverse plane. This connection between the fields of a highly relativistic charge and an associated radiation field is an important one and is used in the *method of virtual quanta*, to be discussed in Chapter 5.

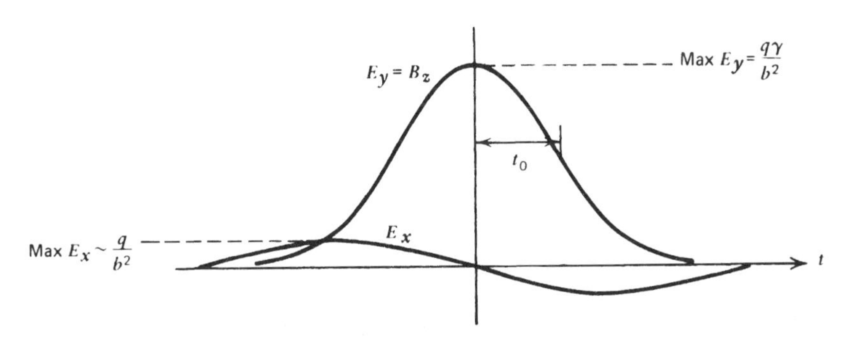

Figure 4.7 Time-dependence of fields from a particle of uniform high velocity.

We can determine the equivalent spectrum of this pulse of virtual radiation. First we must find the transform

$$\hat{E}(\omega)=\frac{1}{2\pi}\int E_2(t)e^{i\omega t}\,dt$$

$$=\frac{q\gamma b}{2\pi}\int_{-\infty}^{\infty}(\gamma^2v^2t^2+b^2)^{-3/2}e^{i\omega t}\,dt. \tag{4.71}$$

This integral can be done in terms of the modified Bessel function of order one, K_1:

$$\hat{E}(\omega)=\frac{q}{\pi bv}\frac{b\omega}{\gamma v}K_1\left(\frac{b\omega}{\gamma v}\right). \tag{4.72a}$$

Thus the spectrum is

$$\frac{dW}{dA\,d\omega}=c|\hat{E}(\omega)|^2=\frac{q^2c}{\pi^2b^2v^2}\left(\frac{b\omega}{\gamma v}\right)^2K_1^2\left(\frac{b\omega}{\gamma v}\right). \tag{4.72b}$$

The spectrum starts to cut off for $\omega>\gamma v/b$, which we could have predicted on the basis of the uncertainty principle, since the pulse is confined roughly to a time interval of order $b/\gamma v$. In fact, the complete behavior of $\hat{E}(\omega)$ can be estimated to within a factor of ~ 2 just by analysis of the picture of $E(t)$: $E(t)$ has a maximum $q\gamma/b^2$ for a time interval $\sim b/\gamma v$. Thus we approximate

$$\hat{E}_{\max}(\omega)\sim E_{\max}(t)\,\Delta t\sim\left(\frac{q\gamma}{b^2}\right)\left(\frac{b}{\gamma v}\right),$$

$$\Delta\omega\sim\frac{1}{\Delta t}\sim\frac{\gamma v}{b}.$$

We have found the spectrum per unit area at a distance b from the line of the charge's motion. To find the total energy per unit frequency range, we must integrate this over $dA=2\pi b\,db$ (see Fig. 4.8):

$$\frac{dW}{d\omega}=2\pi\int_{b_{\min}}^{b_{\max}}\frac{dW}{dA\,d\omega}b\,db. \tag{4.73}$$

The lower limit has been chosen not as zero but as some minimum distance $b_{\min}$, such that the approximation of the field by means of classical electrodynamics and a point charge is valid. Two possible choices

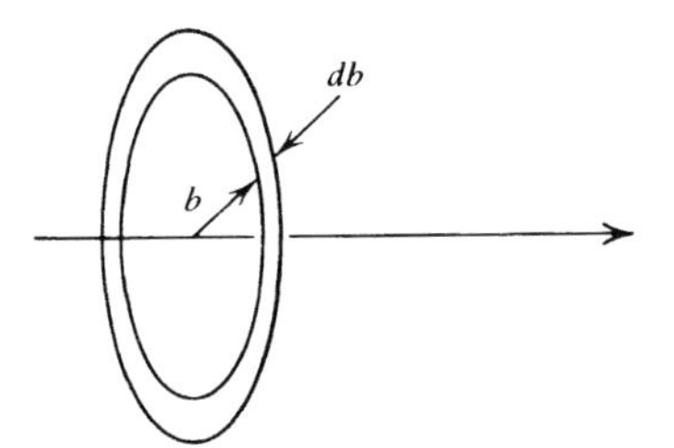

Figure 4.8 ***Area element perpendicular to the velocity of a moving particle.***

are (1) b_{min} = radius of ion, if field is that of an ion and (2) $b_{\text{min}} \sim \hbar/mc$ = Compton wavelength of particle. The integral is now

$$\frac{dW}{d\omega}=\frac{2q^2c}{\pi v^2}\int_x^\infty yK_1^2(y)\,dy, \tag{4.74a}$$

where

$$y\equiv\frac{\omega b}{\gamma v},\qquad x\equiv\frac{\omega b_{\text{min}}}{\gamma v}.$$

This integral can be done in terms of Bessel functions

$$\frac{dW}{d\omega}=\frac{2q^2c}{\pi v^2}\left[xK_0(x)K_1(x)-\tfrac{1}{2}x^2\left(K_1^2(x)-K_0^2(x)\right)\right]. \tag{4.74b}$$

Two limiting forms occur when ω is small, $\omega \ll \gamma v/b_{\text{min}}$, and when ω is large, $\omega \gg \gamma v/b_{\text{min}}$:

$$\frac{dW}{d\omega}=\frac{2q^2c}{\pi v^2}\ln\left(\frac{0.68\gamma v}{\omega b_{\text{min}}}\right),\qquad \omega\ll\frac{\gamma v}{b_{\text{min}}} \tag{4.75a}$$

$$\frac{dW}{d\omega}=\frac{q^2c}{2v^2}\exp\left(-\frac{2\omega b_{\text{min}}}{\gamma v}\right),\qquad \omega\gg\frac{\gamma v}{b_{\text{min}}}. \tag{4.75b}$$

These forms can be derived approximately by direct integration of $xK_1^2(x)$, using the asymptotic results $K_1(x) \sim 1/x$, $x \ll 1$, and $K_1(x) \sim (\pi/2x)^{1/2}e^{-x}$, $x \gg 1$.

4.7 RELATIVISTIC MECHANICS AND THE LORENTZ FOUR-FORCE

The equations of electrodynamics came to us in the already covariant form of Maxwell's equations. Unfortunately, the equations of dynamics as given by Newton are not in convariant form; this is clear since they obey Galilean not Lorentz invariance. Therefore, we must find new equations that reduce to the Newtonian ones for low velocities but that obey the principles of relativity. To do this we are guided by the requirement that these equations be cast in covariant, tensor form.

The rest mass of a particle m_0 is a scalar by definition, since it can be invariantly defined (go to a frame in which the particle is at rest and measure it). Then the *four-momentum* of a particle, P^μ is defined by

$$P^\mu \equiv m_0 U^\mu. \tag{4.76}$$

In the nonrelativistic limit, the spatial components of the four-momentum are just the components of the ordinary three-momentum, $m_0\mathbf{v}$. To interpret all the components relativistically, consider the expansion of $P^0 c$ for $v \ll c$:

$$P^0 c = m_0 c U^0 = m_0 c^2 \left(1 - \frac{v^2}{c^2}\right)^{-1/2} = m_0 c^2 + \tfrac{1}{2} m_0 v^2 + \cdots. \tag{4.77}$$

The second term in (4.77) is the nonrelativistic expression for the kinetic energy of the particle; therefore, we interpret $E = P^0 c$ as the total energy of the particle. The quantity $m_0 c^2$, being independent of v, is interpreted as the rest energy of the particle. If the relativistic expression for the spatial momentum is then defined as $\mathbf{p} = \gamma_v m_0 \mathbf{v}$, then $P^\mu = (E/c, \mathbf{p})$. Then from Eqs. (4.40), (4.76) and (4.77):

$$\vec{P}^2 = -m_0^2 c^2 = -\frac{E^2}{c^2} + |\mathbf{p}|^2,$$

$$E^2 = m_0^2 c^4 + c^2 |\mathbf{p}|^2. \tag{4.78}$$

Since photons are massless and travel at the speed of light, the four-momentum cannot be defined by Eq. (4.76). In this case we still define $P^\mu = (E/c, \mathbf{p})$, but we use the quantum relations $E = \hbar\omega$ and $\mathbf{p} = \hbar\mathbf{k}$. From Eq. (4.43) we then have

$$P^\mu = \hbar k^\mu = \begin{pmatrix} \hbar\omega/c \\ \hbar\mathbf{k} \end{pmatrix}. \tag{4.79}$$

The momentum four-vector for photons is null, $\vec{P}^2 = 0$, since $E = |\mathbf{p}|c$.

Now, in exactly the same way as we obtained the four-velocity from the displacement four-vector, we can define a *four-acceleration* a^μ by taking another derivative, with respect to the scalar interval, of the four-velocity:

$$a^\mu \equiv \frac{dU^\mu}{d\tau}. \tag{4.80}$$

In the nonrelativistic limit, in which $\gamma_u \approx 1$, the spatial components of the four-velocity and four-acceleration are approximately equal to their nonrelativistic, three-vector counterparts.

Note that the four-acceleration and four-velocity are *orthogonal* (their dot product vanishes):

$$\begin{aligned}\vec{a}\cdot\vec{U} &= \frac{dU^\mu}{d\tau}U_\mu = \tfrac{1}{2}\frac{d}{d\tau}(U^\mu U_\mu)\\ &= \tfrac{1}{2}\frac{d}{d\tau}(-c^2) = 0.\end{aligned} \tag{4.81}$$

Having defined the four-acceleration, we can define another four-vector, the *four-force* F^μ, so as to obtain a relativistic form of Newton's equation "$F = ma$":

$$F^\mu \equiv m_0 a^\mu = \frac{dP^\mu}{d\tau}. \tag{4.82}$$

In the case of electromagnetism, we can explicitly evaluate F^μ from the known Lorentz force,

$$\mathbf{F}_{\text{Lorentz force}} = e\left[\mathbf{E} + \frac{1}{c}(\mathbf{v}\times\mathbf{B})\right].$$

Our Lorentz four-force should involve the electromagnetic fields embodied in the tensor $F_{\mu\nu}$ and the particle velocity embodied in the four-velocity U^μ and should also be a four-vector and proportional to the (scalar) charge of the body. The simplest possibility is

$$F^\mu = \frac{e}{c}F^\mu{}_\nu U^\nu. \tag{4.83}$$

Substituting Eq. (4.83) into Eq. (4.82), we have the tensor equation of motion of a charged particle:

$$a^\mu = \frac{e}{m_0 c}F^\mu{}_\nu U^\nu. \tag{4.84}$$

Let us check the components of Eq. (4.84) to see if it is indeed what we want. The $\mu=0$ component is, using Eqs. (4.59) and (4.76):

$$\frac{dW}{dt}=e\mathbf{E}\cdot\mathbf{v}. \tag{4.85a}$$

Equation (4.85a) is just the conservation of energy: the rate of change of particle energy W is the mechanical work done on the particle by the field, $e\mathbf{E}\cdot\mathbf{v}$. Each spatial component (say, $\mu=1$) of Eq. (4.84) is

$$\frac{dP_x}{dt}=e\left[E_x+\frac{1}{c}(\mathbf{v}\times\mathbf{B})_x\right], \tag{4.85b}$$

agreeing with the desired expression for the three-Lorentz force.

Note from Eq. (4.81) and Eq. (4.82) that the four-force, *regardless of its origin*, is always orthogonal to the four-velocity:

$$\vec{F}\cdot\vec{U}=m_0(\vec{a}\cdot\vec{U})=0. \tag{4.86}$$

Equation (4.86) is a general property of any covariant formulation of mechanics in four-dimensional spacetime. It implies that every four-force must have some velocity dependence, although this dependence might become negligible in the nonrelativistic limit. For the Lorentz four-force, in particular, we find

$$\vec{F}_{\text{Lorentz force}}\cdot\vec{U}=\frac{e}{c}F_{\mu\nu}U^{\mu}U^{\nu}=0,$$

because $F_{\mu\nu}$ is antisymmetric and $U^{\mu}U^{\nu}$ is symmetric.

4.8 EMISSION FROM RELATIVISTIC PARTICLES

Total Emission

We would now like to use relativistic transformations to find the radiation emitted by a particle moving at relativistic speeds. The idea is to move into an *instantaneous rest frame K'*, such that the particle has zero velocity at a certain time. The particle will not remain at rest in this frame (since it can accelerate), but at least for infinitesimally neighboring times the particle moves nonrelativistically. We can therefore calculate the radiation emitted by use of the dipole (Larmor) formula. Suppose a total amount of energy dW' is emitted in this frame in time dt'. The momentum of this radiation is zero, $d\mathbf{p}'=0$, because the emission is symmetrical with respect to any direction and its opposite direction. The energy in a frame K moving with velocity $-v$ with respect to the particle is therefore

$$dW = \gamma \, dW',$$

from the transformation properties of the four-momentum. The time interval dt is simply

$$dt = \gamma \, dt',$$

since dt' is the proper time of the particle. The total power emitted in frames K and K' are given by

$$P = \frac{dW}{dt}, \qquad P' = \frac{dW'}{dt'}.$$

From above we see

$$P = P'. \tag{4.87}$$

Thus the total emitted power is a Lorentz invariant for any emitter that emits with front-back symmetry in its instantaneous rest frame. Knowing this, we would like to express the power in covariant form. Now, from the Larmor formula, we have [cf. Eq. (3.19)]

$$P' = \frac{2q^2}{3c^3} |\mathbf{a}'|^2. \tag{4.88}$$

Recall, however, that because $\vec{a} \cdot \vec{U} = 0$ [cf. Eq. (4.81)], and because $U^\mu = (c, \mathbf{0})$ in the instantaneous rest frame of the emitting particle, [cf. Eq. (4.41)], we have

$$a'_0 = 0.$$

Thus

$$|\mathbf{a}'|^2 = a'_k a'^k = a'_0 a'^0 + a'_k a'^k = a'^\alpha a'_\alpha = \vec{a} \cdot \vec{a}. \tag{4.89}$$

So, we can write Eq. (4.88) in manifestly covariant form:

$$P = \frac{2q^2}{3c^3} \vec{a} \cdot \vec{a}. \tag{4.90}$$

The power can thus be evaluated in any frame just by computing $\vec{a}$ in that particular frame and squaring it.

It is convenient to express P in terms of the three-vector acceleration $d^2\mathbf{x}/dt^2$ rather than in terms of the four-vector acceleration $d^2x^\mu/d\tau^2$. It can easily be shown (see Problem 4.3) that if K' is an instantaneous rest frame of a particle, then

$$a'_\parallel = \gamma^3 a_\parallel, \tag{4.91a}$$

$$a'_\perp = \gamma^2 a_\perp. \tag{4.91b}$$

Thus we can write

$$\begin{aligned} P &= \frac{2q^2}{3c^3}\mathbf{a}'\cdot\mathbf{a}' = \frac{2q^2}{3c^3}\left(a_\parallel'^2 + a_\perp'^2\right) \\ &= \frac{2q^2}{3c^3}\gamma^4\left(a_\perp^2 + \gamma^2 a_\parallel^2\right). \end{aligned} \tag{4.92}$$

Angular Distribution of Emitted and Received Power

In the instantaneous rest frame of the particle, let us consider an amount of energy dW' that is emitted into the solid angle $d\Omega' = \sin\theta'\,d\theta'\,d\phi'$ about the direction at angle θ' to the x' axis (see Fig. 4.9). It is convenient to introduce the notations

$$\mu = \cos\theta, \qquad \mu' = \cos\theta',$$

so that

$$d\Omega = d\mu\,d\phi, \qquad d\Omega' = d\mu'\,d\phi'.$$

Since energy and momentum form a four-vector, the transformation of the energy of the radiation is,

$$dW = \gamma(dW' + v\,dP_x') = \gamma(1 + \beta\mu')\,dW'. \tag{4.93}$$

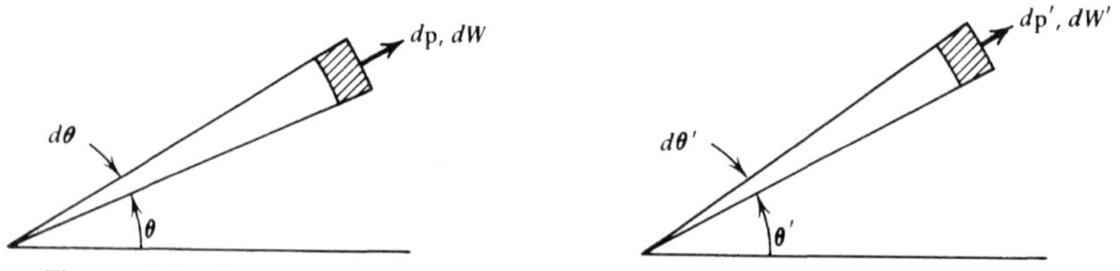

***Figure 4.9** Lorentz transformation of the angular distribution of emitted power.*

We also have from Eq. (4.8b),

$$\mu = \frac{\mu' + \beta}{1 + \beta\mu'}. \tag{4.94}$$

Differentiating this yields

$$d\mu = \frac{d\mu'}{\gamma^2(1+\beta\mu')^2},$$

and since $d\phi = d\phi'$,

$$d\Omega = \frac{d\Omega'}{\gamma^2(1+\beta\mu')^2}. \tag{4.95}$$

Thus we have the result

$$\frac{dW}{d\Omega} = \gamma^3(1+\beta\mu')^3 \frac{dW'}{d\Omega'}. \tag{4.96}$$

The power emitted in the rest frame P' is found simply by dividing dW' by the time interval dt'. However, in frame K there are *two* possible choices for the time interval used to divide dW:

1—$dt = \gamma\, dt'$. This is the time interval during which the emission occurs in frame K [cf. Eq. (4.4)]. With this choice we obtain the *emitted* power in frame $K: P_e$.

2—$dt_A = \gamma(1 - \beta\mu)\, dt'$. This is the time interval of the radiation as received by a stationary observer in K. The extra factor is the retardation effect due to the moving source [cf. Eq. (4.11) and (4.12b)]. With this choice we obtain the *received* power in frame $K: P_r$.

Thus we obtain the two results:

$$\frac{dP_e}{d\Omega} = \gamma^2(1+\beta\mu')^3 \frac{dP'}{d\Omega'} = \frac{1}{\gamma^4(1-\beta\mu)^3}\frac{dP'}{d\Omega'}, \tag{4.97a}$$

$$\frac{dP_r}{d\Omega} = \gamma^4(1+\beta\mu')^4 \frac{dP'}{d\Omega'} = \frac{1}{\gamma^4(1-\beta\mu)^4}\frac{dP'}{d\Omega'}. \tag{4.97b}$$

The alternate forms follow from the equivalence of Eqs. (4.12a) and (4.12b).

Which of these two should we use? P_r is the power actually measured by an observer and so would seem to be the natural one. Also in favor of P_r is that Eq. (4.97b) has the expected symmetry property of yielding the inverse transformation by interchanging primed and unprimed variables, along with a change of sign of β. For these reasons we deal with P_r for the rest of this section, calling it simply P.

It should be pointed out, however, that P_e does have its uses (c.f. Jackson's Sect. 14.3; also our discussion of emission coefficient, §4.9). In practice, the distinction between emitted and received power is often not important, since P_r and P_e are equal in an average sense for stationary distributions of particles. We discuss this further in the context of synchrotron emission in §6.7.

Let us now return to Eq. (4.97b). If the radiation is isotropic in the particle's frame (or nearly isotropic), then the angular distribution in the observer's frame will be highly peaked in the forward direction for highly relativistic velocities ($\beta \sim 1$). In fact, let us write

$$\mu = \cos\theta \approx 1 - \frac{\theta^2}{2}, \tag{4.98a}$$

$$\beta = \left(1 - \frac{1}{\gamma^2}\right)^{1/2} \approx 1 - \frac{1}{2\gamma^2}. \tag{4.98b}$$

It follows by expansion that

$$\frac{1}{\gamma^4(1-\beta\mu)^4} \approx \left(\frac{2\gamma}{1+\gamma^2\theta^2}\right)^4. \tag{4.98c}$$

This latter factor is sharply peaked near $\theta \simeq 0$ with an angular scale of order $1/\gamma$, in agreement with our previous discussion.

Let us now apply these formulas to the case of an emitting particle. In the instantaneous rest frame of the particle the angular distribution is given by [cf. Eq. (3.18)]

$$\frac{dP'}{d\Omega'} = \frac{q^2 a'^2}{4\pi c^3} \sin^2\Theta',$$

where Θ' is the angle between the acceleration and the direction of emission (see Fig. 4.10). Writing $\mathbf{a}' = \mathbf{a}'_\parallel + \mathbf{a}'_\perp$ and using the results (4.91a) and (4.91b), we obtain

$$\frac{dP}{d\Omega} = \frac{q^2}{4\pi c^3} \frac{\left(\gamma^2 a_\parallel^2 + a_\perp^2\right)}{(1-\beta\mu)^4} \sin^2\Theta'. \tag{4.99}$$

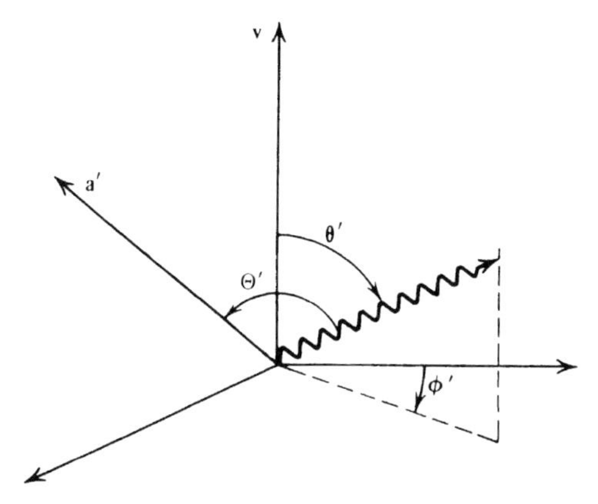

***Figure 4.10** Geometry for dipole emission from a particle instantaneously at rest.*

To use this formula we must relate Θ' to the angles in the frame K. This is difficult in the general case, so we work out the angular distribution of the received power for special cases:

1—Acceleration $\parallel$ to Velocity. Here $\Theta'=\theta'$ so that

$$\sin^2\Theta'=\frac{\sin^2\theta}{\gamma^2(1-\beta\mu)^2} \tag{4.100}$$

where we have used Eq. (4.94). Substituting Eq. (4.100) into Eq. (4.99) with $a_\perp=0$, we obtain

$$\frac{dP_\parallel}{d\Omega}=\frac{q^2}{4\pi c^3}a_\parallel^2\frac{\sin^2\theta}{(1-\beta\mu)^6}. \tag{4.101}$$

2—Acceleration $\perp$ to Velocity. Here $\cos\Theta'=\sin\theta'\cos\phi'$, so that

$$\sin^2\Theta'=1-\frac{\sin^2\theta\cos^2\phi}{\gamma^2(1-\beta\mu)^2}. \tag{4.102}$$

Thus we have the result

$$\frac{dP_\perp}{d\Omega}=\frac{q^2a_\perp^2}{4\pi c^3}\frac{1}{(1-\beta\mu)^4}\left[1-\frac{\sin^2\theta\cos^2\phi}{\gamma^2(1-\beta\mu)^2}\right]. \tag{4.103}$$

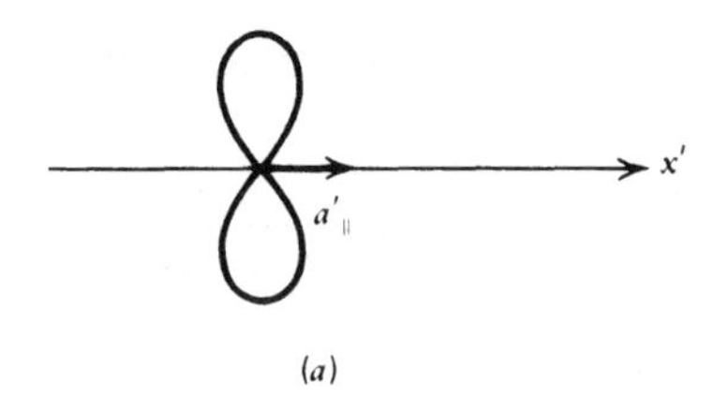

***Figure 4.11a** Dipole radiation pattern for particle at rest.*

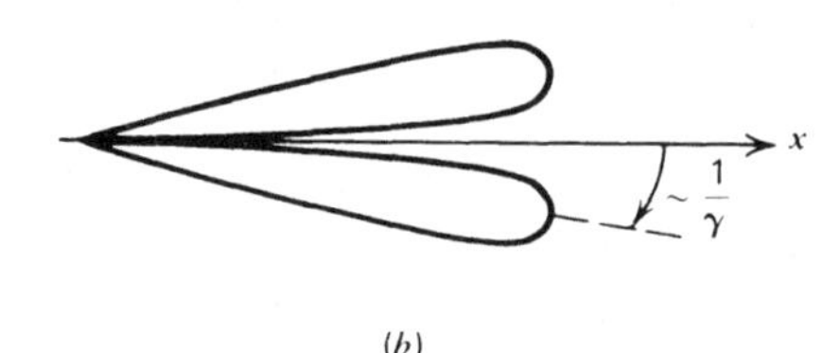

***Figure 4.11b** Angular distribution of radiation emitted by a particle with parallel acceleration and velocity.*

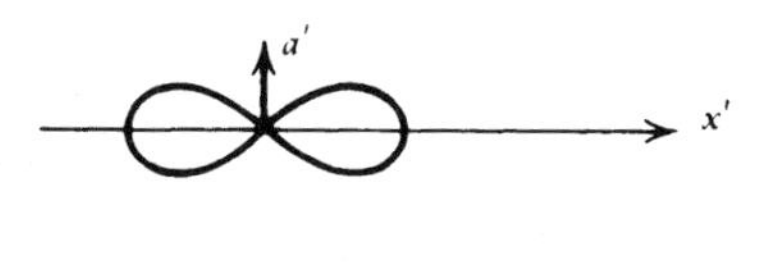

***Figure 4.11c** Same as a.*

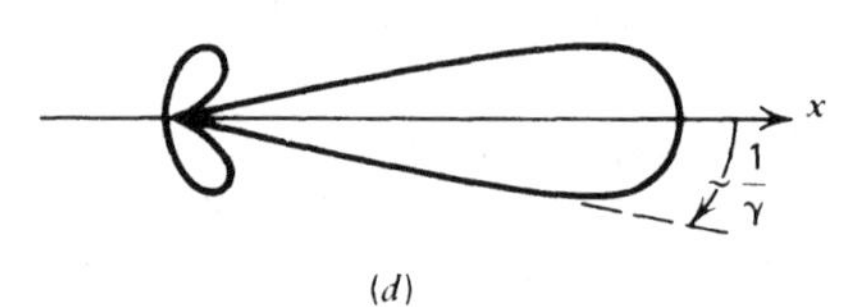

***Figure 4.11d** Angular distribution of radiation emitted by a particle with perpendicular acceleration and velocity.*

3—Extreme Relativistic Limit. When $\gamma \gg 1$, the quantity $(1-\beta\mu)$ in the denominators becomes small in the forward direction, and the radiation becomes strongly peaked in this direction. Using the same arguments as before, we obtain

$$(1-\beta\mu) \approx \frac{1+\gamma^2\theta^2}{2\gamma^2}.$$

For the parallel-acceleration case the received radiation pattern is

$$\frac{dP_{\|}}{d\Omega} \approx \frac{16q^2a_{\|}^2}{\pi c^3}\gamma^{10}\frac{\gamma^2\theta^2}{(1+\gamma^2\theta^2)^6}, \tag{4.104}$$

while for perpendicular acceleration,

$$\frac{dP_{\perp}}{d\Omega} \approx \frac{4q^2a_{\perp}^2}{\pi c^3}\gamma^{8}\frac{1-2\gamma^2\theta^2\cos 2\phi+\gamma^4\theta^4}{(1+\gamma^2\theta^2)^6}. \tag{4.105}$$

Both of these expressions depend on θ solely through the combination $\gamma\theta$. Therefore, the peaking is for angles $\theta \sim 1/\gamma$, which can be seen in Fig. 4.11, where polar diagrams of the radiation patterns are given.

4.9 INVARIANT PHASE VOLUMES AND SPECIFIC INTENSITY

Consider a group of particles that occupy a slight spread in position and in momentum at a particular time. In a frame comoving with the particles they occupy a spatial volume element $d^3\mathbf{x}' = dx'\,dy'\,dz'$ and a momentum volume element $d^3\mathbf{p}' = dP_x'\,dP_y'\,dP_z'$, but no spread in energy, $dW' = -dP_0' = 0$. This is because the contribution to the energy from the space momentum in the rest frame is quadratic and thus vanishes to the first order. The group thus occupies an element of phase space $d\mathcal{V}' = d^3\mathbf{p}'\,d^3\mathbf{x}'$. We now wish to show that any observer not comoving with the particles will conclude that they occupy the same amount of phase space in his frame $d\mathcal{V} = d^3\mathbf{p}\,d^3\mathbf{x}$. Thus a phase space element is Lorentz invariant.

Let the observer have velocity parameter β with respect to the comoving K' frame and orient axes so that he moves along the x axis. Consider first the spatial volume element $d^3\mathbf{x}$ occupied by the particles, as measured by K. Since perpendicular distances are unaffected, $dy = dy'$ and $dz = dz'$. But there is a length contraction in the x direction [cf. Eq. (4.3)], $dx = \gamma^{-1}dx'$, thus yielding the relation

$$d^3\mathbf{x} = \gamma^{-1}d^3\mathbf{x}'. \tag{4.106a}$$

Now consider the momentum volume element measured by the observer, $d^3\mathbf{p}$. The components of momentum transform as components of a four-vector, yielding $dP_y'\,dP_z' = dP_y\,dP_z$, $dP_x = \gamma(dP_x' + \beta dP_0')$. But since the particles have the same energy in the comoving frame, $dP_x = \gamma\,dP_x'$, and we obtain

$$d^3\mathbf{p} = \gamma d^3\mathbf{p}'. \tag{4.106b}$$

Combining Eqs. (4.106a) and (4.106b), we see that

$$d\mathcal{V} = d\mathcal{V}'. \tag{4.107a}$$

Since frames K and K' have arbitrary relative velocity, we have the result

$$d\mathcal{V} = \text{Lorentz invariant.} \tag{4.107b}$$

Equation (4.107a) was strictly derived only for particles of finite mass, so that frame K' could be a rest frame. However, no reference to particle mass occurs in Eq. (4.107b), and therefore it has applicability to the limiting case of photons.

From Eq. (4.107b), it follows simply that the phase space density

$$f = \frac{dN}{d\mathcal{V}} \tag{4.108}$$

is an invariant, since the number of particles within the phase volume element, dN, is a countable quantity and therefore itself invariant.

It is easy to relate the phase space density of photons to the specific intensity I_ν and thus determine the transformation properties of I_ν. This is done by evaluating the energy density per unit solid angle per frequency range in two ways, using f and also the quantity $u_\nu(\Omega)$, defined in §1.3:

$$h\nu f p^2\, dp\, d\Omega = U_\nu(\Omega)\, d\Omega\, d\nu. \tag{4.109}$$

Since $U_\nu(\Omega) = I_\nu / c$ and $p = h\nu / c$ we find that I_ν / ν^3 is simply proportional to the Lorentz invariant f, so that

$$\frac{I_\nu}{\nu^3} = \text{Lorentz invariant.} \tag{4.110}$$

Having determined the Lorentz transformation properties of the specific intensity, we should now like to determine the transformation properties of other transfer quantities. Because the source function occurs in the transfer equation as the difference $I_\nu - S_\nu$, it is clear that S_ν must have the same transformation properties as I_ν, namely,

$$\frac{S_\nu}{\nu^3} = \text{Lorentz invariant.} \tag{4.111}$$

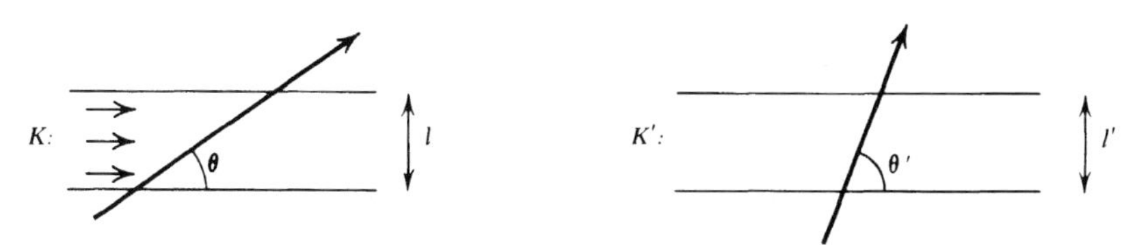

Figure 4.12 Transformation of a moving, absorbing medium.

To find the transformation of absorption coefficient we imagine material in frame K streaming with velocity v between two planes parallel to the x axis. Let K' be the rest frame of the material. (See Fig. 4.12). The optical depth τ along the ray must be an invariant, since $e^{-\tau}$ gives the *fraction of photons* passing through the material, and this involves simple counting. Thus we have the result

$$\tau = \frac{l\alpha_\nu}{\sin\theta} = \frac{l}{\nu\sin\theta}\nu\alpha_\nu = \text{Lorentz invariant.}$$

The transformation of $\sin\theta$ can be found by noting that $\nu\sin\theta$ is simply proportional to the y component of the photon four-momentum k_y. But both k_y and l are the same in both frames, being perpendicular to the motion. Therefore

$$\nu\alpha_\nu = \text{Lorentz invariant.} \tag{4.112}$$

Finally we find the transformation of the emission coefficient $j_\nu = \alpha_\nu S_\nu$ from Eqs. (4.111) and (4.112):

$$\frac{j_\nu}{\nu^2} = \text{Lorentz invariant.} \tag{4.113}$$

Another derivation of Eq. (4.113) can be based on Eq. (4.97a). The emission coefficient can be written as

$$j_\nu = n\frac{dP_e}{d\Omega\, d\nu}, \tag{4.114}$$

where n is the density of emitters (particles/cm^3). Now, from Eq. (4.12b) we have $d\nu = d\nu'\gamma(1+\beta\mu')$, and also $n = \gamma n'$ by Lorentz contraction along the motion. Thus we have

$$j_\nu = \gamma^2(1+\beta\mu')^2 n'\frac{dP'}{d\Omega'\, d\nu'} = \left(\frac{\nu}{\nu'}\right)^2 j'_{\nu'}.$$

and Eq. (4.113) follows. Notice that here it is essential to define the emission coefficient in terms of *emitted* rather than *received* power.

It is often convenient to determine the quantities α_ν, j_ν, S_ν and the like in the rest frame of the material. By the above results we can then find them in any frame. Because the transformation of ν involves the direction θ of the ray, these quantities will not, in general, be isotropic, even when they are isotropic in the rest frame. The observed nonisotropy of the cosmic microwave background can be used to find the velocity of the earth through the background (c.f. Problem 4.13).

PROBLEMS

4.1—In astrophysics it is frequently argued that a source of radiation which undergoes a fluctuation of duration Δt must have a physical diameter of order $D \lesssim c\,\Delta t$. This argument is based on the fact that even if all portions of the source undergo a disturbance at the same instant and for an infinitesimal period of time, the resulting signal at the observer will be smeared out over a time interval $\Delta t_{\min} \sim D/c$ because of the finite light travel time across the source. Suppose, however, that the source is an optically thick spherical shell of radius $R(t)$ that is expanding with relativistic velocity $\beta \sim 1, \gamma \gg 1$ and energized by a stationary point at its center. By consideration of relativistic beaming effects show that if the observer sees a fluctuation from the shell of duration Δt at time t, the source may actually be of radius

$$R < 2\gamma^2 c\,\Delta t,$$

rather than the much smaller limit given by the nonrelativistic considerations. In the rest frame of the shell surface, each surface element may be treated as an isotropic emitter.

This latter argument has been used to show that the active regions in quasars may be much larger than $c\,\Delta t \sim 1$ light month across, and thus avoids much energy being crammed into so small a volume.

4.2—Suppose that an observer at rest with respect to the fixed distant stars sees an isotropic distribution of stars. That is, in any solid angle $d\Omega$ he sees $dN = N(d\Omega/4\pi)$ stars, where N is the total number of stars he can see.

Suppose now that another observer (whose rest frame is K') is moving at a relativistic velocity β in the x direction. What is the distribution of stars seen by this observer? Specifically, what is the distribution function

$P(\theta',\phi')$ such that the number of stars seen by this observer in his solid angle $d\Omega'$ is $P(\theta',\phi')d\Omega'$? Check to see that $\int P(\theta',\phi')d\Omega'=N$, and check that $P(\theta',\phi')=N/4\pi$ for $\beta=0$. In what direction will the stars "bunch up," according to the moving observer?

4.3

a. Show that the transformation of acceleration is

$$a_x=\frac{a'_x}{\gamma^3\sigma^3},$$

$$a_y=\frac{a'_y}{\gamma^2\sigma^2}-\frac{u'_y v}{c^2}\frac{a'_x}{\gamma^2\sigma^3},$$

$$a_z=\frac{a'_z}{\gamma^2\sigma^2}-\frac{u'_z v}{c^2}\frac{a'_x}{\gamma^2\sigma^3},$$

where

$$\sigma\equiv 1+\frac{vu'_x}{c^2}.$$

b. If K' is the instantaneous rest frame of the particle, show that

$$a'_{\parallel}=\gamma^3 a_{\parallel},$$

$$a'_{\perp}=\gamma^2 a_{\perp},$$

where $a_{\parallel}$ and $a_{\perp}$ are the components parallel and perpendicular to the direction of v, respectively.

4.4—A rocket starts out from earth with a constant acceleration of $1g$ in its own frame. After 10 years of its own (proper) time it reverses the acceleration, and in 10 more years it is again at rest with respect to the earth. After a brief time for exploring, the spacemen retrace their journey back to earth, completing the entire trip in 40 years of their own time.

a. Let t be earth time and x be the position of the rocket as measured from earth. Let τ be the proper time of the rocket and let $\beta=c^{-1}dx/dt$. Show that the equation of motion of the rocket during the first phase of positive acceleration is

$$\gamma^3\frac{d^2x}{dt^2}=g.$$

b. Integrate this equation to show that

$$\beta = \frac{gt/c}{\sqrt{(gt/c)^2+1}}.$$

c. Integrating again, show that

$$x = \frac{c^2}{g}\left[\sqrt{(gt/c)^2+1} - 1\right].$$

d. Show that the proper time is related to earth time by

$$\frac{gt}{c} = \sinh\left(\frac{g\tau}{c}\right)$$

so that

$$x = \frac{c^2}{g}\left[\cosh\left(\frac{g\tau}{c}\right) - 1\right].$$

e. How far away do the spacemen get?

f. How long does their journey last from the point of view of an earth observer? Will friends be there to greet them when they return?

Hint: In answering parts (**e**) and (**f**) you need only the results for the first positive phase of acceleration plus simple arguments concerning the other phases.

g. Answer parts (**e**) and (**f**) if the spacemen can tolerate an acceleration of $2g$ rather than $1g$.

4.5—Show that $A^\alpha B^\alpha$ is not in general a scalar, where A^α and B^α are four-vector components.

4.6—Suppose in some inertial frame K a photon has four-momentum components

$$P_\mu = (-E, E, 0, 0).$$

(We use units where $c=1$). There is a special class of Lorentz transformations—called the "little group of P"—which leave the components of P unchanged, for example, a pure rotation through an angle α in the y-z

plane,

$$\begin{bmatrix} 1 & 0 & 0 & 0 \\ 0 & 1 & 0 & 0 \\ 0 & 0 & \cos\alpha & -\sin\alpha \\ 0 & 0 & \sin\alpha & \cos\alpha \end{bmatrix} \begin{bmatrix} -E \\ E \\ 0 \\ 0 \end{bmatrix} = \begin{bmatrix} -E \\ E \\ 0 \\ 0 \end{bmatrix},$$

is such a transformation. Find a sequence of pure boosts and pure rotations whose product is *not* a pure rotation in the y-z plane, but *is* in the little group of P.

4.7—An object emits a blob of material at speed v at an angle θ to the line-of-sight of a distant observer (see Fig. 4.13).

a. Show that the apparent transverse velocity inferred by the observer (i.e., the angular velocity on the sky times the distance to the object) is

$$v_{\text{app}} = \frac{v \sin\theta}{1 - (v/c)\cos\theta}.$$

b. Show that v_{app} can exceed c; find the angle for which v_{app} is maximum, and show that this maximum is $v_{\max} = \gamma v$.

4.8—Let two different uniformly moving observers have velocities $\mathbf{v}_1$ and $\mathbf{v}_2$, in units where $c = 1$. Show that their relative velocity, as measured

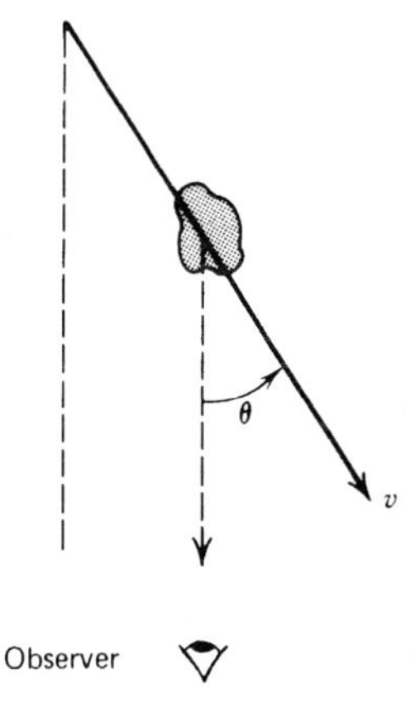

Figure 4.13 ***Emitting blob traveling at angle θ with respect to the line of sight.***

by one of the observers, satisfies

$$v^2 = \frac{(1-\mathbf{v}_1\cdot\mathbf{v}_2)^2-(1-v_1^2)(1-v_2^2)}{(1-\mathbf{v}_1\cdot\mathbf{v}_2)^2}.$$

A straight application of velocity transformations is painfully tedious, but an application of 4-vector invariants is trivial!

4.9—In ordinary three-space, Ohm's law is $\mathbf{j}=\sigma\mathbf{E}$ where $\mathbf{j}$ is the current, $\mathbf{E}$ the electric field, and σ the conductivity. Assuming σ is a scalar, write a four-tensor form of Ohm's law using the four-current j_μ, the Maxwell field tensor $F_{\mu v}$ and the four-velocity of the conducting element U_μ. Remember, a tensor equation that reduces to the correct expression in any frame (e.g., the rest frame of the conducting element) is correct in all frames.

4.10—A particle of rest mass m moves with velocity v in frame K. In its rest frame K' the particle emits some of its internal energy W' in the form of isotropic radiation.

a. Argue that there is no net reaction force on the particle and it remains at rest in K'.

b. What is the total momentum of the emitted radiation as seen in frame K?

c. Since this momentum is emitted into the forward direction, does the particle slow down as a result? If so, how can this be reconciled with the fact that the particle remains at rest in K'? If not, how can this be reconciled with the conservation of momentum?

4.11—A particle (rest mass m) initially at rest absorbs a photon of energy $h\nu$ and converts this energy into increased internal energy (say, heat). The particle has increased its rest mass to m' and moves with some velocity v'.

a. Setting up the conservation of energy and momentum, show that

$$\frac{m}{m'} = \left(1+\frac{2h\nu}{mc^2}\right)^{-1/2}.$$

b. By considering the appropriate Lorentz transformations, show that if the particle had been moving initially and absorbed a photon of energy $h\nu$, this same equation for the ratio of the initial and final rest masses holds with ν' replacing ν, where ν' is given by the Doppler formula.

4.12—Consider a particle of dust orbiting a star in a circular orbit, with velocity v. This particle absorbs stellar photons, heats up, and then emits the excess energy isotropically in its rest frame.

a. Show that in absorbing a photon the angular momentum of the particle about the star does not change. (Assume the photons are traveling radially outward from the star.)

b. When the particle emits its radiation, show that the velocity and its direction do not change, but that the angular momentum now decreases by the ratio m/m' of the rest mass after and before emission. Denoting the angular momenta before and after by l_0 and l, show that

$$l = l_0\left(1 + \frac{2h\gamma\nu}{mc^2}\right)^{-1/2}.$$

c. Having obtained this general result, let us now assume $v \ll c$ and $h\nu \ll mc^2$. By expanding, show that to lowest order the change in angular momentum caused by one photon is

$$\Delta l = -\frac{l_0 h\nu}{mc^2}.$$

Historical note: This result, although now for nonrelativistic particles, apparently cannot be derived classically. Attempts to do so by Poynting and others led to results differing from the correct answer by various numerical factors. Robertson resolved the problem in 1937 (*Mon. Not. Roy. Astron. Soc.* **97**, 423), showing that it is a relativistic effect even to lowest order. The above phenomenon is called the *Poynting–Robertson effect*.

d. A dust grain having a mass $m \sim 10^{-11}$g and cross section $\sigma \sim 10^{-8}$ cm^2 orbits the Sun at 1 A.U. Assuming that it always keeps a circular orbit, find the time for it to fall into the Sun.

4.13

a. Show that an observer moving with respect to a blackbody field of temperature T will see blackbody radiation with a temperature that depends on angle according to

$$T' = \frac{(1 - v^2/c^2)^{1/2}}{1 + (v/c)\cos\theta'}T.$$

b. The isotropy of the 2.7 K universal blackbody radiation at $\lambda = 3$ cm has been established to about one part in 10^3. What is the maximum velocity that the earth can have with respect to the frame in which this radiation is isotropic? [Isotropy is measured by the ratio $(I_{\text{max}} - I_{\text{min}})/(I_{\text{max}} + I_{\text{min}})$.] A positive result of this magnitude has recently been obtained.

4.14—A particle is accelerated by a force having components $F_{\parallel}$ and $F_{\perp}$ with respect to the particle's velocity. Show that the radiated power is

$$P = (2e^2/3m^2c^3)\left(F_{\parallel}^2 + \gamma^2 F_{\perp}^2\right).$$

Thus the perpendicular component has more effect in producing radiation that the parallel component by a factor γ^2.

4.15—Show that $U_{\text{em}}^2 - c^{-2}\mathbf{S}^2$ is a Lorentz scalar, where U_{em} is the free-space electromagnetic energy density and $\mathbf{S}$ is the Poynting vector.

4.16—Consider the *stress-energy tensor* for an electromagnetic field

$$T^{\mu\nu} \equiv \frac{1}{4\pi}\left(F^{\mu\alpha}F^{\nu}{}_{\alpha} - \frac{1}{4}\eta^{\mu\nu}F^{\alpha\beta}F_{\alpha\beta}\right)$$

where $F^{\alpha\beta}$ and $\eta^{\mu\nu}$ are the electromagnetic field tensor and Minkowski metric, respectively.

a. Show that $T^{\mu\nu}$ is traceless: $T^{\mu}{}_{\mu} = 0$.

b. Show that in free space $T^{\mu\nu}$ is divergenceless: $T^{\mu\nu}{}_{,\nu} = 0$.

REFERENCES

Jackson, J. D., 1975, *Classical Electrodynamics*, (Wiley, New York).

Taylor, E. F., and Wheeler, J. A., 1963, *Spacetime Physics* (Freeman, San Francisco).

5

BREMSSTRAHLUNG

Radiation due to the acceleration of a charge in the Coulomb field of another charge is called *bremsstrahlung* or *free-free emission*. A full understanding of this process requires a quantum treatment, since photons of energies comparable to that of the emitting particle can be produced. However, a classical treatment is justified in some regimes, and the formulas so obtained have the correct functional dependence for most of the physical parameters. Therefore, we first give a classical treatment and then state the quantum results as corrections (Gaunt factors) to the classical formulas.

First of all we shall treat nonrelativistic bremsstrahlung. Relativistic corrections are treated in §5.4. We note that bremsstrahlung due to the collision of like particles (electron-electron, proton-proton) is zero in the dipole approximation, because the dipole moment $\Sigma e_i \mathbf{r}_i$ is simply proportional to the center of mass $\Sigma m_i \mathbf{r}_i$, a constant of the motion. We therefore must consider two different particles. In electron-ion bremsstrahlung the electrons are the primary radiators, since the relative accelerations are inversely proportional to the masses, and the charges are roughly equal. Since the ion is comparatively massive, it is permissible to treat the electron as moving in a fixed Coulomb field of the ion.

5.1 EMISSION FROM SINGLE-SPEED ELECTRONS

Let us assume that the electron moves rapidly enough so that the deviation of its path from a straight line is negligible. This is the *small-angle scattering* regime. This approximation is not necessary, but it does simplify the analysis and leads to equations of the correct form. Consider an electron of charge $-e$ moving past an ion of charge Ze with impact parameter b (see Fig. 5.1). The dipole moment is $\mathbf{d}=-e\mathbf{R}$, and its second derivative is

$$\ddot{\mathbf{d}}=-e\dot{\mathbf{v}}, \tag{5.1}$$

where $\mathbf{v}$ is the velocity of the electron. Taking the Fourier transform of this equation, noting that the Fourier transform of $\ddot{\mathbf{d}}$ is $-\omega^2\hat{\mathbf{d}}(\omega)$, [cf. Eq. (3.25a)], we have the result:

$$-\omega^2\hat{\mathbf{d}}(\omega)=-\frac{e}{2\pi}\int_{-\infty}^{\infty}\dot{\mathbf{v}}e^{i\omega t}dt. \tag{5.2}$$

It is easy to derive expressions for $\hat{\mathbf{d}}(\omega)$ in the asymptotic limits of large and small frequencies. First we note that the electron is in close interaction with the ion over a time interval, called the *collision time*, which is of order

$$\tau=\frac{b}{v}. \tag{5.3}$$

For $\omega\tau\gg1$ the exponential in the integral oscillates rapidly, and the integral is small. For $\omega\tau\ll1$ the exponential is essentially unity, so we may write

$$\hat{\mathbf{d}}(\omega)\sim\begin{cases}\dfrac{e}{2\pi\omega^2}\Delta\mathbf{v}, & \omega\tau\ll1\\ 0, & \omega\tau\gg1,\end{cases} \tag{5.4}$$

where $\Delta\mathbf{v}$ is the change of velocity during the collision. Referring to Eq.

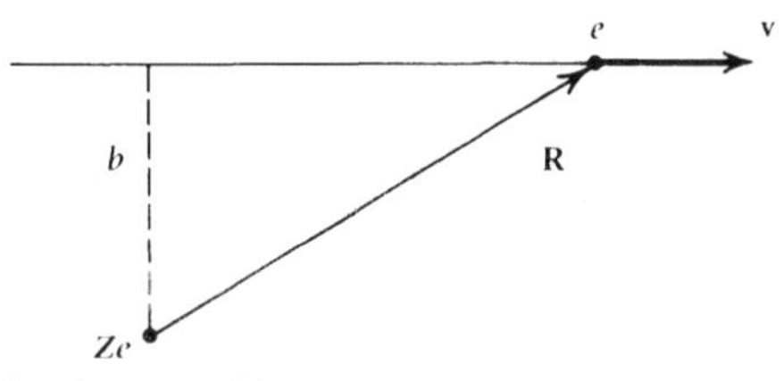

***Figure 5.1** An electron of charge e moving past an ion of charge Ze.*

(3.26b) and using Eq. (5.4), we have

$$\frac{dW}{d\omega}=\begin{cases}\dfrac{2e^2}{3\pi c^3}|\Delta\mathbf{v}|^2, & \omega\tau\ll 1\\ 0, & \omega\tau\gg 1.\end{cases} \tag{5.5}$$

Let us now estimate $\Delta\mathbf{v}$. Since the path is almost linear, the change in velocity is predominantly normal to the path. Thus we merely integrate that component of the acceleration normal to the path:

$$\Delta v=\frac{Ze^2}{m}\int_{-\infty}^{\infty}\frac{b\,dt}{(b^2+v^2t^2)^{3/2}}=\frac{2Ze^2}{mbv},$$

the integral being elementary. Thus for small angle scatterings, the emission from a single collision is

$$\frac{dW(b)}{d\omega}=\begin{cases}\dfrac{8Z^2e^6}{3\pi c^3m^2v^2b^2}, & b\ll v/\omega\\ 0, & b\gg v/\omega.\end{cases} \tag{5.6}$$

We now wish to determine the total spectrum for a medium with ion density n_i, electron density n_e and for a fixed electron speed v. Note that the flux of electrons (electrons per unit area per unit time) incident on one ion is simply $n_e v$. The element of area is $2\pi b\,db$ about a single ion. The total emission per unit time per unit volume per unit frequency range is then

$$\frac{dW}{d\omega\,dV\,dt}=n_e n_i 2\pi v\int_{b_{\min}}^{\infty}\frac{dW(b)}{d\omega}b\,db, \tag{5.7}$$

where $b_{\min}$ is some minimum value of impact parameter; its choice is discussed below.

It would seem that the asymptotic limits (5.6) are insufficient to evaluate the integral in Eq. (5.7), which requires values of $dW(b)/d\omega$ for a full range of impact parameters. However, it turns out that a very good approximation can be achieved using only its low frequency asymptotic form. To see this, substitute the $b\ll v/\omega$ result of Eq. (5.6) into Eq. (5.7). This gives

$$\frac{dW}{d\omega\,dV\,dt}=\frac{16e^6}{3c^3m^2v}n_e n_i Z^2\int_{b_{\min}}^{b_{\max}}\frac{db}{b}=\frac{16e^6}{3c^3m^2v}n_e n_i Z^2\ln\left(\frac{b_{\max}}{b_{\min}}\right), \tag{5.8}$$

where $b_{\max}$ is some value of b beyond which the $b \ll v/\omega$ asymptotic result is inapplicable and the contribution to the integral becomes negligible. The value of $b_{\max}$ is uncertain, but it is of order v/ω. Since $b_{\max}$ occurs inside the logarithm, its precise value is not very important, so we simply take

$$b_{\max} \equiv \frac{v}{\omega}, \tag{5.9}$$

and make a small error. It can now be seen that the use of the asymptotic forms (5.6) is justified, because equal intervals in the logarithm of b contribute equally to the emission, and over most of these intervals the emission is determined by its low frequency asymptotic limit.

The value of $b_{\min}$ can be estimated in two ways. First we can take the value at which the straight-line approximation ceases to be valid. Since this occurs when $\Delta v \sim v$, we take

$$b_{\min}^{(1)} = \frac{4Ze^2}{\pi m v^2}. \tag{5.10a}$$

A second value for $b_{\min}$ is quantum in nature and concerns the possibility of treating the collision process in terms of classical orbits, as we have done. By the uncertainty principle $\Delta x \Delta p \gtrsim \hbar$; and taking $\Delta x \sim b$ and $\Delta p \sim mv$ we have

$$b_{\min}^{(2)} = \frac{h}{mv}. \tag{5.10b}$$

When $b_{\min}^{(1)} \gg b_{\min}^{(2)}$ a classical description of the scattering process is valid, and we use $b_{\min} = b_{\min}^{(1)}$. This occurs when $\frac{1}{2}mv^2 \ll Z^2 Ry$, where $Ry = me^4/(2\hbar^2)$ is the Rydberg energy for the hydrogen atom. When $b_{\min}^{(1)} \ll b_{\min}^{(2)}$, or, equivalently, $\frac{1}{2}mv^2 \gg Z^2 Ry$, the uncertainty principle plays an important role, and the classical calculation cannot strictly be used. Nonetheless, results of the correct order of magnitude are obtained by simply setting $b_{\min} = b_{\min}^{(2)}$.

For any regime the exact results are conveniently stated in terms of a correction factor or *Gaunt factor* $g_{ff}(v,\omega)$ such that

$$\frac{dW}{d\omega\, dV\, dt} = \frac{16\pi e^6}{3\sqrt{3}\, c^3 m^2 v} n_e n_i Z^2 g_{ff}(v,\omega). \tag{5.11}$$

Comparison of Eqs. (5.8) and (5.11) gives g_{ff} in terms of an effective logarithm

$$g_{ff}(v,\omega)=\frac{\sqrt{3}}{\pi}\ln\left(\frac{b_{\max}}{b_{\min}}\right). \tag{5.12}$$

The Gaunt factor is a certain function of the energy of the electron and of the frequency of the emission. Extensive tables and graphs of it exist in the literature. See, for instance, the review article by Bressaard and van de Hulst, (1962) and the article by Karzas and Latter (1961).

5.2 THERMAL BREMSSTRAHLUNG EMISSION

The most interesting use of these formulas is their application to *thermal bremsstrahlung*; that is, we average the above single-speed expression over a thermal distribution of speeds. The probability dP that a particle has velocity in the velocity range $d^3\mathbf{v}$ is

$$dP\propto e^{-E/kT}d^3\mathbf{v}=\exp\left(-\frac{mv^2}{2kT}\right)d^3\mathbf{v}$$

Since $d^3\mathbf{v}=4\pi v^2dv$ for an isotropic distribution of velocities, the probability that a particle has a speed in the speed range dv is

$$dP\propto v^2\exp\left(-\frac{mv^2}{2kT}\right)dv. \tag{5.13}$$

Now we want to integrate Eq. (5.11) over this function. What are the limits of integration? At first guess, one would choose $0\leqslant v<\infty$. But at frequency ν, the incident velocity must be at least such that

$$h\nu\leqslant\tfrac{1}{2}mv^2$$

because otherwise a photon of energy $h\nu$ could not be created. This cutoff in the lower limit of the integration over electron velocities is called a *photon discreteness effect*. Performing the integral

$$\frac{dW(T,\omega)}{dV\,dt\,d\omega}=\frac{\displaystyle\int_{v_{\min}}^{\infty}\frac{dW(v,\omega)}{d\omega\,dV\,dt}v^2\exp(-mv^2/2kT)\,dv}{\displaystyle\int_0^{\infty}v^2\exp(-mv^2/2kT)\,dv},$$

where $v_{min} \equiv (2h\nu/m)^{1/2}$, and using $d\omega = 2\pi d\nu$, we obtain

$$\frac{dW}{dV\,dt\,d\nu} = \frac{2^5 \pi e^6}{3mc^3}\left(\frac{2\pi}{3km}\right)^{1/2} T^{-1/2} Z^2 n_e n_i e^{-h\nu/kT} \bar{g}_{ff}. \tag{5.14a}$$

Evaluating eq. (5.14) in CGS units, we have for the emission (erg s^{-1} cm^{-3} Hz^{-1})

$$\varepsilon_\nu^{ff} \equiv \frac{dW}{dV\,dt\,d\nu} = 6.8 \times 10^{-38} Z^2 n_e n_i T^{-1/2} e^{-h\nu/kT} \bar{g}_{ff}. \tag{5.14b}$$

Here $\bar{g}_{ff}(T,\nu)$ is a *velocity averaged Gaunt factor*. The factor $T^{-1/2}$ in Eq. (5.14) comes from the fact that $dW/dV\,dt\,d\omega \propto v^{-1}$ [cf. Eq. (5.11) and $\langle v \rangle \propto T^{1/2}$. The factor $e^{-h\nu/kT}$ comes from the lower-limit cutoff in the velocity integration due to photon discreteness and the Maxwellian shape for the velocity distribution.

Approximate analytic formulas for $\bar{g}_{ff}$ in the various regimes in which large-angle scatterings and small-angle scatterings are dominant, in which

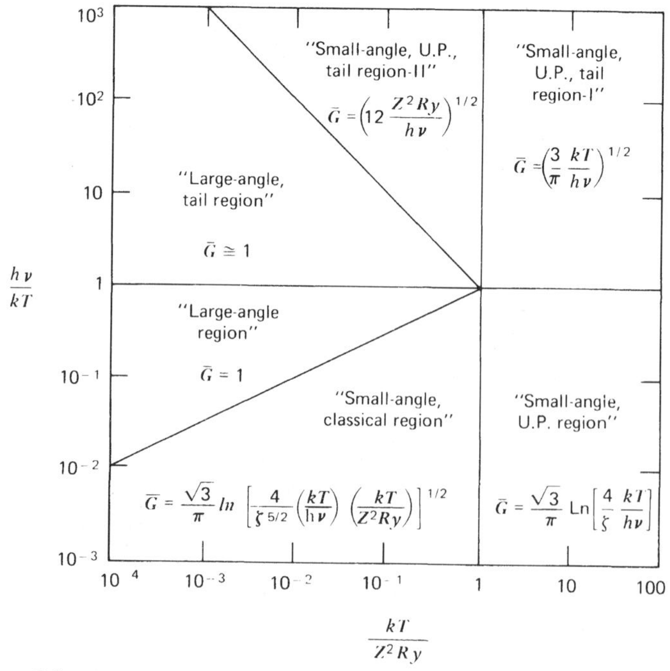

***Figure 5.2** Approximate analytic formulae for the gaunt factor $\bar{g}_{ff}(\nu, T)$ for thermal bremsstrahlung. Here $\bar{g}_{ff}$ is denoted by $\bar{G}$ and the energy unit Ry = 13.6 eV. (Taken from Novikov, I. D. and Thorne, K. S. 1973 in Black Holes, Les Houches, Eds. C. DeWitt and B. DeWitt, Gordon and Breach, New York.)*

the uncertainty principle (U. P.) is important in the minimum impact parameter, and so on are indicated in Fig. 5.2. Figure 5.3 gives numerical graphs of $\bar{g}_{ff}$. The values of $\bar{g}_{ff}$ for $u \equiv h\nu/kT \gg 1$ are not important, since the spectrum cuts off for these values. Thus $\bar{g}_{ff}$ is of order unity for $u \sim 1$ and is in the range 1 to 5 for $10^{-4} < u < 1$. We see that good order of magnitude estimates can be made by setting $\bar{g}_{ff}$ to unity.

We also see that bremsstrahlung has a rather "flat spectrum" in a log-log plot up to its cutoff at about $h\nu \sim kT$. (This is true only for optically thin sources. We have not yet considered *absorption* of photons by free electrons.)

To obtain the formulas for *nonthermal bremsstrahlung*, one needs to know the actual distributions of velocities, and the formula for emission from a single-speed electron must be averaged over that distribution. To do this one also must have the appropriate Gaunt factors.

Let us now give formulas for the total power per unit volume emitted by thermal bremsstrahlung. This is obtained from the spectral results by integrating Eq. (5.14) over frequency. The result may be stated as

$$\frac{dW}{dt\,dV} = \left(\frac{2\pi kT}{3m}\right)^{1/2} \frac{2^5 \pi e^6}{3hmc^3} Z^2 n_e n_i \bar{g}_B, \tag{5.15a}$$

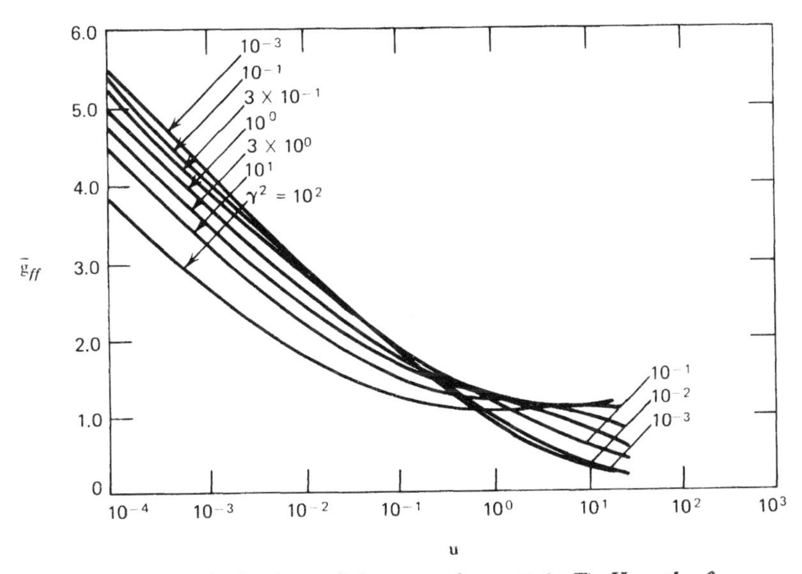

Figure 5.3 *Numerical values of the gaunt factor $\bar{g}_{ff}(\nu, T)$. Here the frequency variable is $u = 4.8 \times 10^{11}\nu/T$ and the temperature variable is $\gamma^2 = 1.58 \times 10^5 Z^2/T$. (Taken from Karzas, W. and Latter, R. 1961, Astrophys. J. Suppl., 6, 167.)*

or numerically, again in CGS units, the emission (erg s^{-1} cm^{-3}) is

$$\epsilon^{ff} \equiv \frac{dW}{dt\,dV} = 1.4 \times 10^{-27} T^{1/2} n_e n_i Z^2 \bar{g}_B. \tag{5.15b}$$

Here $\bar{g}_B(T)$ is a frequency average of the velocity averaged Gaunt factor, which is in the range 1.1 to 1.5. Choosing a value of 1.2 will give an accuracy to within about 20%.

5.3 THERMAL BREMSSTRAHLUNG (FREE-FREE) ABSORPTION

It is possible to relate the absorption of radiation by an electron moving in the field of an ion to the preceding bremsstrahlung emission process. The most interesting case is thermal free-free absorption. In that case we have Kirchhoff's law [cf. Eq. (1.37)]

$$j_\nu^{ff} = \alpha_\nu^{ff} B_\nu(T). \tag{5.16}$$

Here α_ν^{ff} is the free-free absorption coefficient, and j_ν^{ff} is related to the preceding emission formula by

$$\frac{dW}{dt\,dV\,d\nu} = 4\pi j_\nu^{ff}. \tag{5.17}$$

With the form for the Planck function [Eq. (1.51)], we have then

$$\alpha_\nu^{ff} = \frac{4e^6}{3mhc}\left(\frac{2\pi}{3km}\right)^{1/2} T^{-1/2} Z^2 n_e n_i \nu^{-3}\left(1 - e^{-h\nu/kT}\right)\bar{g}_{ff}. \tag{5.18a}$$

Evaluating Eq. (5.18a) in CGS units, we have for $\alpha_\nu^{ff}(\mathrm{cm}^{-1})$:

$$\alpha_\nu^{ff} = 3.7 \times 10^8 T^{-1/2} Z^2 n_e n_i \nu^{-3}\left(1 - e^{-h\nu/kT}\right)\bar{g}_{ff}. \tag{5.18b}$$

For $h\nu \gg kT$ the exponential is negligible, and α_ν is proportional to ν^{-3}. For $h\nu \ll kT$, we are in the Rayleigh–Jeans regime, and Eq. (5.18a) becomes

$$\alpha_\nu^{ff} = \frac{4e^6}{3mkc}\left(\frac{2\pi}{3km}\right)^{1/2} T^{-3/2} Z^2 n_e n_i \nu^{-2}\bar{g}_{ff}, \tag{5.19a}$$

or, numerically,

$$\alpha_\nu^{ff} = 0.018\, T^{-3/2} Z^2 n_e n_i \nu^{-2}\bar{g}_{ff}. \tag{5.19b}$$

The Rosseland mean of the free-free absorption coefficient [Eq. (1.109)] is, in CGS units,

$$\alpha_R^{ff} = 1.7 \times 10^{-25} T^{-7/2} Z^2 n_e n_i \bar{g}_R, \tag{5.20}$$

where $\bar{g}_R$ is an appropriately weighted frequency average of $\bar{g}_{ff}$, and is of order unity.

5.4 RELATIVISTIC BREMSSTRAHLUNG

Our previous discussion of bremsstrahlung was for nonrelativistic particles. We now show how the relativistic case can be treated by an interesting and physically picturesque method called the *method of virtual quanta.* A classical treatment provides useful insight, even though a full understanding would require quantum electrodynamics.

We consider the collision between an electron and a heavy ion of charge Ze. Normally, the ions move rather slowly in comparison to the electrons (in the rest frame of the medium as a whole), but it is possible to view the process in a frame of reference in which the electron is initially at rest. In that case the ion appears to move rapidly toward the electron. With no loss of generality we can assume that the ion moves along the x axis with velocity v while the electron is initially at rest on the y axis, a distance b from the origin. From the discussion of §4.6 we recall that the electrostatic field of the ion is transformed into an essentially transverse pulse with $|\mathbf{E}| \sim |\mathbf{B}|$, which appears to the electron to be a pulse of electromagnetic radiation (see Fig. 5.4). This radiation then Compton scatters off the electron to produce emitted radiation. Transforming back to the rest frame

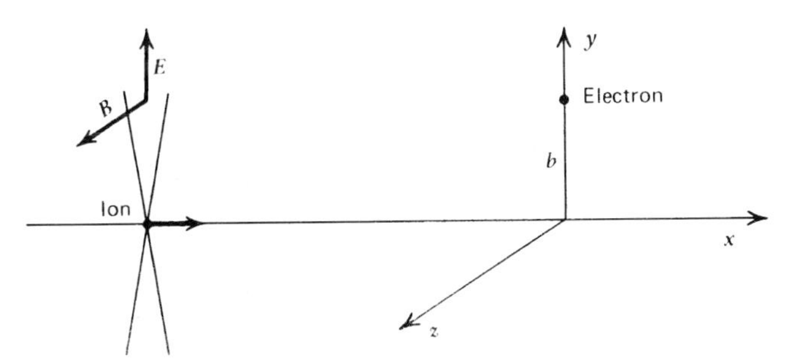

***Figure 5.4** Electric and magnetic fields of an ion moving rapidly past an electron.*

of the ion (or lab frame) we obtain the bremsstrahlung emission of the electron. Thus relativistic bremsstrahlung can be regarded as the Compton scattering of the *virtual quanta* of the ion's electrostatic field as seen in the electron's frame.

In the (primed) electron rest frame, the spectrum of the pulse of virtual quanta has the form, [cf. Eq. (4.72b)]

$$\frac{dW'}{dA'\,d\omega'}(\text{erg cm}^{-2}\text{ Hz}^{-1})=\frac{(Ze)^2}{\pi^2 b'^2 c}\left(\frac{b'\omega'}{\gamma c}\right)^2 K_1^2\left(\frac{b'\omega'}{\gamma c}\right), \tag{5.21}$$

where we have set $v=c$ in the ultrarelativistic limit. Now, in this frame the virtual quanta are scattered by the electron according to the Thomson cross section for $\hbar\omega'\lesssim mc^2$, and according to the Klein-Nishina cross section for $\hbar\omega'>mc^2$ [see Chapter 7]. In the low-frequency limit, the scattered radiation is

$$\frac{dW'}{d\omega'}=\sigma_T\frac{dW'}{dA'\,d\omega'}, \tag{5.22}$$

where σ_T is the Thomson cross section. Now, since energy and frequency transform identically under Lorentz transformations, we have for the energy emitted per frequency in the lab frame, $dW/d\omega=dW'/d\omega'$. To write $dW/d\omega$ as a function of b and ω, rather than b' and ω', we note that transverse lengths are unchanged, $b=b'$, and that $\omega=\gamma\omega'(1+\beta\cos\theta')$, [cf. Eq. (4.12b), where θ' is the scattering angle in the electron rest frame]. Because such scattering is forward-backward symmetric, we have the averaged relation $\omega=\gamma\omega'$. Thus the emission in the lab frame is

$$\frac{dW}{d\omega}=\frac{8Z^2e^6}{3\pi b^2c^5m^2}\left(\frac{b\omega}{\gamma^2 c}\right)^2 K_1\left(\frac{b\omega}{\gamma^2 c}\right). \tag{5.23}$$

Equation (5.23) is the energy per unit frequency emitted by the collision of an ion and a relativistic electron at impact parameter b. For a plasma with electron and ion densities n_e and n_i, respectively, we can repeat the arguments leading to Eq. (5.7), where v is replaced by c and where $b_{\min}\sim h/mc$ according to the uncertainty principle. The integral in Eqs. (5.7) and (5.23) is identical to that in Eq. (4.74a), except for an additional factor of γ in the argument. Thus we have the low-frequency limit, $\hbar\omega\ll\gamma mc^2$,

$$\frac{dW}{dt\,dV\,d\omega}\sim\frac{16Z^2e^6n_en_i}{3c^4m^2}\ln\left(\frac{0.68\gamma^2c}{\omega b_{\min}}\right). \tag{5.24}$$

At higher frequencies Klein–Nishina corrections must be used.

For a thermal distribution of electrons, a useful approximate expression for the frequency integrated power (erg s^{-1} cm^{-3}) in CGS units is [see Novikov and Thorne 1973]

$$\frac{dW}{dV\,dt} = 1.4\times 10^{-27} T^{1/2} Z^2 n_e n_i \bar{g}_B (1 + 4.4\times 10^{-10} T). \tag{5.25}$$

The second term in brackets is a relativistic correction to Eq. (5.15b).

PROBLEMS

5.1—Consider a sphere of ionized hydrogen plasma that is undergoing spherical gravitational collapse. The sphere is held at constant isothermal temperature T_0, uniform density and constant mass M_0 during the collapse, and has decreasing radius $R(t)$. The sphere cools by emission of bremsstrahlung radiation in its interior. At $t = t_0$ the sphere is optically thin.

a. What is the total luminosity of the sphere as a function of M_0, $R(t)$ and T_0 while the sphere is optically thin?

b. What is the luminosity of the sphere as a function of time after it becomes optically thick?

c. Give an implicit relation, in terms of $R(t)$, for the time t_1 when the sphere becomes optically thick.

d. Draw a qualitative curve of the luminosity as a function of time.

5.2—Suppose X-rays are received from a source of known distance L with a flux F (erg cm^{-2} s^{-1}). The X-ray spectrum has the form of Fig. 5.5 It is conjectured that these X-rays are due to bremsstrahlung from an optically thin, hot, plasma cloud, which is in hydrostatic equilibrium around a central mass M. Assume that the cloud thickness ΔR is roughly its radius R, $\Delta R \sim R$. Find R and the density of the cloud, ρ, in terms of the known observations and conjectured mass M. If $F = 10^{-8}$ erg cm^{-2} s^{-1}, $L = 10$ kpc, what are the constraints on M such that the source would indeed be effectively thin (for self-consistency)? Does electron scattering play any role? Here 1 kpc$\equiv$one kiloparsec, a unit of distance$\approx 3.1\times 10^{21}$ cm.

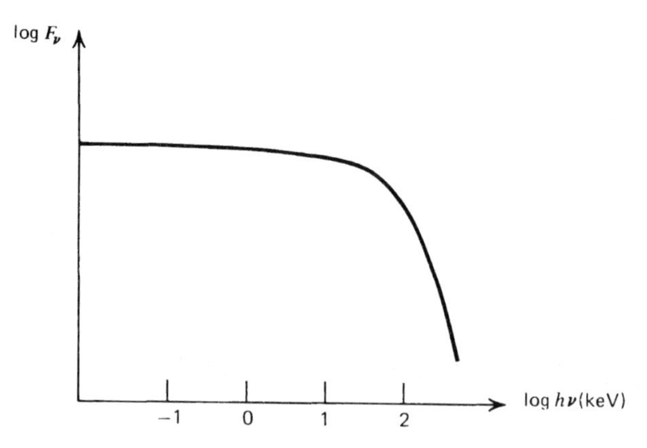

Figure 5.5 Detected spectrum from an X-ray source.

REFERENCES

Bressaard, P. J. and van de Hulst, H. C., 1962, *Rev. Mod. Phys.* **34**, 507.

Ginzburg, V. L. 1967, in *High Energy Astrophysics*, Vol. 1, C. DeWitt, E. Schatzman, P. Veron, Eds. (Gordon and Breach, New York).

Jackson, J. D., op. cit. 1975, *Classical Electrodynamics*, (Wiley, New York).

Karzas W. and Latter, R., 1961, *Astrophys. Journ. Suppl.* **6**, 167.

Novikov, I. D. and Thorne, K. S., 1973, in *Black Holes, Les Houches*, C. DeWitt and B. DeWitt, Eds. (Gordon and Breach, New York).

6

SYNCHROTRON RADIATION

Particles accelerated by a magnetic field **B** will radiate. For nonrelativistic velocities the complete nature of the radiation is rather simple and is called *cyclotron radiation*. The frequency of emission is simply the frequency of gyration in the magnetic field.

However, for extreme relativistic particles the frequency spectrum is much more complex and can extend to many times the gyration frequency. This radiation is known as *synchrotron radiation*.

6.1 TOTAL EMITTED POWER

Let us start by finding the motion of a particle of mass m and charge q in a magnetic field using the correct relativistic equations [cf. Eqs. (4.84)].

$$\frac{d}{dt}(\gamma m\mathbf{v}) = \frac{q}{c}\mathbf{v}\times\mathbf{B} \tag{6.1a}$$

$$\frac{d}{dt}(\gamma mc^2) = q\mathbf{v}\cdot\mathbf{E} = 0. \tag{6.1b}$$

This last equation implies that γ=constant or that $|\mathbf{v}|$=constant. Therefore, it follows that

$$m\gamma\frac{d\mathbf{v}}{dt}=\frac{q}{c}\mathbf{v}\times\mathbf{B}. \tag{6.2}$$

Separating the velocity components along the field $\mathbf{v}_{\parallel}$ and in a plane normal to the field $\mathbf{v}_{\perp}$ we have

$$\frac{d\mathbf{v}_{\parallel}}{dt}=0, \qquad \frac{d\mathbf{v}_{\perp}}{dt}=\frac{q}{\gamma mc}\mathbf{v}_{\perp}\times\mathbf{B}. \tag{6.3}$$

It follows that $\mathbf{v}_{\parallel}$=constant, and, since the total $|\mathbf{v}|$=constant, also $|\mathbf{v}_{\perp}|$= constant. The solution to this equation is clearly uniform circular motion of the projected motion on the normal plane, since the acceleration in this plane is normal to the velocity and of constant magnitude. The combination of this circular motion and the uniform motion along the field is a *helical* motion of the particle (Fig. 6.1). The frequency of the rotation, or gyration, is

$$\omega_B=\frac{qB}{\gamma mc}. \tag{6.4}$$

The acceleration is perpendicular to the velocity, with magnitude $a_{\perp}=\omega_B v_{\perp}$, so that the total emitted radiation is, [cf. Eq. (4.92)].

$$P=\frac{2q^2}{3c^3}\gamma^4\frac{q^2B^2}{\gamma^2m^2c^2}v_{\perp}^2, \tag{6.5a}$$

or

$$P=\frac{2}{3}r_0^2c\beta_{\perp}^2\gamma^2B^2. \tag{6.5b}$$

For an isotropic distribution of velocities it is necessary to average this formula over all angles for a given speed β. Let α be the *pitch angle*, which is the angle between field and velocity. Then we obtain

$$\langle\beta_{\perp}^2\rangle=\frac{\beta^2}{4\pi}\int\sin^2\alpha\,d\Omega=\frac{2\beta^2}{3}, \tag{6.6}$$

and the result

$$P=\left(\tfrac{2}{3}\right)^2r_0^2c\beta^2\gamma^2B^2, \tag{6.7a}$$

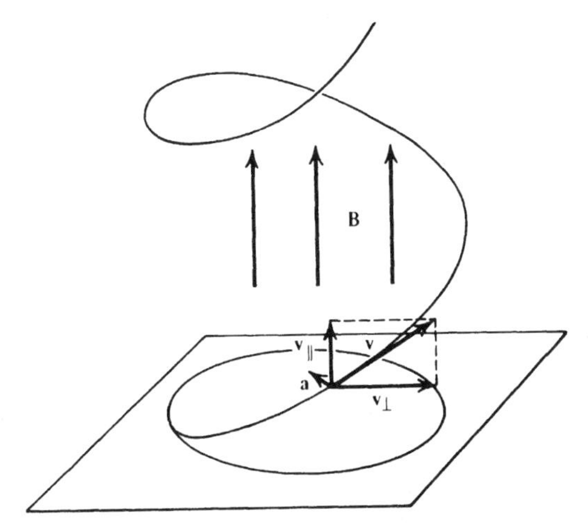

Figure 6.1 Helical motion of a particle in a uniform magnetic field.

which may be written

$$P=\frac{4}{3}\sigma_T c\beta^2\gamma^2 U_B. \tag{6.7b}$$

Here $\sigma_T=8\pi r_0^2/3$ is the Thomson cross section, and U_B is the magnetic energy density, $U_B=B^2/8\pi$.

6.2 SPECTRUM OF SYNCHROTRON RADIATION: A QUALITATIVE DISCUSSION

The spectrum of synchrotron radiation must be related to the detailed variation of the electric field as seen by an observer. Because of beaming effects the emitted radiation fields appear to be concentrated in a narrow set of directions about the particle's velocity. Since the velocity and acceleration are perpendicular, the appropriate diagram is like the one in Fig. 4.11d.

The observer will see a pulse of radiation confined to a time interval much smaller than the gyration period. The spectrum will thus be spread over a much broader region than one of order $\omega_B/2\pi$. This is an essential feature of synchrotron radiation.

We can find orders of magnitude by reference to Fig. 6.2. The observer will see the pulse from points 1 and 2 along the particle's path, where these points are such that the cone of emission of angular width $\sim 1/\gamma$ includes

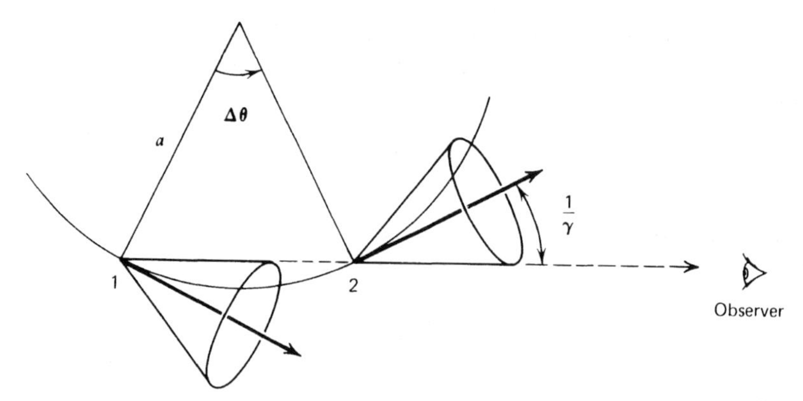

Figure 6.2 ***Emission cones at various points of an accelerated particle's trajectory.***

the direction of observation. The distance Δs along the path can be computed from the radius of curvature of the path, $a=\Delta s/\Delta\theta$.

From the geometry we have $\Delta\theta=2/\gamma$, so that $\Delta s=2a/\gamma$. But the radius of curvature of the path follows from the equation of motion

$$\gamma m\frac{\Delta \mathbf{v}}{\Delta t}=\frac{q}{c}\mathbf{v}\times\mathbf{B},$$

Since $|\Delta\mathbf{v}|=v\,\Delta\theta$ and $\Delta s=v\,\Delta t$, we have

$$\frac{\Delta\theta}{\Delta s}=\frac{qB\sin\alpha}{\gamma mcv}, \tag{6.8a}$$

$$a=\frac{v}{\omega_B\sin\alpha}. \tag{6.8b}$$

Note that this differs by a factor $\sin\alpha$ from the radius of the circle of the projected motion in a plane normal to the field. Thus Δs is given by

$$\Delta s\approx\frac{2v}{\gamma\omega_B\sin\alpha}. \tag{6.8c}$$

The times t_1 and t_2 at which the particle passes points 1 and 2 are such that $\Delta s=v(t_2-t_1)$ so that

$$t_2-t_1\approx\frac{2}{\gamma\omega_B\sin\alpha}. \tag{6.9}$$

Let t_1^A and t_2^A be the *arrival times* of radiation at the point of observation from points 1 and 2. The difference $t_2^A - t_1^A$ is less than $t_2 - t_1$ by an amount $\Delta s/c$, which is the time for the radiation to move a distance Δs. Thus we have

$$\Delta t^A = t_2^A - t_1^A = \frac{2}{\gamma \omega_B \sin \alpha}\left(1 - \frac{v}{c}\right). \tag{6.10a}$$

It should be noted that the factor $(1 - v/c)$ is the same one that enters the Doppler effect [cf. §4.1]. Since $\gamma \gg 1$, we have

$$1 - \frac{v}{c} \approx \frac{1}{2\gamma^2},$$

so that

$$\Delta t^A \approx (\gamma^3 \omega_B \sin \alpha)^{-1}. \tag{6.10b}$$

Therefore, the width of the observed pulses is smaller than the gyration period by a factor γ^3. The pulse is shown in Fig. 6.3. From our general discussion of spectra associated with particular pulses, §2.3, we expect that the spectrum will be fairly broad, cutting off at frequencies like $1/\Delta t^A$. If we define a critical frequency

$$\omega_c \equiv \frac{3}{2} \gamma^3 \omega_B \sin \alpha \tag{6.11a}$$

or

$$\nu_c = \frac{3}{4\pi} \gamma^3 \omega_B \sin \alpha, \tag{6.11b}$$

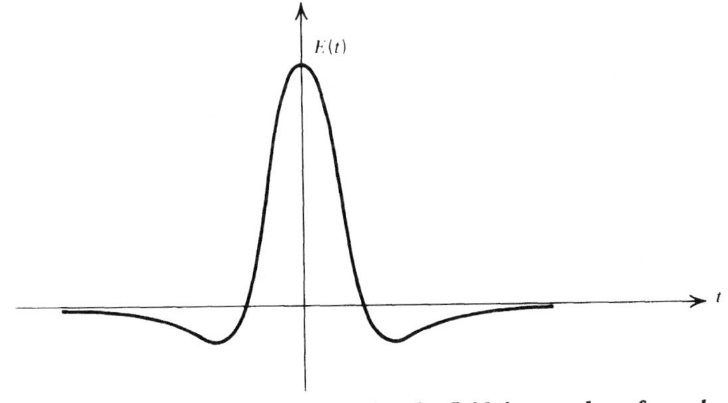

Figure 6.3 ***Time-dependence of the electric field in a pulse of synchrotron radiation.***

then we expect the spectrum to extend to something of order ω_c before falling away. We can actually derive quite a lot about the spectrum, simply using the fact that the electric field is a function of θ solely through the combination $\gamma\theta$, (see, e.g., §4.8) where θ is a polar angle about the direction of motion. This is a manifestation of the *beaming effect*. Let us write

$$E(t) \propto F(\gamma\theta), \tag{6.12}$$

where t here refers to time measured in the observer's frame. We set the zero of time and the path length s to be when the pulse is centered on the observer. Using arguments similar to those used to find Δs, we find $\theta \approx s/a$ and $t \approx (s/v)(1-v/c)$. Then the relationship of θ to t is found to be

$$\gamma\theta \approx 2\gamma(\gamma^2\omega_B \sin\alpha)t \propto \omega_c t. \tag{6.13}$$

Therefore, we write the time dependence of the electric field as

$$E(t) \propto g(\omega_c t). \tag{6.14}$$

The proportionality constant here is not yet known, and it may depend on any physical parameters except time t. This is still sufficient for us to derive the general dependence of the spectrum on ω. The Fourier transform of the electric field is

$$\hat{E}(\omega) \propto \int_{-\infty}^{\infty} g(\omega_c t) e^{i\omega t} dt. \tag{6.15a}$$

Changing variables of integration to $\xi \equiv \omega_c t$, we have

$$\hat{E}(\omega) \propto \int_{-\infty}^{\infty} g(\xi) e^{i\omega\xi/\omega_c} d\xi. \tag{6.15b}$$

The spectrum $dW/d\omega d\Omega$ is proportional to the square of $\hat{E}(\omega)$ [cf. Eqs. (2.33) and (3.11a)]. Integrating this over solid angle and dividing by the orbital period, both independent of frequency, then gives for the time-averaged power per unit frequency, [cf. Eq. (2.34)],

$$\frac{dW}{dt\,d\omega} = T^{-1}\frac{dW}{d\omega} \equiv P(\omega) = C_1 F\left(\frac{\omega}{\omega_c}\right), \tag{6.16}$$

where F is a dimensionless function and C_1 is a constant of proportionality. We may now evaluate C_1 by the simple trick of comparing the total power as evaluated by the integral over ω to the previous result in Eq. (6.5):

$$P=\int_0^\infty P(\omega)\,d\omega=C_1\int_0^\infty F\left(\frac{\omega}{\omega_c}\right)d\omega=\omega_c C_1\int_0^\infty F(x)\,dx, \qquad (6.17a)$$

where we have set $x\equiv\omega/\omega_c$. We do not know what $\int F(x)\,dx$ is until we specify $F(x)$. However, we can regard its nondimensional value as arbitrary, merely setting a convention for the normalization of $F(x)$. We can still find the dependence of the constant C_1 on all the physical parameters. From our previous discussion, we have

$$P=\frac{2q^4B^2\gamma^2\beta^2\sin^2\alpha}{3m^2c^3}, \qquad (6.17b)$$

and

$$\omega_c=\frac{3\gamma^2qB\sin\alpha}{2mc}. \qquad (6.17c)$$

We thus conclude that for the highly relativistic case ($\beta\approx1$), the power per unit frequency emitted by each electron is

$$P(\omega)=\frac{\sqrt{3}}{2\pi}\frac{q^3B\sin\alpha}{mc^2}F\left(\frac{\omega}{\omega_c}\right). \qquad (6.18)$$

The choice $\sqrt{3}/2\pi$ for the nondimensional constant has been made to anticipate the conventional choice for the normalization of F, discussed below. If the power per frequency interval $d\nu$ is desired, one can use the relation $P(\nu)=2\pi P(\omega)$.

6.3 SPECTRAL INDEX FOR POWER-LAW ELECTRON DISTRIBUTION

From the formula for $P(\omega)$ given above, it is clear that no factor of γ appears, except for that contained in ω_c. From this fact alone it is possible to derive an extremely important result concerning synchrotron spectra. Often the spectrum can be approximated by a power law over a limited range of frequency. When this is so, one defines the *spectral index* as the constant s in the expression

$$P(\omega) \propto \omega^{-s}. \tag{6.19}$$

This is the negative slope on a log $P(\omega) - \log\omega$ plot. Often the spectra of astronomical radiation has a spectral index that is constant over a fairly wide range of frequencies: for example, the Rayleigh–Jeans portion of the blackbody law has $s = -2$.

An analogous result sometimes holds for the particle distribution law of relativistic electrons. Often the number density of particles with energies between E and $E + dE$ (or γ and $\gamma + d\gamma$) can be approximately expressed in the form

$$N(E)dE = CE^{-p}dE, \qquad E_1 < E < E_2, \tag{6.20a}$$

or

$$N(\gamma)d\gamma = C\gamma^{-p}d\gamma, \qquad \gamma_1 < \gamma < \gamma_2. \tag{6.20b}$$

The quantity C can vary with pitch angle and the like. The total power radiated per unit volume per unit frequency by such a distribution is given by the integral of $N(\gamma)d\gamma$ times the single particle radiation formula over all energies or γ. Thus, we have

$$P_{\text{tot}}(\omega) = C\int_{\gamma_1}^{\gamma_2} P(\omega)\gamma^{-p}d\gamma \propto \int_{\gamma_1}^{\gamma_2} F\left(\frac{\omega}{\omega_c}\right)\gamma^{-p}d\gamma. \tag{6.21a}$$

Let us change variables of integration to $x \equiv \omega/\omega_c$, noting $\omega_c \propto \gamma^2$;

$$P_{\text{tot}}(\omega) \propto \omega^{-(p-1)/2}\int_{x_1}^{x_2} F(x)x^{(p-3)/2}dx. \tag{6.21b}$$

The limits x_1 and x_2 correspond to the limits γ_1 and γ_2 and depend on ω. However, if the energy limits are sufficiently wide we can approximate $x_1 \approx 0$, $x_2 \approx \infty$, so that the integral is approximately constant. In that case, we have

$$P_{\text{tot}}(\omega) \propto \omega^{-(p-1)/2} \tag{6.22a}$$

so that the spectral index s is related to the particle distribution index p by

$$s = \frac{p-1}{2}. \tag{6.22b}$$

Let us summarize the results of this simplified treatment of synchrotron radiation: We have shown that

1. The angular distribution from a single radiating particle lies close (within $1/\gamma$) to the cone with half-angle equal to the pitch angle.
2. The single-particle spectrum extends up to something of the order of a critical frequency ω_c. More precisely, the spectrum is a function of ω/ω_c alone.
3. For power law distribution of particle energies with index p over a sufficiently broad energy range, the spectral index of the radiation is $s=(p-1)/2$.

6.4 SPECTRUM AND POLARIZATION OF SYNCHROTRON RADIATION: A DETAILED DISCUSSION

Consider the orbital trajectory in Fig. 6.4, where the origin of the coordinates is the location of the particle at the origin of retarded time $t'=0$, and a is the radius of curvature of the trajectory. The coordinate system has been chosen so that the particle has velocity $\mathbf{v}$ along the x axis at time $t'=0$; $\boldsymbol{\epsilon}_\perp$ is a unit vector along the y axis in the orbital (x-y) plane, and

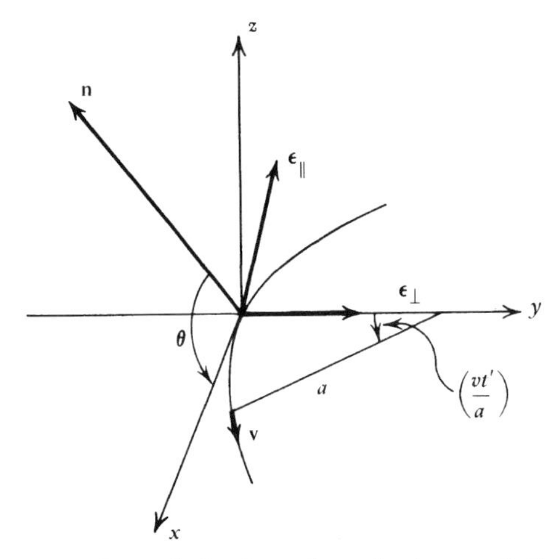

Figure 6.4 Geometry for polarization of synchrotron radiation. At $t=0$, the particle velocity is along the x axis, and a is the radius of curvature of the trajectory.

$\boldsymbol{\epsilon}_{\|}=\mathbf{n}\times\boldsymbol{\epsilon}_{\perp}$. Using Fig. 6.4, we have

$$\mathbf{n}\times(\mathbf{n}\times\boldsymbol{\beta})=-\boldsymbol{\epsilon}_{\perp}\sin\left(\frac{vt'}{a}\right)+\boldsymbol{\epsilon}_{\|}\cos\left(\frac{vt'}{a}\right)\sin\theta, \tag{6.23}$$

where we have set $|\boldsymbol{\beta}|=1$. This gives the first factor in Eq. (3.13) for $dW/d\omega\, d\Omega$. For the second factor in that equation, $\exp i\omega[t'-\mathbf{n}\cdot\mathbf{r}(t')/c]$, we note that

$$t'-\frac{\mathbf{n}\cdot\mathbf{r}(t')}{c}=t'-\frac{a}{c}\cos\theta\sin\left(\frac{vt'}{a}\right)$$
$$\approx(2\gamma^2)^{-1}\left[(1+\gamma^2\theta^2)t'+\frac{c^2\gamma^2t'^3}{3a^2}\right], \tag{6.24}$$

where we have expanded the sine and cosine functions for small arguments, used the approximation $(1-v/c)\approx1/2\gamma^2$, and set $v=c$ elsewhere. Note how the argument of the exponential in Eq. (6.24) is large and the integral is small unless $\gamma\theta\lesssim1$, $c\gamma t'/a\lesssim1$, in accordance with our qualitative discussion in 6.2 above.

An expression for the spectrum in the two polarizations states, that is, the intensity along $\boldsymbol{\epsilon}_{\|}$ and intensity along $\boldsymbol{\epsilon}_{\perp}$, may now be obtained from Eq. (3.13) and Eqs. (6.23) and (6.24) above. Expanding the sine and cosine functions again in Eq. (6.23), we obtain

$$\frac{dW}{d\omega\, d\Omega}\equiv\frac{dW_{\|}}{d\omega\, d\Omega}+\frac{dW_{\perp}}{d\omega\, d\Omega} \tag{6.25a}$$

$$\frac{dW_{\perp}}{d\omega\, d\Omega}=\frac{q^2\omega^2}{4\pi^2c}\left|\int\frac{ct'}{a}\exp\left[\frac{i\omega}{2\gamma^2}\left(\theta_\gamma^2t'+\frac{c^2\gamma^2t'^3}{3a^2}\right)\right]dt'\right|^2, \tag{6.25b}$$

$$\frac{dW_{\|}}{d\omega\, d\Omega}=\frac{q^2\omega^2\theta^2}{4\pi^2c}\left|\int\exp\left[\frac{i\omega}{2\gamma^2}\left(\theta_\gamma^2t'+\frac{c^2\gamma^2t'^3}{3a^2}\right)\right]dt'\right|^2, \tag{6.25c}$$

where

$$\theta_\gamma^2\equiv1+\gamma^2\theta^2. \tag{6.26a}$$

Now, making the changes of variables

$$y\equiv\gamma\frac{ct'}{a\theta_\gamma}, \tag{6.26b}$$

$$\eta\equiv\frac{\omega a\theta_\gamma^3}{3c\gamma^3}, \tag{6.26c}$$

Eqs. (6.25) become

$$\frac{dW_\perp}{d\omega\, d\Omega}=\frac{q^2\omega^2}{4\pi^2 c}\left(\frac{a\theta_\gamma^2}{\gamma^2 c}\right)^2\left|\int_{-\infty}^{\infty} y\exp\left[\frac{3}{2}i\eta\left(y+\frac{1}{3}y^3\right)\right]dy\right|^2, \tag{6.27a}$$

$$\frac{dW_\parallel}{d\omega\, d\Omega}=\frac{q^2\omega^2\theta^2}{4\pi^2 c}\left(\frac{a\theta_\gamma}{\gamma c}\right)^2\left|\int_{-\infty}^{\infty}\exp\left[\frac{3}{2}i\eta\left(y+\frac{1}{3}y^3\right)\right]dy\right|^2, \tag{6.27b}$$

where little error is made in extending the limits of integration from $-\infty$ to ∞. The integrals in Eqs. (6.27a) and (6.27b) are functions only of the parameter η. Since most of the radiation occurs at angles $\theta\approx 0$, η can be written as

$$\eta\approx\eta(\theta=0)=\frac{\omega}{2\omega_c}, \tag{6.28}$$

where we have used Eqs. (6.8b) and (6.11a). Thus the frequency dependence of the spectrum depends on ω only through ω/ω_c, as found in our qualitative discussion. It should also be clear that the angular dependence uses θ only through the combination $\gamma\theta$.

To make further progress, we note that the integrals in Eq. (6.27) may be expressed in terms of the modified Bessel functions of 1/3 and 2/3 order, for example, formulas: 10.4.26, 10.4.31, and 10.4.32 of Abramovitz and Stegun (1965). Therefore we can write

$$\frac{dW_\perp}{d\omega\, d\Omega}=\frac{q^2\omega^2}{3\pi^2 c}\left(\frac{a\theta_\gamma^2}{\gamma^2 c}\right)^2 K_{\frac{2}{3}}^2(\eta), \tag{6.29a}$$

$$\frac{dW_\parallel}{d\omega\, d\Omega}=\frac{q^2\omega^2\theta^2}{3\pi^2 c}\left(\frac{a\theta_\gamma}{\gamma c}\right)^2 K_{\frac{1}{3}}^2(\eta). \tag{6.29b}$$

These formulas can now be integrated over solid angle to give the energy per frequency range radiated by the particle per complete orbit in the projected normal plane. During one such orbit the emitted radiation is almost completely confined to the solid angle shown shaded in Fig. 6.5, which lies within an angle $1/\gamma$ of a cone of half-angle α. Thus it is permissible to take the element of solid angle to be $d\Omega=2\pi\sin\alpha\, d\theta$, and we can write

$$\frac{dW_\perp}{d\omega}=\frac{2q^2\omega^2a^2\sin\alpha}{3\pi c^3\gamma^4}\int_{-\infty}^{\infty}\theta_\gamma^4 K_{\frac{2}{3}}^2(\eta)\,d\theta, \tag{6.30a}$$

$$\frac{dW_\parallel}{d\omega}=\frac{2q^2\omega^2a^2\sin\alpha}{3\pi c^3\gamma^2}\int_{-\infty}^{\infty}\theta_\gamma^2\theta^2 K_{\frac{1}{3}}^2(\eta)\,d\theta. \tag{6.30b}$$

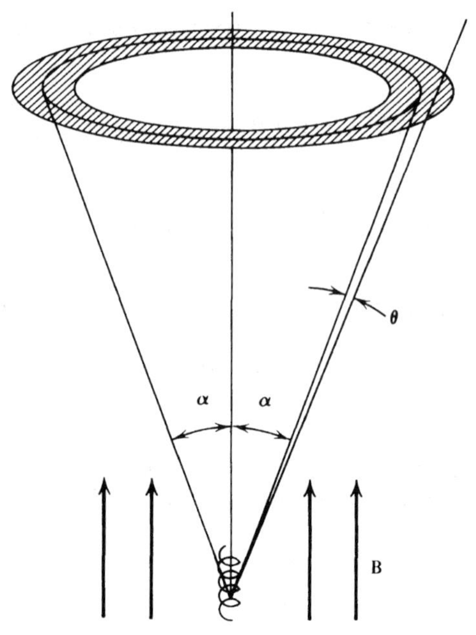

***Figure 6.5** Synchrotron emission from a particle with pitch angle α. Radiation is confined to the shaded solid angle.*

The infinite limits on the integral are convenient and permissible, because the integrand is concentrated to small values of $\Delta\theta$ about α, of order $1/\gamma$. The above integrals can be reduced further (see Westfold, 1959 for details), and we can write

$$\frac{dW_{\perp}}{d\omega}=\frac{\sqrt{3}\,q^2\gamma\sin\alpha}{2c}\left[F(x)+G(x)\right] \tag{6.31a}$$

$$\frac{dW_{\parallel}}{d\omega}=\frac{\sqrt{3}\,q^2\gamma\sin\alpha}{2c}\left[F(x)-G(x)\right], \tag{6.31b}$$

where

$$F(x)\equiv x\int_x^{\infty}K_{\frac{5}{3}}(\xi)\,d\xi, \qquad G(x)\equiv xK_{\frac{2}{3}}(x), \tag{6.31c}$$

and, again $x\equiv\omega/\omega_c$.

To convert this to emitted power per frequency we divide by the orbital period of the charge, $T=2\pi/\omega_B$,

$$P_\perp(\omega)=\frac{\sqrt{3}\,q^3B\sin\alpha}{4\pi mc^2}\left[F(x)+G(x)\right], \tag{6.32a}$$

$$P_\parallel(\omega)=\frac{\sqrt{3}\,q^3B\sin\alpha}{4\pi mc^2}\left[F(x)-G(x)\right]. \tag{6.32b}$$

The total emitted power per frequency is the sum of these:

$$P(\omega)=\frac{\sqrt{3}\,q^3B\sin\alpha}{2\pi mc^2}F(x), \tag{6.33}$$

in agreement with our previous Eq. (6.18). The function $F(x)$ is plotted in Fig. 6.6. Asymptotic forms for small and large values of x are:

$$F(x)\sim\frac{4\pi}{\sqrt{3}\,\Gamma\left(\frac{1}{3}\right)}\left(\frac{x}{2}\right)^{1/3}, \qquad x\ll 1, \tag{6.34a}$$

$$F(x)\sim\left(\frac{\pi}{2}\right)^{1/2}e^{-x}x^{1/2}, \qquad x\gg 1. \tag{6.34b}$$

To obtain frequency-integrated emission, or emission from a power-law distribution of electrons, it is useful to have expressions for integrals over the F and G functions. From Eq. 11.4.22 of Abramowitz and Stegun (1965)

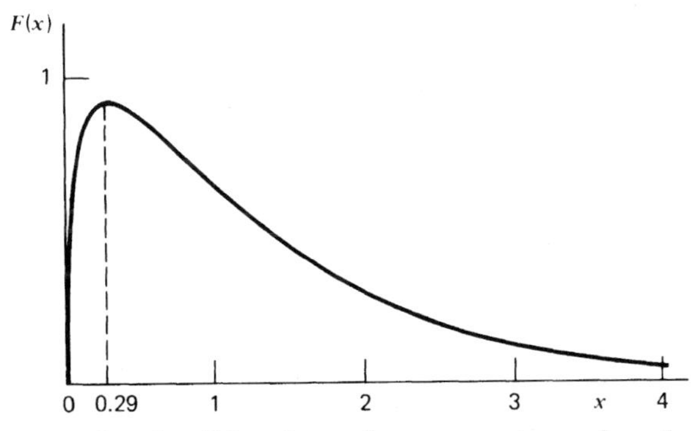

Figure 6.6 ***Function describing the total power spectrum of synchrotron emission. Here $x=\omega/\omega_c$. (Taken from Ginzburg, V. and Syrovatskii, S. 1965, Ann. Rev. Astron. Astrophys., 3, 297.)***

one may derive the following relations:

$$\int_0^\infty x^\mu F(x)\,dx = \frac{2^{\mu+1}}{\mu+2}\Gamma\left(\frac{\mu}{2}+\frac{7}{3}\right)\Gamma\left(\frac{\mu}{2}+\frac{2}{3}\right) \tag{6.35a}$$

$$\int_0^\infty x^\mu G(x)\,dx = 2^\mu\Gamma\left(\frac{\mu}{2}+\frac{4}{3}\right)\Gamma\left(\frac{\mu}{2}+\frac{2}{3}\right) \tag{6.35b}$$

where $\Gamma(y)$ is the gamma function of argument y.

For a *power-law distribution of electrons*, Eq. (6.20b), it can be shown from Eqs. (6.33) and (6.35a) that the total power per unit volume per unit frequency, $P_{\text{tot}}(\omega)$, is

$$P_{\text{tot}}(\omega) = \frac{\sqrt{3}\,q^3 C B \sin\alpha}{2\pi mc^2(p+1)}\Gamma\left(\frac{p}{4}+\frac{19}{12}\right)\Gamma\left(\frac{p}{4}-\frac{1}{12}\right)\left(\frac{mc\omega}{3qB\sin\alpha}\right)^{-(p-1)/2} \tag{6.36}$$

6.5 POLARIZATION OF SYNCHROTRON RADIATION

We can also compute the polarization for synchrotron radiation. The first point to notice is that the radiation from a single charge will be elliptically polarized, the sense of the polarization (right or left handed) being determined by whether the observed line of sight lies just inside or just outside of the cone of maximal radiation (see Fig. 6.5). However, for any reasonable distribution of particles that varies smoothly with pitch angle, the elliptical component will cancel out, as emission cones will contribute equally from both sides of the line of sight. Thus the radiation will be partially linearly polarized, and we can completely characterize the radiation by its powers per unit frequency $P_{\parallel}(\omega)$ and $P_{\perp}(\omega)$, in directions parallel and perpendicular to the projection of the magnetic field on the plane of the sky (see Fig. 6.7). From Eqs. (2.57), (6.32a), and (6.32b) we obtain the degree of linear polarization for particles of a single energy γ:

$$\Pi(\omega) = \frac{P_{\perp}(\omega) - P_{\parallel}(\omega)}{P_{\perp}(\omega) + P_{\parallel}(\omega)} = \frac{G(x)}{F(x)}. \tag{6.37}$$

This polarization is rather high; the polarization of the frequency integrated radiation is 75% (see Problem 6.5b).

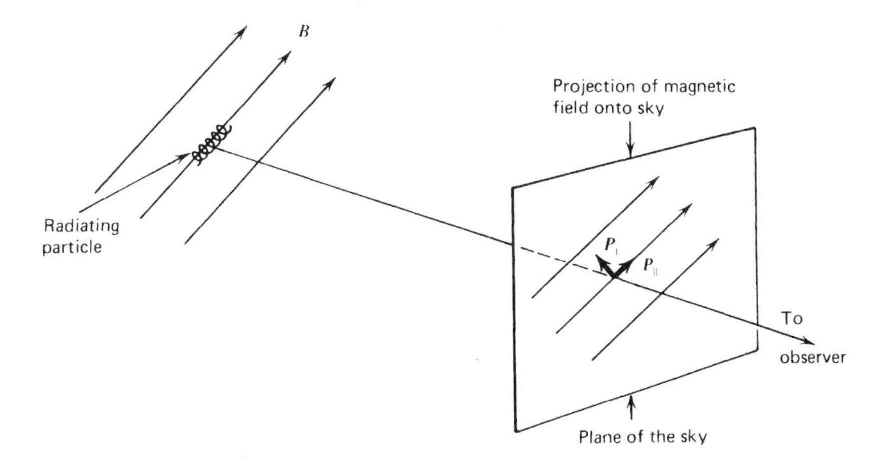

Figure 6.7 Decomposition of synchrotron polarization vectors on the plane of the sky.

For particles with a power law distribution of energies, Eq. (6.20), the degree of polarization can be shown to be (see Problem 6.5a)

$$\Pi = \frac{p+1}{p+\frac{7}{3}}. \tag{6.38}$$

6.6 TRANSITION FROM CYCLOTRON TO SYNCHROTRON EMISSION

It is interesting to follow the development of the typical synchrotron spectrum as the electron's energy is varied from the nonrelativistic through the highly relativistic regimes. Let us consider both the electric field at the observation point and the associated spectrum of radiation. For low energies the electric field components vary sinusoidally with the same frequency as the gyration in the magnetic field, and the spectrum consists of a single line, as shown in Figs. 6.8a and 6.8b (see Problem 3.2).

When v/c increases, higher harmonics of the fundamental frequency, ω_B, begin to contribute. It should be clear that the general spectrum, in fact, must be a superposition of contributions at integer multiples of ω_B, since there is periodicity in time intervals $T=2\pi/\omega_B$. Problem 3.7 demonstrates the general property that a circulating charge produces radiation at harmonics of the fundamental and that increasing harmonics contribute at

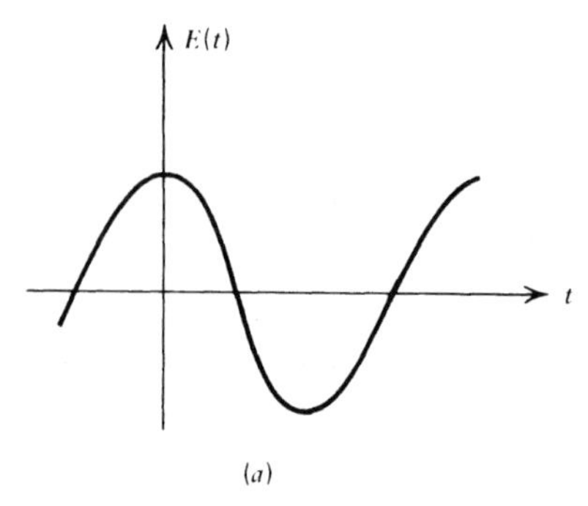

***Figure 6.8a** Time dependence of electric field from slowly moving particle in a magnetic field (cyclotron radiation).*

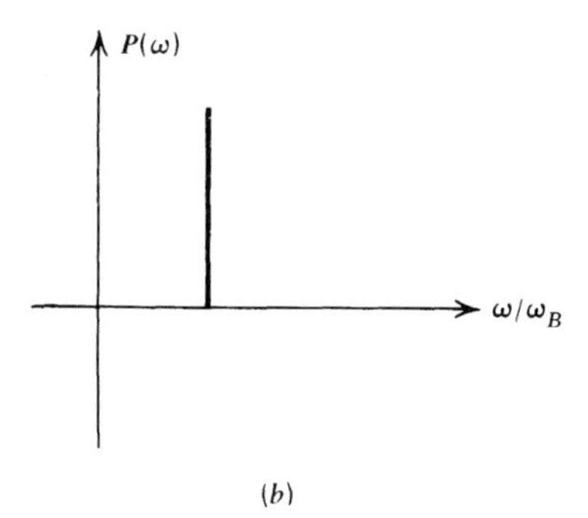

***Figure 6.8b** Power spectrum for a.*

a strength proportional to increasing powers of v/c for $v/c \ll 1$. For example, at slightly relativistic velocities, Fig. 6.8 becomes Fig. 6.9. Here we have adopted the convention that the electric field is positive as the particle approaches the observer. We see that the positive phase of the electric field has become somewhat sharper and more intense relative to the negative phase (Doppler effect). There is now a substantial amount of radiation at the first harmonic of ω_B (i.e., $2\omega_B$).

Finally, for very relativistic velocities, $v \sim c$, we have Fig. 6.10. The originally sinusoidal form of $E(t)$ has now become a series of sharp pulses, which are repeated at time intervals $2\pi/\omega_B$. The spectrum now involves a great number of harmonics, the envelope of which approaches the form of the function $F(x)$. As soon as the frequency resolution becomes large with respect to ω_B, or if other physical broadening mechanisms fill in the spaces between the lines, we approach the results derived earlier. One such

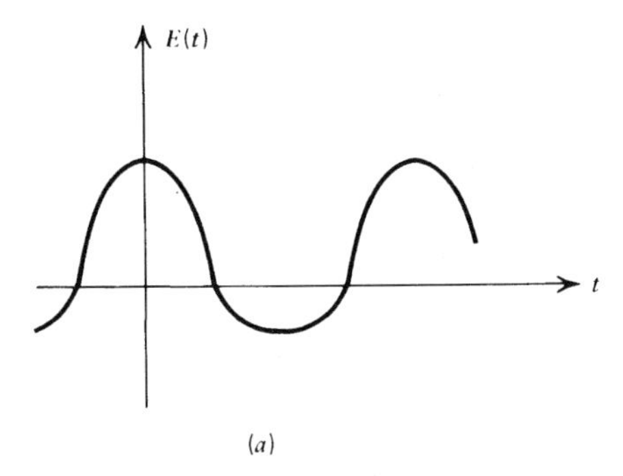

Figure 6.9a Time dependence of electric field from a particle of intermediate velocity in a magnetic field.

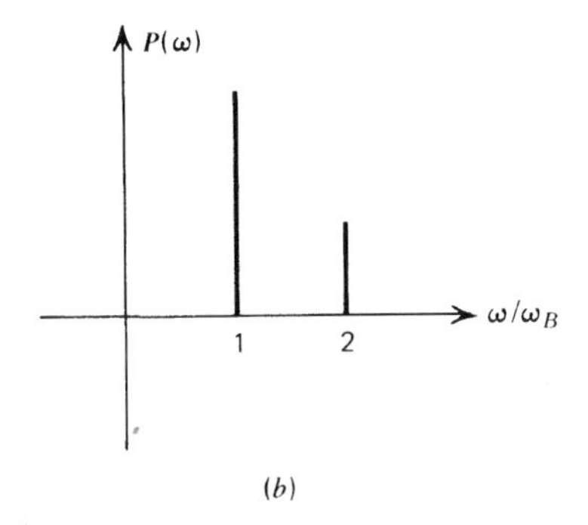

Figure 6.9b Power spectrum for a.

physical broadening mechanism occurs for a distribution of particle energies; then the gyration frequency ω_B is proportional to $1/\gamma$, so that the spectra of the particles do not fall on the same lines. Another effect that will cause the spectrum to become continuous is that emission from different parts of the emitting region may have different values and directions for the magnetic field, so that the harmonics fall at different places in the observed spectrum.

The electric field received by the observer from a distribution of particles consists of a random superposition of many pulses of the kind described here. The net result is a spectrum that is simply the sum of the spectra from the individual pulses (see Problem 3.6).

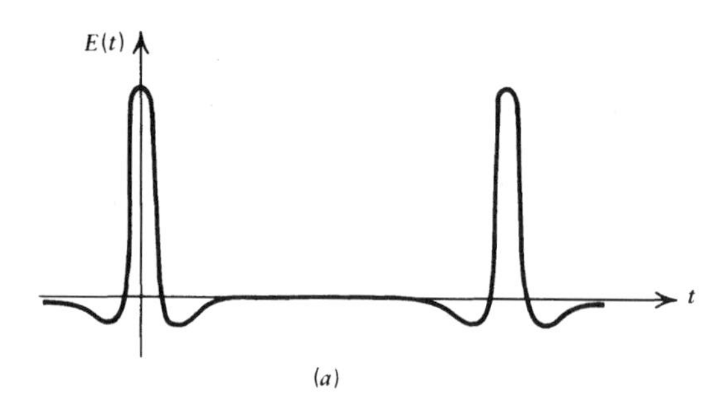

***Figure 6.10a** Time dependence of electric field from a rapidly moving particle in a magnetic field (synchrotron radiation).*

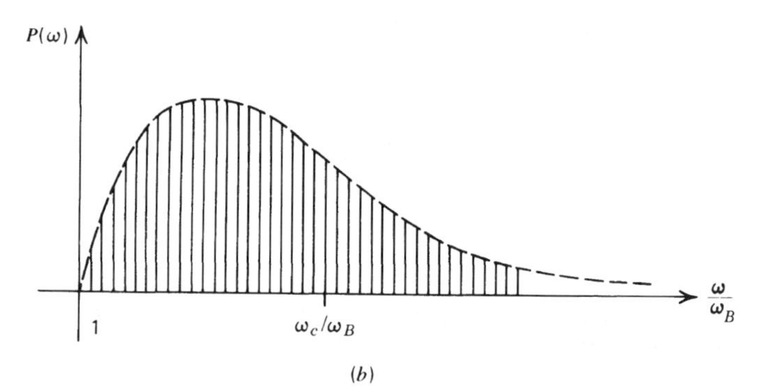

***Figure 6.10b** Power spectrum for a.*

6.7 DISTINCTION BETWEEN RECEIVED AND EMITTED POWER

In about 1968 (e.g., Pacholczyk, 1970; Ginzburg and Syrovatskii, 1969), it was noticed that a proper distinction between received and emitted power had not been made. (In looking at references before then check your formulas carefully.) The problem is that the *received* pulses are not at the frequency ω_B but at an appropriately Doppler-shifted frequency, because of the progressive motion of the particle toward the observer. This can be seen clearly in Fig. 6.11. If $T=2\pi/\omega_B$ is the orbital period of the projected motion, then time-delay effects (cf. §4.1), will give a period between the arrival of pulses T_A satisfying

$$T_A = T\left(1-\frac{v_{\|}}{c}\cos\alpha\right)$$
$$= T\left(1-\frac{v}{c}\cos^2\alpha\right)\approx\frac{2\pi}{\omega_B}\sin^2\alpha. \tag{6.39}$$

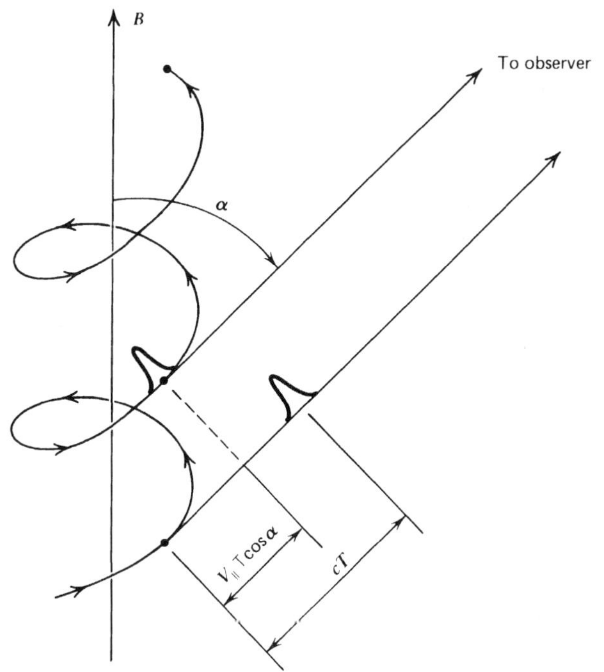

Figure 6.11 Doppler shift of synchrotron radiation emitted by a particle moving toward the observer.

The fundamental *observed* frequency is thus $\omega_B/\sin^2\alpha$ rather than ω_B. This leads to two modifications to the preceding theory, neither of which is serious, fortunately:

1. The first is that the spacing of the harmonics is $\omega_B/\sin^2\alpha$ not ω_B. For extreme relativistic particles this is not important, because one sees a continuum rather than the harmonic structure. In deriving the expression for the pulse width Δt_A and consequently for the critical frequency ω_c, we did take the Doppler compression of the radiation properly into account. Thus the continuum radiation is still a function $F(\omega/\omega_c)$.

2. The second comes from the fact that we found the *emitted power* by dividing the energy by the period T of the gyration. This is correct, but the *received power* must be obtained by dividing the energy by T_A. Thus, we have

$$P_r = \frac{P_e}{\sin^2\alpha}. \tag{6.40}$$

The question arises, should we include the $\sin^2\alpha$ factor in determining the received power? The answer depends on the physical case. Usually one observes a region fairly localized in space with only moderate net velocity toward the receiver. Then any particle that is progressing toward the receiver at one time will at a later time be moving away (and thus not contributing to the power). The average power emitted and received under these circumstances will be the same, because the total number of emitted and received pulses must be the same in the long run (see Problem 6.3). Even over short intervals this will hold when there is a stationary distribution of particles.

We conclude then that for the usual situation encountered in astrophysics one should use the expression for the emitted power to give the proper observed power. Thus the "corrections" due to helical motion are not important for most cases of interest.

6.8 SYNCHROTRON SELF-ABSORPTION

Synchrotron emission is accompanied by absorption, in which a photon interacts with a charge in a magnetic field and is absorbed, giving up its energy to the charge. Another process that can occur is stimulated emission or negative absorption, in which a particle is induced to emit more strongly into a direction and at a frequency where photons already are present. These processes can be interrelated by means of the Einstein coefficients. In our previous discussion of the Einstein coefficients (§1.6) we treated transitions between discrete states, and we must generalize that discussion now to include continuum states. This is easily done by recognizing that the states of an emitting particle are simply the free particle states, defined by its momentum, position, and possibly its internal state. According to the statistical mechanics there is one quantum state associated with the translational degrees of freedom of the particle within a volume of phase space of magnitude h^3. Thus we break up the continuous classical phase space into elements of size h^3, and consider transitions between these states as being between discrete states, for which our previous discussion applies.

A further modification of our previous results is necessary, because for a given energy of a photon $h\nu$ there are many possible transitions possible between states differing in energy by an amount $h\nu$. This means that, in the formula for the absorption coefficient given in Eq. (1.74), we must sum over all upper states 2 and lower states 1:

$$\alpha_\nu = \frac{h\nu}{4\pi}\sum_{E_1}\sum_{E_2}\left[n(E_1)B_{12} - n(E_2)B_{21}\right]\phi_{21}(\nu). \tag{6.41}$$

The profile function $\phi_{21}(\nu)$ is essentially a δ-function that restricts the summations to those states differing by an energy $h\nu = E_2 - E_1$; it will drop out of the final formulas. We have assumed that the emission and absorption are isotropic [as we did for Eq. (1.74)]. For synchrotron emission this requires that the magnetic field be tangled and have no net direction, and that the particle distributions also be isotropic.

It is now our task to reduce Eq. (6.41) to a form depending only on the previously derived formula for synchrotron emission (6.33). It is more convenient here to write the emission in terms of the frequency ν rather than ω, so that we use $P(\nu, E_2) = 2\pi P(\omega)$. We have also explicitly written the argument E_2, the energy of the radiating electron. In terms of the Einstein coefficients we have

$$\begin{aligned} P(\nu, E_2) &= h\nu \sum_{E_1} A_{21}\phi_{21}(\nu) \\ &= (2h\nu^3/c^2)h\nu \sum_{E_1} B_{21}\phi_{21}(\nu) \end{aligned} \tag{6.42}$$

where we have used one of the Einstein relations (1.71b). (Since we are dealing with elementary states, the statistical weights are all unity.)

The parts of the absorption coefficient (6.41) due to stimulated emission can be now written in terms of $P(\nu, E_2)$:

$$\frac{-h\nu}{4\pi} \sum_{E_1} \sum_{E_2} n(E_2) B_{21}\phi_{21} = \frac{-c^2}{8\pi h\nu^3} \sum_{E_2} n(E_2) P(\nu, E_2). \tag{6.43}$$

The true absorption part can be written

$$\frac{h\nu}{4\pi} \sum_{E_1} \sum_{E_2} n(E_1) B_{12}\phi_{21} = \frac{c^2}{8\pi h\nu^3} \sum_{E_2} n(E_2 - h\nu) P(\nu, E_2). \tag{6.44}$$

Here we have used the Einstein relation $B_{12} = B_{21}$. Also we have made use of the *continuous* nature of the problem by moving $n(E_1)$ from under the summation sign and replacing it by $n(E_2 - h\nu)$. This is permissible because $\phi_{21}(\nu)$ acts essentially like a δ function, enforcing the energy relation $E_1 = E_2 - h\nu$. Therefore, we have

$$\alpha_\nu = \frac{c^2}{8\pi h\nu^3} \sum_{E_2} \left[n(E_2 - h\nu) - n(E_2) \right] P(\nu, E_2). \tag{6.45}$$

Let us introduce the isotropic electron distribution function $f(p)$ by $f(p)d^3p$ = number of electrons/volume with momenta in d^3p about p.

According to statistical mechanics, the number of quantum states/volume range d^3p is simply $\tilde{\omega}h^{-3}d^3p$, where $\tilde{\omega}$ is the internal statistical weight of the electron (=2 for spin=1/2 particles). The electron density per quantum state is thus $(h^3/\tilde{\omega})f(p)$. Therefore, we can make the replacements

$$\sum_2 \to \frac{\tilde{\omega}}{h^3}\int d^3p_2, \qquad n(E_2)\to\frac{h^3}{\tilde{\omega}}f(p_2).$$

Then Eq. (6.45) becomes

$$\alpha_\nu = \frac{c^2}{8\pi h\nu^3}\int d^3p_2\big[f(p_2^*)-f(p_2)\big]P(\nu,E_2) \tag{6.46}$$

where p_2^* is the momentum corresponding to energy $E_2 - h\nu$. Before specializing this formula further, let us check that it yields the correct result for a *thermal distribution* of particles, that is

$$f(p) = K\exp\left[-\frac{E(p)}{kT}\right].$$

We note that

$$\begin{aligned} f(p_2^*)-f(p_2) &= K\exp\left(-\frac{E_2-h\nu}{kT}\right) - K\exp\left(-\frac{E_2}{kT}\right) \\ &= f(p_2)(e^{h\nu/kT}-1). \end{aligned}$$

Thus the absorption coefficient is

$$(\alpha_\nu)_{\text{thermal}} = \frac{c^2}{8\pi h\nu^3}(e^{h\nu/kT}-1)\int d^3p_2 f(p_2)P(\nu,E_2). \tag{6.47}$$

But the integral here simply represents the total power per volume per frequency range, which is $4\pi j_\nu$ for isotropic emission. Recognizing the formula for $B_\nu(T)$ this can be written

$$(\alpha_\nu)_{\text{thermal}} = \frac{j_\nu}{B_\nu(T)},$$

which is the correct result for thermal emission (Kirchhoff's Law).

Because the electron distribution is isotropic it is convenient to use the energy rather than the momentum to describe the distribution function, that is, $N(E)$, as in Eq. (6.20). We shall also assume the extreme relativistic relation $E=pc$. Then from the relation

$$N(E)dE=f(p)4\pi p^2dp \tag{6.48}$$

we obtain

$$\alpha_\nu=\frac{c^2}{8\pi h\nu^3}\int dE\,P(\nu,E)E^2\left[\frac{N(E-h\nu)}{(E-h\nu)^2}-\frac{N(E)}{E^2}\right], \tag{6.49}$$

where we now have simply written E instead of E_2.

We now assume that $h\nu \ll E$. This is, in fact, a necessary condition for the application of classical electrodynamics, so is already an implicit restriction on our formula for $P(\nu,E)$. Expanding to first order in $h\nu$ we obtain

$$\alpha_\nu=-\frac{c^2}{8\pi\nu^2}\int dE\,P(\nu,E)E^2\frac{\partial}{\partial E}\left[\frac{N(E)}{E^2}\right]. \tag{6.50}$$

Let us again look at the case of a thermal distribution, which for ultrarelativistic particles is

$$N(E)=KE^2e^{-E/kT}. \tag{6.51}$$

This leads to the result

$$(\alpha_\nu)_{\text{thermal}}=\frac{c^2}{8\pi\nu^2kT}\int N(E)P(\nu,E)dE=\frac{j_\nu c^2}{2\nu^2kT},$$

which is Kirchhoff's law in the Rayleigh–Jeans regime. This is to be expected, because of the assumption $h\nu \ll E$.

For a *power law distribution of particles*, Eq. (6.20), we have

$$-E^2\frac{d}{dE}\left[\frac{N(E)}{E^2}\right]=(p+2)CE^{-(p+1)}=\frac{(p+2)N(E)}{E},$$

and the absorption coefficient (6.50) can be written

$$\alpha_\nu=\frac{(p+2)c^2}{8\pi\nu^2}\int dE\,P(\nu,E)\frac{N(E)}{E}. \tag{6.52}$$

It is straightforward to show, using Eqs. (6.33) and (6.35a), that the integral gives

$$\alpha_\nu = \frac{\sqrt{3}\, q^3}{8\pi m}\left(\frac{3q}{2\pi m^3 c^5}\right)^{p/2} C(B\sin\alpha)^{(p+2)/2}\Gamma\left(\frac{3p+2}{12}\right)\Gamma\left(\frac{3p+22}{12}\right)\nu^{-(p+4)/2}. \tag{6.53}$$

The source function can be found from

$$S_\nu = \frac{j_\nu}{\alpha_\nu} = \frac{P(\nu)}{4\pi\alpha_\nu} \propto \nu^{5/2}, \tag{6.54}$$

using Eq. (6.53). A simple way of deriving this latter result is to note that S_ν can be written as $S_\nu \propto \nu^2 \bar{E}$ where $\bar{E}$ is a mean particle energy [cf. Eqs. (6.52) and (6.54). The appropriate value for $\bar{E}$ is the energy of those electrons whose critical frequency equals ν, that is, $\bar{E}^2 \propto \nu_c = \nu$, so that one obtains the proportionality given in Eq. (6.54). It is of some interest that the source function is a power law with an index $-\frac{5}{2}$, independent of the value of p. It should be particularly noted that this index is not equal to -2, the Rayleigh–Jeans value, because the emission is nonthermal.

For optically thin synchrotron emission, the observed intensity is proportional to the emission function, while for optically thick emission it is proportional to the source function. Since the emission and source functions for a nonthermal power law electron distribution are proportional to $\nu^{-(p-1)/2}$ and $\nu^{5/2}$, respectively, [cf. eqs. (6.22a) and (6.54)] we see that the optically thick region occurs at low frequencies and produces a low-frequency cutoff of the spectrum (see Fig. 6.12).

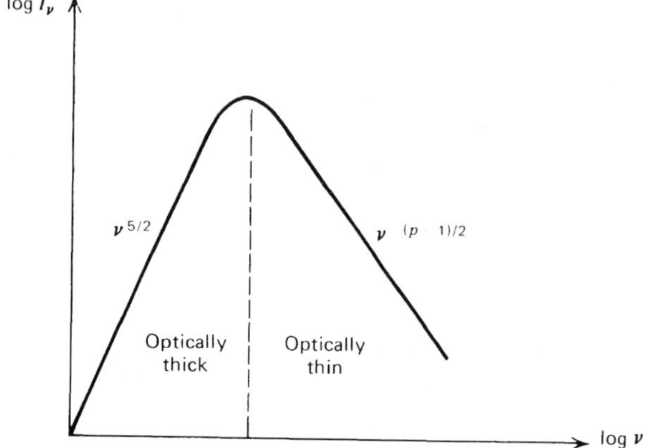

***Figure 6.12** Synchrotron spectrum from a power-law distribution of electrons.*

6.9 THE IMPOSSIBILITY OF A SYNCHROTRON MASER IN VACUUM

It is possible to prove that the absorption coefficient is positive for an arbitrary distribution of particle energies $N(E)$. That is, if we attempted to cause a population inversion by increasing $N(E)$ at a certain energy E_0 so that emission from E_0 to $E_0 - h\nu$ was a maser transition, we would inevitably be making a still stronger positive absorption somewhere else that would more than compensate. To show this analytically we can integrate Eq. (6.50) by parts, noting that $N(E)P(\nu,E)$ vanishes for low and high energies:

$$\alpha_\nu = \frac{c^2}{8\pi\nu^2}\int \frac{N(E)}{E^2}\frac{d}{dE}\left[E^2P(\nu,E)\right]dE.$$

For any fixed ν,

$$E^2P(\nu,E) \propto x^{-1}F(x) = \int_x^\infty K_{\frac{5}{3}}(\eta)\,d\eta.$$

This is clearly a monotonically decreasing function of x, since $K_{\frac{5}{3}}(\eta)$ is positive. Therefore, $E^2P(\nu,E)$ is a monotonically increasing function of E, and α_ν is positive.

We actually should also look at the absorption coefficients for specific polarization states to complete the proof of impossibility of masers. For the two states of polarization

$$P(\nu,E) \propto F(x) \pm G(x).$$

Since $x^{-1}G(x) = K_{\frac{2}{3}}(x)$, which decreases monotonically with x, we need only consider the polarization state in the parallel direction. By use of Eq. 10.1.22 of Abramowitz and Stegun (1965), we obtain the identity

$$\frac{1}{x}\left[F(x) - G(x)\right] = \frac{2}{3}\int_x^\infty K_{\frac{2}{3}}(\eta)\eta^{-1}\,d\eta,$$

which again is clearly monotonically decreasing with x.

Although synchrotron masers cannot exist in vacuum, it is possible to show that in a plasma, where the index of refraction is not unity, such synchrotron maser emission is possible.

PROBLEMS

6.1—An ultrarelativistic electron emits synchrotron radiation. Show that its energy decreases with time according to

$$\gamma=\gamma_0(1+A\gamma_0 t)^{-1}, \qquad A=\frac{2e^4B_\perp^2}{3m^3c^5}.$$

Here γ_0 is the initial value of γ and $B_\perp = B\sin\alpha$. Show that the time for the electron to lose half its energy is

$$t_{\frac{1}{2}}=(A\gamma_0)^{-1}=\frac{5.1\times10^8}{\gamma_0 B_\perp^2}.$$

How does one reconcile the decrease of γ here with the result of constant γ implied by Eqs. (6.1)?

6.2—A region of space contains relativistic electrons and magnetic fields. Let a typical linear scale of this region be l. Suppose the region is compressed (by passage of a shock wave, perhaps). Assume that the compression is the same in all directions. We want to see what effect this compression has on various properties of the electrons and magnetic field.

a. Show that the magnetic field satisfies $B\propto l^{-2}$.

b. If the compression is slow, show that the momentum of an electron satisfies $p\propto l^{-1}$, and that magnetic flux through electron orbits is approximately conserved.

c. Show that the synchrotron emission $P\propto l^{-6}$, that the critical frequency $\nu_c\propto l^{-4}$ and that the half-life for the electron $t_{\frac{1}{2}}\propto l^5$. (This shows that moderate compression can profoundly effect observed emission.)

6.3—Ultrarelativistic electrons are emitting synchrotron radiation in a fairly uniform magnetic field. The observer's line of sight makes an angle α with respect to B. (See Fig. 6.13).

The electrons are confined to the region between points 1 and 2 by constrictions in the magnetic field, which reflect the electrons back and forth along the field lines while maintaining their pitch angles. Show that a given electron, while radiating continually in its own frame, produces observable radiation only for a fraction $\frac{1}{2}\sin^2\alpha$ of the time.

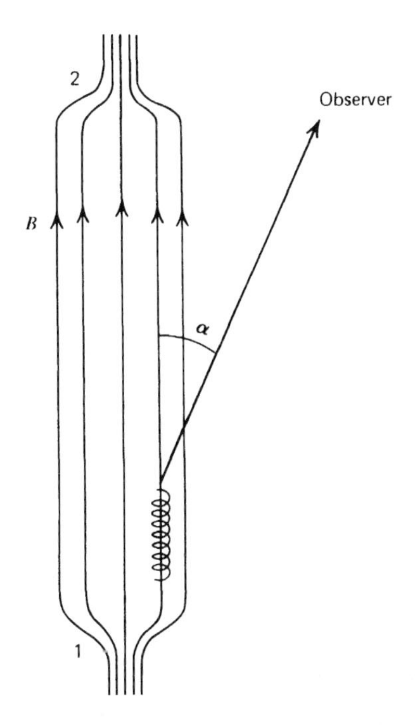

Figure 6.13 Synchrotron emission from electrons confined between positions 1 and 2.

6.4—The spectrum shown in Fig. 6.14 is observed from a point source of unknown distance d. A model for this source is a spherical mass of radius R that is emitting synchrotron radiation in a magnetic field of strength B. The space between us and the source is uniformly filled with a thermal bath of hydrogen that emits and absorbs mainly by bound-free transitions, and it is believed that the hydrogen bath is unimportant compared to the synchrotron source at frequencies where the former is optically thin. The synchrotron source function can be written as

$$S_\nu = A(\mathrm{erg\,cm^{-2}\,s^{-1}\,Hz^{-1}})\left(\frac{B}{B_0}\right)^{-1/2}\left(\frac{\nu}{\nu_0}\right)^{5/2}.$$

The absorption coefficient for synchrotron radiation is

$$\alpha_\nu^s = C(\mathrm{cm^{-1}})\left(\frac{B}{B_0}\right)^{(p+2)/2}\left(\frac{\nu}{\nu_0}\right)^{-(p+4)/2},$$

and that for bound-free transitions is

$$\alpha_\nu^{bf} = D(\mathrm{cm^{-1}})\left(\frac{\nu}{\nu_0}\right)^{-3},$$

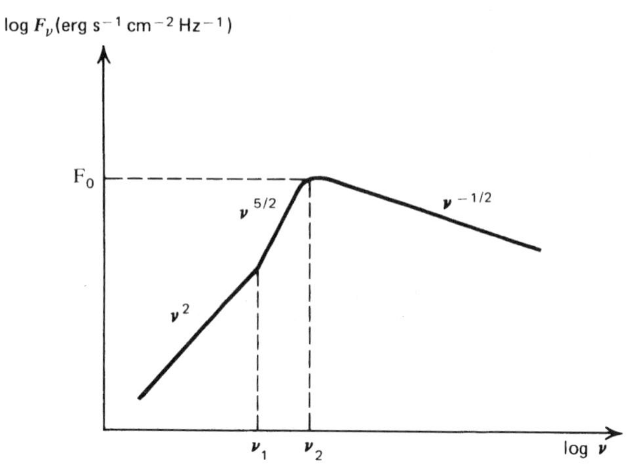

Figure 6.14 Observed spectrum from a point source.

where A, B_0, ν_0, C and D are constants and p is the power law index for the assumed power law distribution of relativistic electrons in the synchrotron source.

a. Find the size of the source R and the magnetic field strength B in terms of the solid angle $\Omega = \pi(R^2/d^2)$ subtended by the source and the constants A, B_0, ν_0, C, D.

b. Now using D and ν_1, in addition to the previous constants, find the solid angle of the source and its distance.

6.5

a. Derive the linear polarization for a power-law distribution of electrons, $N(\gamma) = C\gamma^{-p}$, emitting synchrotron radiation, Eq. (6.38).

b. Show that the linear polarization for the frequency-integrated synchrotron emission of particles of the same γ is 75%.

REFERENCES

Abramowitz, M. and Stegun, I. A., 1965, *Handbook of Mathematical Functions* (National Bureau of Standards, Washington, D.C.).

Ginzburg, V. L. and Syrovatskii, S. I., 1969, *Ann. Rev. Astron. Astrophys.*, 7, 375.

Jackson, J. D., 1975, *Classical Electrodynamics*, (Wiley, New York).

Pacholczyk, A. G. 1970, *Radio Astrophysics*, (Freeman, San Francisco).

Westfold, K. C., 1959, "The Polarization of Synchrotron Radiation," *Astrophys. J.*, **130**, 241.

7

COMPTON SCATTERING

7.1 CROSS SECTION AND ENERGY TRANSFER FOR THE FUNDAMENTAL PROCESS

Scattering from Electrons at Rest

For low photon energies, $h\nu \ll mc^2$, the scattering of radiation from free charges reduces to the classical case of Thomson scattering, discussed in Chapter 4. Recall that for Thomson scattering, when the incident photons are approximated as a continuous electromagnetic wave [cf. Eq. (3.40)],

$$\epsilon = \epsilon_1, \tag{7.1a}$$

$$\frac{d\sigma_T}{d\Omega} = \frac{1}{2} r_0^2 (1 + \cos^2\theta), \tag{7.1b}$$

$$\sigma_T = \frac{8\pi}{3} r_0^2. \tag{7.1c}$$

Here ϵ and ϵ_1 are the incident and scattered photon energy, respectively, $d\sigma_T / d\Omega$ is the differential Thomson cross section for unpolarized incident radiation, and r_0 is the classical electron radius. When $\epsilon = \epsilon_1$, the scattering is called *coherent* or *elastic*.

Quantum effects appear in two ways: First, through the kinematics of the scattering process, and, second, through the alteration of the cross sections. The kinematic effects occur because a photon possesses a momentum $h\nu/c$ as well as an energy $h\nu$. The scattering will no longer be elastic $(\epsilon_1 \neq \epsilon)$ because of the recoil of the charge. Let us set up the conservation of energy and momentum relations. The initial and final four-momenta of the photon are $\vec{P}_{\gamma i}=(\epsilon/c)(1,\mathbf{n}_i)$ and $\vec{P}_{\gamma f}=(\epsilon_1/c)(1,\mathbf{n}_f)$ and the initial and final momenta of the electron are $\vec{P}_{ei}=(mc,\mathbf{0})$ and $\vec{P}_{ef}=(E/c,\mathbf{p})$, where $\mathbf{n}_i$ and $\mathbf{n}_f$ are the initial and final directions of the photons (see Fig. 7.1). Conservation of momentum and energy is expressed by $\vec{P}_{ei}+\vec{P}_{\gamma i}=\vec{P}_{ef}+\vec{P}_{\gamma f}$. Rearranging terms and squaring gives $|\vec{P}_{ef}|^2=|\vec{P}_{ei}+\vec{P}_{\gamma i}-\vec{P}_{\gamma f}|^2$, which eliminates the final electron momentum. We thus finally obtain

$$\epsilon_1=\frac{\epsilon}{1+\dfrac{\epsilon}{mc^2}(1-\cos\theta)}. \tag{7.2}$$

In terms of wavelength, this can be written:

$$\lambda_1-\lambda=\lambda_c(1-\cos\theta) \tag{7.3a}$$

where the *Compton wavelength* is defined by

$$\lambda_c\equiv\frac{h}{mc}$$
$$=0.02426\ \text{Å for electrons.} \tag{7.3b}$$

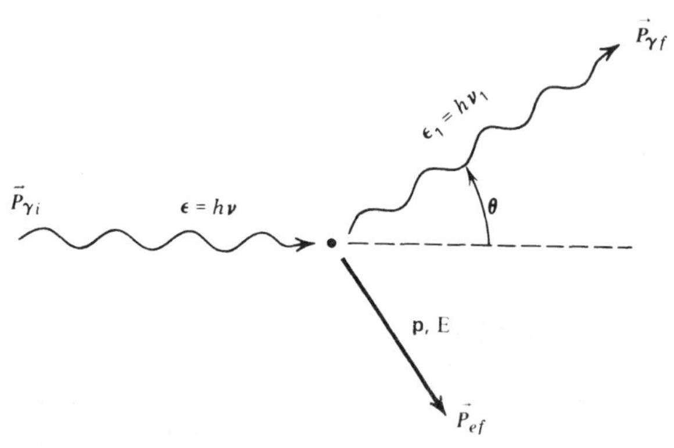

***Figure 7.1** Geometry for scattering of a photon by an electron initially at rest.*

We see that there is a wavelength change of the order of λ_c upon scattering. For long wavelengths $\lambda \gg \lambda_c$ (i.e., $h\nu \ll mc^2$) the scattering is closely elastic. When this condition is satisfied, we can assume that there is no change in photon energy in the rest frame of the electron.

Although a derivation is outside the scope of this text, let us briefly describe the quantum effect on the cross section. The differential cross section for unpolarized radiation is shown in quantum electrodynamics (Heitler, 1954) to be given by the *Klein–Nishina* formula

$$\frac{d\sigma}{d\Omega} = \frac{r_0^2}{2} \frac{\epsilon_1^2}{\epsilon^2} \left(\frac{\epsilon}{\epsilon_1} + \frac{\epsilon_1}{\epsilon} - \sin^2\theta \right). \tag{7.4}$$

Note that for $\epsilon_1 \sim \epsilon$ Eq. (7.4) reduces to the classical expression. The principal effect is to *reduce* the cross section from its classical value as the photon energy becomes large. Thus Compton scattering becomes less efficient at high energies. The total cross section can be shown to be

$$\sigma = \sigma_T \cdot \frac{3}{4} \left[\frac{1+x}{x^3} \left\{ \frac{2x(1+x)}{1+2x} - \ln(1+2x) \right\} + \frac{1}{2x} \ln(1+2x) - \frac{1+3x}{(1+2x)^2} \right] \tag{7.5}$$

where $x \equiv h\nu / mc^2$. In the nonrelativistic regime we have approximately

$$\sigma \approx \sigma_T \left(1 - 2x + \frac{26x^2}{5} + \cdots \right), \qquad x \ll 1, \tag{7.6a}$$

whereas for the extreme relativistic regime we have

$$\sigma = \frac{3}{8} \sigma_T x^{-1} \left(\ln 2x + \frac{1}{2} \right), \qquad x \gg 1. \tag{7.6b}$$

Scattering from Electrons in Motion: Energy Transfer

In the rest of this section we assume that in the rest frame of the electron $h\nu \ll mc^2$, so that the relativistic corrections in the Klein–Nishina formula may be neglected. Whenever the moving electron has sufficient kinetic energy compared to the photon, net energy may be transferred from the electron to the photon, in contrast to the situation indicated in Eq. (7.2). In such a case the scattering process is called *inverse Compton*.

Let us call K the lab or observer's frame, and let K' be the rest frame of the electron. The scattering event as seen in each frame is given in Fig. 7.2.

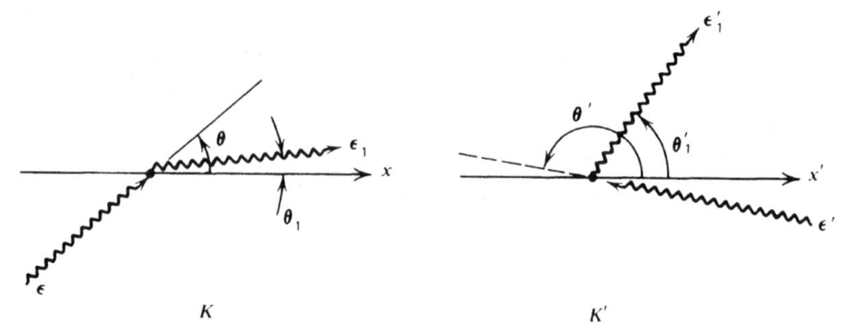

***Figure 7.2** Scattering geometries in the observer's frame K and in the electron rest frame K′.*

Note that our previous formulas for scattering from electrons at rest should now be written in primed notation, since they hold in the electron rest frame. From the Doppler shift formulas, [cf. (4.12)],

$$\epsilon' = \epsilon\gamma(1-\beta\cos\theta), \tag{7.7a}$$

$$\epsilon_1 = \epsilon_1'\gamma(1+\beta\cos\theta_1'). \tag{7.7b}$$

Now, we also know, from Eq. (7.2) that

$$\epsilon_1' \approx \epsilon'\left[1-\frac{\epsilon'}{mc^2}(1-\cos\Theta)\right] \tag{7.8a}$$

$$\cos\Theta = \cos\theta_1'\cos\theta' + \sin\theta'\sin\theta_1'\cos(\phi'-\phi_1'), \tag{7.8b}$$

where ϕ_1' and ϕ' are the azimuthal angles of the scattered photon and incident photon in the rest frame.

In the case of relativistic electrons, $\gamma^2 - 1 \gg h\nu/mc^2$, the energies of the photon before scattering, in the rest frame of the electron, and after scattering are in the approximate ratios

$$1:\gamma:\gamma^2,$$

providing that the condition for Thomson scattering in the rest frame $\gamma\epsilon \ll mc^2$ is met. This follows from Eqs. (7.7), since θ and θ_1' are characteristically of order $\pi/2$.

This process therefore converts a low-energy photon to a high-energy one by a factor of order γ^2. Since the intermediate photon energy can be as high as, say, 100 keV and still be in the Thomson limit, it can be seen that photons of enormous energies ($\gamma \times 100$ keV) can be produced. If the intermediate energy is too high, then both quantum effects mentioned in the previous section act to reduce the effectiveness of the process, by making $\epsilon_1' < \epsilon$ and by reducing the probability of scattering. Kinematical effects alone limit the energy attainable: From the conservation of energy we can write $\epsilon_1 < \gamma mc^2 + \epsilon$. Fixing ϵ and letting γ become large, we see that photon energies larger than $\sim \gamma mc^2$ cannot be obtained.

7.2 INVERSE COMPTON POWER FOR SINGLE SCATTERING

In the preceding section the formulas referred to Compton scattering of a single photon off a single electron. Now we want to derive average formulas for the case of a given isotropic distribution of photons scattering off a given isotropic distribution of electrons. An elegant way to average Eqs. (7.7) and (7.8) over angles, due to Blumenthal and Gould (1970), is sketched below. Let the photon phase space distribution function be $n(p)$, which is a Lorentz invariant. Let $v\,d\epsilon$ be the density of photons having energy in range $d\epsilon$. Then v and n are related by

$$v\,d\epsilon = n\,d^3p. \tag{7.9}$$

Recall that d^3p transforms in the same way as energy under Lorentz transformations [cf. Eq. (4.106b)]. Thus $v\,d\epsilon/\epsilon$ is a Lorentz invariant:

$$\frac{v\,d\epsilon}{\epsilon} = \frac{v'\,d\epsilon'}{\epsilon'}. \tag{7.10}$$

The total power emitted (i.e., scattered) in the electron's rest frame can be found from

$$\frac{dE_1'}{dt'} = c\sigma_T \int \epsilon_1' v'\,d\epsilon', \tag{7.11}$$

where $v'\,d\epsilon'$ is the number density of incident photons. We now assume that the change in energy of the photon in the rest frame is negligible compared to the energy change in the lab frame, $\gamma^2 - 1 \gg \epsilon/mc^2$; thus we can equate $\epsilon_1' = \epsilon'$. Now, we also know

$$\frac{dE_1}{dt} = \frac{dE_1'}{dt'} \tag{7.12}$$

by the invariance of emitted power. Thus we have the result

$$\frac{dE_1}{dt} = c\sigma_T \int \epsilon'^2 \frac{v'\,d\epsilon'}{\epsilon'} = c\sigma_T \int \epsilon'^2 \frac{v\,d\epsilon}{\epsilon}. \tag{7.13}$$

In Eqs. (7.12) and (7.13) we have again made the assumption that $\gamma\epsilon \ll mc^2$, so that the Thomson cross section is applicable. As is seen in Problem 7.3, a variety of scattering processes might be expected to satisfy this criterion.

Now, since $\epsilon' = \epsilon\gamma(1-\beta\cos\theta)$, Eq. (7.13) becomes

$$\frac{dE_1}{dt} = c\sigma_T\gamma^2 \int (1-\beta\cos\theta)^2 \epsilon v\,d\epsilon, \tag{7.14}$$

which now referes solely to quantities in frame K. For an isotropic distribution of photons we have

$$\langle (1-\beta\cos\theta)^2 \rangle = 1 + \frac{1}{3}\beta^2,$$

since $\langle \cos\theta \rangle = 0$ and $\langle \cos^2\theta \rangle = \frac{1}{3}$. Thus we obtain

$$\frac{dE_1}{dt} = c\sigma_T\gamma^2\left(1+\frac{1}{3}\beta^2\right)U_{\mathrm{ph}}, \tag{7.15a}$$

where

$$U_{\mathrm{ph}} \equiv \int \epsilon v\,d\epsilon, \tag{7.15b}$$

is the initial photon energy density. The rate of decrease of the total initial photon energy is

$$\frac{dE_1}{dt} = -c\sigma_T \int \epsilon v\,d\epsilon = -\sigma_T c U_{\mathrm{ph}}.$$

Thus the net power lost by the electron, and thereby converted into increased radiation, is

$$\frac{dE_{\mathrm{rad}}}{dt} = c\sigma_T U_{\mathrm{ph}}\left[\gamma^2\left(1+\frac{1}{3}\beta^2\right) - 1\right]$$

Since $\gamma^2 - 1 = \gamma^2\beta^2$, we finally have

$$P_{\mathrm{compt}} = \frac{dE_{\mathrm{rad}}}{dt} = \tfrac{4}{3}\sigma_T c\gamma^2\beta^2 U_{\mathrm{ph}}. \tag{7.16a}$$

When the energy transfer in the electron rest frame is not neglected, Eq. (7.16a) becomes (cf. Blumenthal and Gould, 1970, but subtract out incoming energy)

$$P_{\text{compt}} = \frac{4}{3}\sigma_T c\gamma^2\beta^2 U_{\text{ph}}\left[1 - \frac{63}{10}\frac{\gamma\langle\epsilon^2\rangle}{mc^2\langle\epsilon\rangle}\right], \tag{7.16b}$$

where $\langle\epsilon^2\rangle$ and $\langle\epsilon\rangle$ are mean values integrated over U_{ph}. Note that Eq. (7.16b) allows energy to be either given or taken from the photons.

Recall that the formula for the synchrotron power emitted by each electron is [cf. Eq. (6.7b)]

$$P_{\text{synch}} = \frac{4}{3}\sigma_T c\gamma^2\beta^2 U_B. \tag{7.17}$$

Using Eq. (7.16a), we have the general result:

$$\frac{P_{\text{synch}}}{P_{\text{compt}}} = \frac{U_B}{U_{\text{ph}}}, \tag{7.18}$$

that is, the radiation losses due to synchrotron emission and to inverse Compton effect are in the same ratio as the magnetic field energy density and photon energy density. Note that this result also holds for *arbitrary* values of the electron's velocity, not just for ultrarelativistic values. It does, however, depend on the validity of Thomson scattering in the rest frame so that $\gamma\epsilon \ll mc^2$.

From Eq. (7.16) one can compute the total Compton power, per unit volume, from a medium of relativistic electrons. Let $N(\gamma)\,d\gamma$ be the number of electrons per unit volume with γ in the range γ to $\gamma + d\gamma$. Then

$$P_{\text{tot}}(\text{erg s}^{-1}\text{ cm}^{-3}) = \int P_{\text{compt}} N(\gamma)\,d\gamma. \tag{7.19}$$

For example, if

$$N(\gamma) = \begin{cases} C\gamma^{-p}, & \gamma_{\min} \leqslant \gamma \leqslant \gamma_{\max} \\ 0, & \text{otherwise,} \end{cases} \tag{7.20}$$

then, with $\beta \sim 1$, we obtain

$$P_{\text{tot}}(\text{erg s}^{-1}\text{ cm}^{-3}) = \frac{4}{3}\sigma_T c U_{\text{ph}} C(3-p)^{-1}\left(\gamma_{\max}^{3-p} - \gamma_{\min}^{3-p}\right). \tag{7.21}$$

From Eq. (7.16a) we can also compute the total power from a thermal distribution of nonrelativistic electrons of number density n_e. Taking $\gamma \approx 1$, $\langle \beta^2 \rangle = \langle v^2/c^2 \rangle = 3kT/mc^2$, we obtain

$$P_{\text{tot}}(\text{erg s}^{-1}\ \text{cm}^{-3}) = \left(\frac{4kT}{mc^2}\right) c\sigma_T n_e U_{\text{ph}}. \tag{7.22}$$

We show below, in Eq. (7.36), that the factor in parentheses is the fractional photon energy gain per scattering, when $\epsilon \ll 4kT$.

7.3 INVERSE COMPTON SPECTRA FOR SINGLE SCATTERING

The spectrum of inverse Compton scattering depends on both the incident spectrum and the energy distribution of the electrons. However, it is only necessary to determine the spectrum for the scattering of photons of a given energy ϵ_0 off electrons of a given energy γmc^2, because the general spectrum can then be found by averaging over the actual distributions of photons and electrons. We consider here cases in which both the photons and electrons have isotropic distributions; the scattered photons are then also isotropically distributed, and it only remains to find their energy spectrum.

To demonstrate the techniques involved without being burdened by excessive detail, we treat the case $\gamma\epsilon_0 \ll mc^2$, implying Thomson scattering in the rest frame. The small energy shift given by Eq. (7.2) is also ignored. In addition, we make the assumption that the scattering in the rest frame is *isotropic*, that is, we assume that

$$\frac{d\sigma'}{d\Omega'} = \frac{1}{4\pi}\sigma_T = \frac{2}{3}r_0^2,$$

instead of the more exact Eq. (7.1b). This will give the correct qualitative behavior of the results.

It is convenient when dealing with such problems of scattering to use an intensity I based on photon number rather than energy. The number of photons crossing area dA in time dt within solid angle $d\Omega$ and energy range $d\epsilon$ is, then, $I\,dA\,dt\,d\Omega\,d\epsilon$. This intensity can be found from the monochromatic specific intensity by dividing by the energy. A similar definition holds for the emission functions.

Suppose that the isotropic incident photon field is monoenergetic:

$$I(\epsilon) = F_0\delta(\epsilon - \epsilon_0),$$

where F_0 is the number of photons per unit area, per unit time per steradian. Let us determine the scattering off a beam of electrons of density N and energy γmc^2 traveling along the x axis (see Fig. 7.2). The incident intensity field in the rest frame K' is

$$I'(\epsilon',\mu') = F_0\left(\frac{\epsilon'}{\epsilon}\right)^2 \delta(\epsilon-\epsilon_0),$$

using Eq. (4.110) and remembering the extra factor of ϵ implied by the present definition of I. From the Doppler formulas (4.12) we have

$$\begin{aligned} I'(\epsilon',\mu') &= \left(\frac{\epsilon'}{\epsilon_0}\right)^2 F_0\delta(\gamma\epsilon'(1+\beta\mu')-\epsilon_0) \\ &= \left(\frac{\epsilon'}{\epsilon_0}\right)^2 \frac{F_0}{\gamma\beta\epsilon'}\delta\left(\mu' - \frac{\epsilon_0-\gamma\epsilon'}{\gamma\beta\epsilon'}\right), \end{aligned}$$

where μ' is the cosine of the angle between the photon direction in the rest frame and the x axis. The emission function in K' is given by Eqs. (1.84) and (1.85):

$$j'(\epsilon_1') = N'\sigma_T \frac{1}{2}\int_{-1}^{+1} I'(\epsilon_1',\mu')\,d\mu',$$

where j' is the number of emitted photons per unit volume per unit per steradian. We have here introduced the elastic scattering assumption that the scattered photon energy ϵ_1' equals the incident energy ϵ'. It follows that

$$\begin{aligned} j'(\epsilon_1') &= \frac{N'\sigma_T\epsilon_1' F_0}{2\epsilon_0^2\gamma\beta}, \quad \text{if } \frac{\epsilon_0}{\gamma(1+\beta)} < \epsilon_1' < \frac{\epsilon_0}{\gamma(1-\beta)} \\ &= 0, \quad \text{otherwise.} \end{aligned}$$

The emission function in frame K can be found from Eq. (4.113)

$$\begin{aligned} j(\epsilon_1,\mu_1) &= \frac{\epsilon_1}{\epsilon_1'} j'(\epsilon_1'), \\ &= \frac{N\sigma_T\epsilon_1 F_0}{2\epsilon_0^2\gamma^2\beta}, \quad \text{if } \frac{\epsilon_0}{\gamma^2(1+\beta)(1-\beta\mu_1)} < \epsilon_1 < \frac{\epsilon_0}{\gamma(1-\beta)(1-\beta\mu_1)} \\ &= 0, \quad \text{otherwise.} \end{aligned} \qquad (7.23)$$

Here we have used $N=\gamma N'$, relating the densities in the two frames, and also Eq. (4.12).

The above results hold for a beam of electrons. To obtain the results for an isotropic distribution of electrons we must average over the angle between the electron and emitted photon:

$$j(\epsilon_1)=\frac{1}{2}\int_{-1}^{+1} j(\epsilon_1,\mu_1)\,d\mu_1.$$

The quantity $j(\epsilon_1,\mu_1)$ is nonzero only for a certain interval of μ_1:

$$\frac{1}{\beta}\left[1-\frac{\epsilon_0}{\epsilon_1}(1+\beta)\right]<\mu_1<\frac{1}{\beta}\left[1-\frac{\epsilon_0}{\epsilon_1}(1-\beta)\right],$$

which follows from the restriction on Eq. (7.23). When ϵ_1/ϵ_0 is less than $(1-\beta)/(1+\beta)$ or greater than $(1+\beta)/(1-\beta)$, there is no overlap between this interval and $(-1,1)$, so $j(\epsilon_1)$ vanishes. The other cases for the limits of the μ_1 integral are:

$$-1<\mu_1<\frac{1}{\beta}\left[1-\frac{\epsilon_0}{\epsilon_1}(1-\beta)\right], \quad \text{for } \frac{1-\beta}{1+\beta}<\frac{\epsilon_1}{\epsilon_0}<1,$$

$$\frac{1}{\beta}\left[1-\frac{\epsilon_0}{\epsilon_1}(1+\beta)\right]<\mu_1<1, \quad \text{for } 1<\frac{\epsilon_1}{\epsilon_0}<\frac{1+\beta}{1-\beta}.$$

Therefore, we obtain the result:

$$j(\epsilon_1)=\frac{N\sigma_T F_0}{4\epsilon_0\gamma^2\beta^2}\begin{cases}(1+\beta)\dfrac{\epsilon_1}{\epsilon_0}-(1-\beta), & \dfrac{1-\beta}{1+\beta}<\dfrac{\epsilon_1}{\epsilon_0}<1 \quad (7.24a)\\ (1+\beta)-\dfrac{\epsilon_1}{\epsilon_0}(1-\beta), & 1<\dfrac{\epsilon_1}{\epsilon_0}<\dfrac{1+\beta}{1-\beta}, \quad (7.24b)\\ 0, & \text{otherwise.} \quad (7.24c)\end{cases}$$

It may easily be checked that

$$\int_0^\infty j(\epsilon_1)\,d\epsilon_1=N\sigma_T F_0,$$

$$\int_0^\infty j(\epsilon_1)(\epsilon_1-\epsilon_0)\,d\epsilon_1=N\sigma_T\frac{4}{3}\gamma^2\beta^2\epsilon_0 F_0.$$

Since $N\sigma_T F_0$ is the rate of photon scattering per unit volume, per unit solid angle, the first of these simply expresses the conservation of number of photons upon scattering. The second expresses the average increase in photon energy per scattering [cf. Eq. (7.16a)].

The function $j(\epsilon_1)$ is plotted for several values of β in Fig. 7.3*a*. For small β the curves are symmetrical about the initial photon energy ϵ_0. As β increases, the portion of the curve for $\epsilon \gg \epsilon_0$ becomes more and more dominant, expressing the upward shift of average energy of the scattered photon.

For values of β near unity ($\gamma \gg 1$) it is convenient to rescale the energy variable and write

$$x \equiv \frac{\epsilon_1}{4\gamma^2 \epsilon_0}. \tag{7.25}$$

The emission function, in our isotropic approximation, is dominated by Eq. (7.24b) and can be written as

$$j(\epsilon_1) = \frac{3N\sigma_T F_0}{4\gamma^2 \epsilon_0} f_{\text{iso}}(x), \tag{7.26a}$$

where

$$f_{\text{iso}}(x) \equiv \frac{2}{3}(1-x), \qquad 0<x<1, \tag{7.26b}$$

and zero otherwise. Note that the vanishing of $f_{\text{iso}}(x)$ for $x>1$ comes about from the restriction $\epsilon_1/\epsilon_0 < (1+\beta)/(1-\beta)$ on Eq. (7.24), which for $\gamma \gg 1$ becomes $\epsilon_1/\epsilon_0 < 4\gamma^2$.

When the exact angular dependence in $d\sigma'/d\Omega'$ is included, the expression for $f(x)$ in the limit $\gamma \gg 1$ is given by (see Blumenthal and Gold, 1970):

$$f(x) = 2x \ln x + x + 1 - 2x^2, \qquad 0<x<1. \tag{7.27}$$

A comparison of these two forms for $f(x)$ is given in Fig. 7.3*b*. Notice that most qualitative features of the exact result are preserved by the approximate one.

The spectrum resulting from the scattering of an arbitrary initial spectrum off a power law distribution (Eq. 7.20) of relativistic electrons can now be found. Let us use $v(\epsilon)$, the initial photon number density

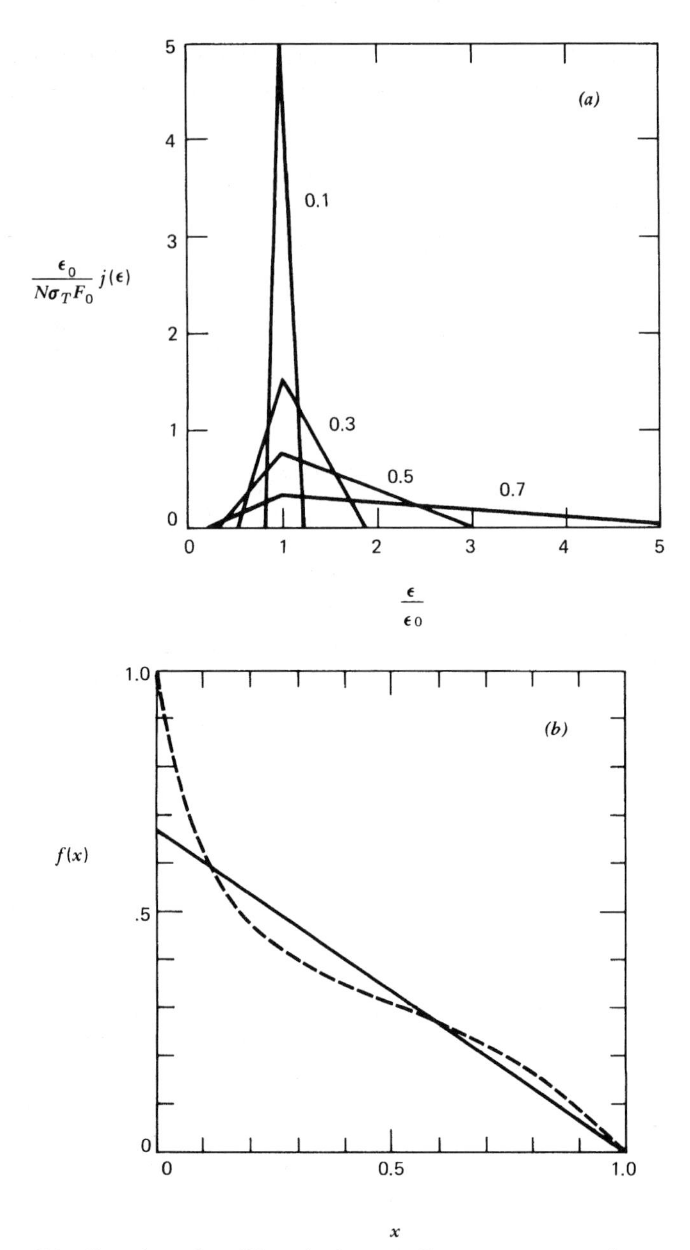

Figure 7.3 ***Functions describing the inverse Compton spectrum from a single scattering. (a) Emission function for various values of β within isotropic approximation (b) Comparisons of exact (dashed) and isotropic (solid) approximations for $f(x)$.***

introduced in Eq. (7.9), related to the isotropic intensity by $v(\epsilon)=4\pi c^{-1}I(\epsilon)$. Then the total scattered power per volume per energy is

$$\frac{dE}{dV\,dt\,d\epsilon_1}=4\pi\epsilon_1 j(\epsilon_1)$$

$$=\frac{3}{4}c\sigma_T C\int d\epsilon\left(\frac{\epsilon_1}{\epsilon}\right)v(\epsilon)\int_{\gamma_1}^{\gamma_2}d\gamma\,\gamma^{-p-2}f\left(\frac{\epsilon_1}{4\gamma^2\epsilon}\right). \tag{7.28a}$$

Changing the variable of integration from γ to x in the second integral yields

$$\frac{dE}{dV\,dt\,d\epsilon_1}=3\sigma_T cC2^{p-2}\epsilon_1^{-(p-1)/2}\int d\epsilon\,\epsilon^{(p-1)/2}v(\epsilon)\int_{x_1}^{x_2}dx\,x^{(p-1)/2}f(x), \tag{7.28b}$$

where $x_1\equiv\epsilon_1/(4\gamma_1^2\epsilon)$ and $x_2\equiv\epsilon_1/(4\gamma_2^2\epsilon)$. Now, suppose that $\gamma_2\gg\gamma_1$ and that $v(\epsilon)$ peaks at some value $\bar{\epsilon}$. The second integral in Eq. (7.28b) is then independent of ϵ_1 and can be removed. The final result is then

$$\frac{dE}{dV\,d\epsilon_1\,dt}=\pi c r_0^2 CA(p)\epsilon_1^{-(p-1)/2}\int d\epsilon\,\epsilon^{(p-1)/2}v(\epsilon) \tag{7.29a}$$

where

$$A(p)\equiv 2^{p+1}\int_0^{\infty}dx\,x^{(p-1)/2}f(x)=2^{p+3}\frac{p^2+4p+11}{(p+3)^2(p+5)(p+1)}. \tag{7.29b}$$

We point out that Eq. (7.29) is valid only over a range in ϵ_1 such that the upper and lower limits in the integral of Eq. (7.28b) can be extended to zero and infinity. If $\bar{\epsilon}$ is the typical energy of a photon in the distribution of incident photons, then this range is approximately given by $4\gamma_1^2\bar{\epsilon}\ll\epsilon_1\ll 4\gamma_2^2\bar{\epsilon}$. In particular, Eq. (7.29) cannot be integrated over all ϵ_1 to obtain the total power—instead, one must return to Eqs. (7.28a) or (7.28b) in their exact forms. The spectral index is seen to be

$$s=\frac{p-1}{2} \tag{7.30}$$

identical to the case of synchrotron emission (cf. §6.3).

When $v(\epsilon)$ is the blackbody distribution, that is,

$$v(\epsilon)=\frac{8\pi\epsilon^2}{h^3c^3}\frac{1}{\exp(\epsilon/kT)-1},$$

we obtain from (7.29), and 23.2.7 of Abramovitz and Stegun (1965),

$$\frac{dE}{dV\,dt\,d\epsilon_1}=\frac{C8\pi^2r_0^2}{h^3c^2}(kT)^{(p+5)/2}F(p)\epsilon_1^{-(p-1)/2} \tag{7.31}$$

where

$$F(p)\equiv A(p)\Gamma\left(\frac{p+5}{2}\right)\zeta\left(\frac{p+5}{2}\right)$$

and ζ denotes the Riemann zeta function, defined by

$$\zeta(s)\equiv\sum_{n=1}^{\infty}n^{-s}.$$

The general problem of scattering of an isotropic photon field from an isotropic electron distribution, including the Compton effect and Klein–Nishina cross section, has been solved by Jones (1968), and the interested reader should look there for details.

7.4 ENERGY TRANSFER FOR REPEATED SCATTERINGS IN A FINITE, THERMAL MEDIUM: THE COMPTON *Y* PARAMETER

Before discussing in some detail the effect of repeated Compton scattering on the spectrum and total energy of the photon distribution, it is useful to determine the conditions under which the scattering process significantly alters the total photon energy. We restrict our considerations to situations in which the Thomson limit applies: $\gamma\epsilon\ll mc^2$.

In finite media one may define a Compton y parameter, to determine whether a photon will significantly change its energy in traversing the medium:

$$y\equiv\begin{bmatrix}\text{average fractional}\\ \text{energy change per}\\ \text{scattering}\end{bmatrix}\times\begin{pmatrix}\text{mean number of}\\ \text{scatterings}\end{pmatrix}. \tag{7.32}$$

The quantities in parentheses are evaluated below. In general when $y \gtrsim 1$, the total photon energy and spectrum will be significantly altered; whereas for $y \ll 1$, the total energy is not much changed.

It is convenient to evaluate the first term in Eq. (7.32) for a thermal distribution of electrons. Consider first the nonrelativistic limit. Averaging Eq. (7.8a) over angles, we obtain

$$\frac{\Delta\epsilon'}{\epsilon'} \equiv \frac{\epsilon_1' - \epsilon'}{\epsilon'} = -\frac{\epsilon'}{mc^2}. \tag{7.33}$$

Now, in the lab frame to lowest order in the two small parameters ϵ/mc^2 and kT/mc^2, this must be of the form

$$\frac{\Delta\epsilon}{\epsilon} = -\frac{\epsilon}{mc^2} + \frac{\alpha kT}{mc^2}, \tag{7.34}$$

where α is some coefficient to be determined. To calculate α, imagine that the photons and electrons are in complete equilibrium but interact only through scattering. We assume that the photon density is sufficiently small that stimulated processes can be neglected. The photons thus have a Bose–Einstein distribution with a chemical potential rather than a Planck distribution because photons cannot be created or destroyed by scattering. In the nondegenerate limit (where stimulated effects are negligible) the appropriate distribution is Eq. (6.51), and we have the averages

$$\langle\epsilon\rangle = \int \epsilon \frac{dN}{d\epsilon} d\epsilon \Big/ \int \frac{dN}{d\epsilon} d\epsilon = 3kT, \tag{7.35a}$$

$$\langle\epsilon^2\rangle = 12(kT)^2. \tag{7.35b}$$

For this hypothetical case no net energy can be transferred from photons to electrons, so

$$\begin{aligned}\langle\Delta\epsilon\rangle = 0 &= \frac{\alpha kT}{mc^2}\langle\epsilon\rangle - \frac{\langle\epsilon^2\rangle}{mc^2}\\ &= \frac{3kT}{mc^2}(\alpha - 4)kT,\end{aligned}$$

giving the result $\alpha = 4$. Thus for nonrelativistic electrons in thermal equilibrium, the expression for the energy transfer per scattering is

$$(\Delta\epsilon)_{NR} = \frac{\epsilon}{mc^2}(4kT - \epsilon). \tag{7.36}$$

Note that if the electrons have high enough temperature relative to incident photons, the photons may gain energy. This is called *inverse Compton scattering*. If $\epsilon > 4kT$, on the other hand, energy is transferred from photons to electrons.

In the ultrarelativistic limit, $\gamma \gg 1$, ignoring the energy transfer in the electron rest frame, Eqs. (7.7) show that

$$(\Delta\epsilon)_R \sim \frac{4}{3}\gamma^2\epsilon, \tag{7.37}$$

where the 4/3 results from angle averaging Eqs. (7.37) and is derived in §7.2. For a thermal distribution of ultrarelativistic electrons, we have, using arguments analogous to those leading to Eq. (7.35),

$$\langle\gamma^2\rangle = \frac{\langle E^2\rangle}{(mc^2)^2} = 12\left(\frac{kT}{mc^2}\right)^2.$$

Thus Eq. (7.37) becomes

$$(\Delta\epsilon)_R \sim 16\epsilon\left(\frac{kT}{mc^2}\right)^2, \tag{7.38}$$

Now, the second term in Eq. (7.32) may be evaluated using Eqs. (1.89a) and (1.89b). For a pure scattering medium we have

$$\left(\begin{matrix}\text{mean number of}\\ \text{scatterings}\end{matrix}\right) \approx \text{Max}(\tau_{es}, \tau_{es}^2), \tag{7.39a}$$

where τ_{es} is given by

$$\tau_{es} \sim \rho\kappa_{es}R. \tag{7.39b}$$

Here κ_{es} is the electron scattering opacity, which for ionized hydrogen is

$$\kappa_{es} = \frac{\sigma_T}{m_p} = 0.40\ \text{cm}^2\ \text{g}^{-1} \tag{7.40}$$

and where R is the size of the finite medium. Combining Eqs. (7.32), (7.36), (7.37), and (7.39), we then obtain expressions for the Compton y parameter for relativistic and nonrelativistic thermal distributions of electrons:

$$y_{NR} = \frac{4kT}{mc^2}\text{Max}(\tau_{es}, \tau_{es}^2), \tag{7.41a}$$

$$y_R = 16\left(\frac{kT}{mc^2}\right)^2 \text{Max}(\tau_{es}, \tau_{es}^2). \tag{7.41b}$$

We have assumed that the energy transfer in the electron rest frame is negligible, that is, $4kT \gg \epsilon$ in the nonrelativistic case. The importance of the y parameter is illustrated in Problem 7.1. There it is shown that input photons of initial energy ϵ_i emerge with average energy $\epsilon_f \sim \epsilon_i e^y$ after scattering in a cloud of nonrelativistic electrons (as long as $\epsilon_f \ll 4kT$).

In media in which absorption is important, it is convenient to define a frequency-dependent Compton parameter, $y(\nu)$. For this parameter the relevant $\tau_{es}(\nu)$ must be measured from an effective absorption optical depth, $\tau_*(\nu)$, of order unity. Thus $\tau_{es}(\nu) = \rho\kappa_{es} l_*(\nu)$ (cf. §1.7), and using Eqs. (1.96), we obtain

$$\tau_{es}(\nu) \sim \left(\frac{\kappa_{es}/\kappa_a(\nu)}{1+\kappa_a(\nu)/\kappa_{es}} \right)^{1/2} \tag{7.42}$$

where $\kappa_a(\nu)$ is the absorption opacity. Equation (7.42) gives the scattering optical depth to the surface from the characteristic point of emission of a photon of frequency ν. The definitions for $y_{NR}(\nu)$ and $y_R(\nu)$ are identical to Eqs. (7.41a) and (7.41b) with τ_{es} replaced by $\tau_{es}(\nu)$ of Eq. (7.42).

7.5 INVERSE COMPTON SPECTRA AND POWER FOR REPEATED SCATTERINGS BY RELATIVISTIC ELECTRONS OF SMALL OPTICAL DEPTH

In §7.3 it has been shown that a power-law spectrum results from inverse Compton scattering off a power-law distribution of relativistic electrons. This is not surprising, since any quantity scaled by a factor that has a power-law distribution will itself have a power-law distribution. However, as we now show here, for relativistic electrons, and below for nonrelativistic electrons, a power-law photon distribution can also be produced from *repeated* scatterings off a nonpower-law electron distribution of small scattering depth.

Let A be the mean amplification of photon energy per scattering, that is,

$$\begin{aligned} A &\equiv \frac{\epsilon_1}{\epsilon} \\ &\sim \frac{4}{3}\langle\gamma^2\rangle = 16\left(\frac{kT}{mc^2}\right)^2, \end{aligned} \tag{7.43}$$

where the second equation follows for a thermal electron distribution, (cf. §7.4). Consider an initial photon distribution of mean photon energy ϵ_i, such that $\epsilon_i \ll \langle\gamma^2\rangle^{-1/2}mc^2$, and intensity $I(\epsilon_i)$ at ϵ_i. Then, after k scatterings, the energy of a mean initial photon will be

$$\epsilon_k \sim \epsilon_i A^k. \tag{7.44}$$

If the medium is of small scattering optical depth (and much smaller absorption depth), then the probability $p_k(\tau_{es})$ of a photon undergoing k scatterings before escaping the medium is approximately $p_k(\tau_{es}) \sim \tau_{es}^k$. The intensity of emergent radiation at energy ϵ_k is roughly proportional to $p_k(\tau_{es})$, since the bandwidth of the Compton produced spectrum is comparable to the frequency. Thus the emergent intensity at energy ϵ_k has the power-law shape

$$I(\epsilon_k) \sim I(\epsilon_i)\tau_{es}^k \sim I(\epsilon_i)\left(\frac{\epsilon_k}{\epsilon_i}\right)^{-\alpha}, \tag{7.45a}$$

where

$$\alpha \equiv \frac{-\ln \tau_{es}}{\ln A}. \tag{7.45b}$$

The above qualitative derivation of Eq. (7.45) was first given by Ya. B. Zeldovich and has been verified in numerical Monte Carlo calculations by L. A. Pozdnyakov, I. M. Sobol, and R. A. Sunyaev (1976).

Equation (7.45) only holds for emergent photons satisfying $\epsilon_k/\langle\gamma^2\rangle^{1/2} \lesssim mc^2$, so that the energy amplification at the last scattering is correctly described by Eq. (7.43). Note, however, that such photons are just those that emerge at energies $\sim kT$ in a thermal distribution of relativistic electrons.

The total Compton power in the output spectrum is given by

$$P \propto \int_{\epsilon_i}^{A^{1/2}mc^2} I(\epsilon_k)\,d\epsilon_k = I(\epsilon_i)\epsilon_i\left[\int_1^{A^{1/2}mc^2/\epsilon_i} x^{-\alpha}\,dx\right]. \tag{7.46}$$

The factor in square brackets is approximately the factor by which the initial power $\propto I(\epsilon_i)\epsilon_i$ is amplified in energy. Clearly, this amplification will be important if $\alpha \lessdot 1$. From Eq. (7.45b) we conclude that energy amplification of a soft photon input spectrum is therefore important when

$$A\tau_{es} \sim 16(kT/mc^2)^2\tau_{es} \gtrsim 1, \tag{7.47}$$

where the intermediate step holds if the electrons are thermal. Note that Eq. (7.47) is equivalent to $y_R \gtrsim 1$ [cf. Eq. (7.41b)] for $\tau_{es} \lesssim 1$.

7.6 REPEATED SCATTERINGS BY NONRELATIVISTIC ELECTRONS: THE KOMPANEETS EQUATION

Consider now the evolution of the photon phase space density $n(\omega)$ due to scattering from electrons. We assume that $n(\omega)$ is isotropic. If $f_e(\mathbf{p})$ is the phase density of electrons of momentum $\mathbf{p}$, then the *Boltzmann equation* for $n(\omega)$ is

$$\frac{\partial n(\omega)}{\partial t} = c\int d^3p \int \frac{d\sigma}{d\Omega} d\Omega\left[f_e(\mathbf{p}_1)n(\omega_1)(1+n(\omega)) - f_e(\mathbf{p})n(\omega)(1+n(\omega_1))\right] \tag{7.48}$$

where we consider the scattering events

$$p+\omega \leftrightarrows p_1+\omega_1.$$

The first term in Eq. (7.48) represents scattering into frequency ω by photons of frequency ω_1, whereas the second term represents scattering out of frequency ω into frequencies ω_1. The relationship between ω and ω_1 is given by Eqs. (7.50), (7.53) and Problem 7.4 and is a function of the scattering angles. The dependence on angles disappears after integration over $d\Omega$. The factors $1+n(\omega)$ and $1+n(\omega_1)$ take into account stimulated scattering effects; that is, the probability of scattering from frequency ω_1 to ω is increased by the factor $1+n(\omega)$ because photons obey Bose–Einstein statistics and tend toward mutual occupation of the same quantum state [cf. Eqs. (1.68) and (1.74) and §1.5]. Aside from these quantum mechanical correction factors, Eq. (7.48) is a standard form in kinetic theory. In general, the Boltzmann equation can be solved only for special cases or with approximations. We give approximate solutions in the nonrelativistic limit below.

A detailed analysis of the evolution of the spectrum in the presence of repeated scatterings off relativistic electrons is difficult because the energy transfer per scattering is large and one must solve the full integrodifferential equation, (7.48). However, when the electrons are nonrelativistic, the fractional energy transfer per scattering is small. In particular, the Boltzmann equation may be expanded to second order in this small quantity, yielding an approximation called the *Fokker–Planck equation*. For photons scattering off a nonrelativistic, thermal distribution of electrons, the Fokker–Planck equation was first derived by A. S. Kompaneets (1957) and is known as the Kompaneets equation.

For a thermal distribution of nonrelativistic electrons, the phase space density $f_e(E)$, where $E=p^2/2m$, is given by

$$f_e(E)=n_e(2\pi mkT)^{-3/2}e^{-E/kT}, \tag{7.49}$$

where n_e is the electron space density. We define the dimensionless energy transfer to the photons as

$$\Delta \equiv \frac{\hbar(\omega_1 - \omega)}{kT}. \tag{7.50}$$

We now consider situations in which the energy transfer is small, $\Delta \ll 1$, and expand $f_e(E_1)$ and $n(\omega_1)$ for this regime. For example, for $n(\omega_1)$ this expansion, *to second order*, is

$$n(\omega_1) = n(\omega) + (\omega_1 - \omega)\frac{\partial n(\omega)}{\partial \omega} + \frac{1}{2}(\omega_1 - \omega)^2 \frac{\partial^2 n(\omega)}{\partial \omega^2} + \cdots \tag{7.51}$$

Now letting

$$x \equiv \frac{\hbar\omega}{kT},$$

we obtain, to second order in Δ,

$$\begin{aligned} c^{-1}\frac{\partial n}{\partial t} &= [n' + n(1+n)] \int\int d^3p \frac{d\sigma}{d\Omega} d\Omega f_e \Delta \\ &+ \left[\frac{1}{2}n'' + n'(1+n) + \frac{1}{2}n(1+n)\right] \int\int d^3p \frac{d\sigma}{d\Omega} d\Omega f_e \Delta^2, \end{aligned} \tag{7.52}$$

where $n' \equiv \partial n / \partial x$ and so on. The term in Δ gives the "secular" shift in energy, and the term in Δ^2 gives the "random walk" change in energy.

Let us first compute the second integral, I_2, in Eq. (7.52), which gives the random walk contribution to $\partial n / \partial t$. Using a derivation completely analogous to that leading to Eq. (7.2) but with the electron not initially at rest, one finds (Problem 7.4),

$$\Delta = \frac{x\mathbf{p}\cdot(\mathbf{n}_1 - \mathbf{n})}{mc} + O\left(\frac{kT}{mc^2}\right), \tag{7.53}$$

where $\mathbf{p}$ is the electron momentum before collision and $\mathbf{n}$ and $\mathbf{n}_1$ are unit vectors along the photon direction before and after collision, respectively. Now, using the formula for $d\sigma / d\Omega$, Eq. (7.1b), and the above equations, one obtains (Problem 7.4),

$$I_2 = 2x^2 n_e \sigma_T \left(\frac{kT}{mc^2}\right) + O\left(\frac{kT}{mc^2}\right)^2. \tag{7.54}$$

We can similarly evaluate the integral I_1, but this is more difficult than I_2. A simpler method uses photon conservation and detailed balancing. Since n is the photon phase space density and x is proportional to momentum, then the change in number of photons per unit volume, which must vanish, is proportional to

$$\frac{d}{dt}\int nx^2\,dx = \int \frac{\partial n}{\partial t}x^2\,dx = 0.$$

It is thus clear that $\partial n/\partial t$ must be of the form (Problem 7.4)

$$\frac{\partial n}{\partial t} = -\frac{1}{x^2}\frac{\partial}{\partial x}\left[x^2 j(x)\right]. \tag{7.55a}$$

By comparison with Eq. (7.52), j must be of the form

$$j = g(x)\left[n' + h(n,x)\right], \tag{7.55b}$$

with h and g two functions to be determined. Now, we know that a Bose–Einstein photon distribution with finite chemical potential,

$$n = (e^{\alpha + x} - 1)^{-1}, \tag{7.56}$$

must be in thermal equilibrium with the electrons, with $j=0$. Requiring $n' + h(n,x) = 0$ for n given by Eq. (7.56) then determines

$$h(n,x) = n(1+n). \tag{7.57a}$$

Comparison of Eqs. (7.57a) and (7.55) with (7.54) and (7.52) then yields the two desired results:

$$g(x) = -cx^2 n_e \sigma_T\left(\frac{kT}{mc^2}\right), \tag{7.57b}$$

$$I_1 = n_e \sigma_T x(4-x)\left(\frac{kT}{mc^2}\right). \tag{7.58}$$

Note that the "secular term" of the Fokker–Planck equation, proportional to I_1, states that energy is gained or lost depending on the sign of $4-x$, in agreement with Eq. (7.36).

Substitution of Eqs. (7.57) into (7.55) then yields the Kompaneets equation, describing the evolution of the photon distribution function due to repeated, nonrelativistic, inverse Compton scattering:

$$\frac{\partial n}{\partial t_c} = \left(\frac{kT}{mc^2}\right)\frac{1}{x^2}\frac{\partial}{\partial x}\left[x^4(n' + n + n^2)\right]. \tag{7.59}$$

Here, the quantity

$$t_c \equiv (n_e \sigma_T c) t$$

is the time measured in units of mean time between scatterings.

In general, Eq. (7.59) must be solved by numerical integration. However, several important limiting cases can be pointed out here. First, note that the spectrum reaches equilibrium after photons have been "scattered up" to energies forming the Bose–Einstein distribution, Eq. (7.56). This steady-state, "saturated" spectrum is approximated by a *Wien law* [cf. Eq. (1.54)]

$$n(x) \propto e^{-x} \tag{7.60}$$

when the occupation number is small; that is, $\alpha \gg 1$. Note also that for times short compared to that required to reach saturation, so that the mean $h\nu$ of an initially low energy photon distribution is still small compared to kT, $x \ll 1$, the total energy density of the photons increases with time according to

$$\frac{dE}{dt_c} = \frac{8\pi}{c^3h^3}(kT)^4 \frac{d}{dt_c}\int_0^\infty nx^3\,dx \approx \left(\frac{4kT}{mc^2}\right)E. \tag{7.61a}$$

Here we have neglected the n and n^2 terms compared to the n' on the right-hand side of Eq. (7.59) and have performed two integrations by parts. From Eq. (7.61a) it can be seen that the total energy in a soft input spectrum increases initially as

$$E(t) \approx E(0)\exp\left(\frac{4kT}{mc^2}t_c\right). \tag{7.61b}$$

Note the similarity between this expression and that for the energy gain of a single photon in scattering out of a finite medium, Problem 7.1:

$$\epsilon_f = \epsilon_i e^y, \tag{7.62}$$

where $\mathrm{Max}(\tau_{es}, \tau_{es}^2)$ plays the role of t_c.

7.7 SPECTRAL REGIMES FOR REPEATED SCATTERING BY NONRELATIVISTIC ELECTRONS

A detailed analysis of Compton spectra requires a solution of the Kompaneets equation, Eq. (7.59), with a photon source term. For frequencies where $y \ll 1$ (modified blackbody) or $y \gg 1$ (saturated Comptonization), approximate analyses are usually adequate. For intermediate cases (unsaturated Comptonization) we return to the more detailed treatment required by the Kompaneets equation.

To delineate regimes it is convenient to introduce several characteristic frequencies. We are concerned with thermal media in which absorption and emission arise from free-free (bremsstrahlung) processes, (see §5.3). In such media the relative importance of absorption is greatest at low frequencies. Consider first the frequency, ν_0, at which the scattering and absorption coefficients are equal: From Eqs. (1.22), (5.18), and (7.40), we have

$$\kappa_{es} = \kappa_{ff}(\nu_0), \tag{7.63a}$$

$$\frac{x_0^3}{1-e^{-x_0}} \sim 4\times 10^{25} T^{-7/2} \rho \bar{g}_{ff}(x_0), \tag{7.63b}$$

where $x_0 \equiv h\nu_0/kT$ and $\bar{g}_{ff}(x)$ is the free-free Gaunt factor. In the range of interest, $\bar{g}_{ff}$ is approximated by, (Fig. 5.2)

$$\bar{g}_{ff}(x) \sim 3\pi^{-1/2} \ln\left(\frac{2.25}{x}\right).$$

For $x \equiv h\nu/kT < x_0$, scattering will be unimportant; whereas for $x > x_0$, scattering will modify the spectrum. Note that if $x_0 \gtrsim 1$, scattering is unimportant over most of the spectrum. In all the following discussion we assume $x_0 \ll 1$.

Consider next the frequency ν_t at which the medium becomes effectively thin. From Eqs. (1.97) and (5.18) we have

$$\kappa_{es} = \kappa_{ff}(\nu_t)\tau_{es}^2, \tag{7.64a}$$

$$\frac{x_t^3}{1-e^{-x_t}} \sim 4\times 10^{25} T^{-7/2} \rho \bar{g}_{ff}(x_t)\tau_{es}^2, \tag{7.64b}$$

where $x_t \equiv h\nu_t/kT$ and τ_{es} is the total optical depth to electron scattering, Eq. (7.39b). For values of $x > x_t$, absorption is unimportant. Note that in the range $x_0 < x < x_t$ both scattering and absorption are important.

Finally, we introduce the frequency ν_{coh} for which incoherent scattering (inverse Compton effects) can be important. This frequency is so defined that $y(\nu_{coh}) = 1$, that is, for $\nu > \nu_{coh}$ inverse Compton is important between emission and escape from the medium. Note that this frequency is defined only if the y parameter for the full thickness of the medium, Eqs. (7.39)–(7.41), exceeds unity. Otherwise, inverse Compton scattering is unimportant at all frequencies. From Eqs. (7.41), (7.42), and (5.18), for $x_{coh} \ll 1$,

$$\kappa_{es} = \left(\frac{mc^2}{4kT} \right) \kappa_{ff}(\nu_{coh}), \tag{7.65a}$$

$$x_{coh} \sim 2.4 \times 10^{17} \rho^{1/2} T^{-9/4} \left[\bar{g}_{ff}(x_{coh}) \right]^{1/2}. \tag{7.65b}$$

From Eqs. (7.64) and (7.65), we see that inverse Compton is important, and x_{coh} is defined, only when $x_{coh} < x_t$.

Modified Blackbody Spectra; $y \ll 1$

For $y \ll 1$, only coherent scattering is important. Then, from Problem 1.10, we have for the emergent intensity in a scattering and absorbing medium

$$I_\nu = \frac{2B_\nu}{1 + \sqrt{(\kappa_{ff} + \kappa_{es}) \kappa_{ff}^{-1}}}. \tag{7.66}$$

The functional form of Eq. (7.66), in the limit $\kappa_{es} \gg \kappa_{ff}(\nu)$, may also be derived by the simple random-walk considerations leading to Eq. (1.102). We see that at values of $x \ll x_0$ Eq. (7.66) reduces to the blackbody intensity, whereas at values of $x \gg x_0$ Eq. (7.66) becomes a "modified blackbody spectrum,"

$$I_\nu^{MB} \equiv 2B_\nu \sqrt{\kappa_{ff} / \kappa_{es}} \tag{7.67a}$$

$$= 8.4 \times 10^{-4} T^{5/4} \rho^{1/2} \bar{g}_{ff}^{1/2} x^{3/2} e^{-x/2} (e^x - 1)^{-1/2}$$

$$\times \text{ erg s}^{-1} \text{ cm}^{-2} \text{ Hz}^{-1} \text{ ster}^{-1}. \tag{7.67b}$$

For $x_0 \ll 1$ Eq. (7.63b) gives the approximate equation for x_0:

$$x_0 \sim 6.3 \times 10^{12} T^{-7/4} \rho^{1/2} \left[\bar{g}_{ff}(x_0) \right]^{1/2}. \tag{7.68}$$

Note that at frequencies $x_0 \ll x \ll 1$, $I_\nu^{MB} \propto \nu$ instead of the Rayleigh–Jeans law $I_\nu^{RJ} \propto \nu^2$. The total flux in a modified blackbody spectrum is approximately

$$F^{MB} \sim \sigma T^4 \left(\frac{\kappa_R}{\kappa_{es}} \right)^{1/2}$$

$$\sim 2.3 \times 10^7 T^{9/4} \rho^{1/2} \text{ erg s}^{-1} \text{ cm}^{-2}, \tag{7.69}$$

where we have taken the Rosseland mean, κ_R, for the frequency-averaged κ_{ff} [cf. Eq. (5.20)].

Equation (7.66) actually applies only to a medium that is an infinite half-space. For finite media it is necessary to determine the value of x_t [cf. Eq. (7.64b)]. For $x_t < x_0$ the emission is blackbody at $x < x_t$ and optically thin bremsstrahlung for $x > x_t$, with scattering never important. For $x_0 < x_t < 1$, the emission is correctly described by Eq. (7.66) for $x < x_t$ and is then optically thin bremsstrahlung for $x > x_t$. For $x_t > 1$ the medium behaves as if it were infinite, and Eq. (7.66) may be used for the entire spectrum.

The above relations for the modified blackbody spectrum were first discussed by Felten and Rees (1972) and by Illarionov and Sunyaev (1972).

Wien Spectra; $y \gg 1$

When $y \gg 1$, inverse Compton may be important, depending on whether $x_{coh} \ll 1$ or $x_{coh} \gg 1$. In the latter case, inverse Compton may be neglected, since the majority of the photons and energy, that is, the spectrum in the region $x \lesssim 1$, undergo coherent scattering. The preceding subsection may be used to describe the spectrum. We therefore consider only the case $x_{coh} \ll 1$.

For $x_{coh} \ll 1$, Eqs. (7.63) and (7.65) give

$$x_{coh} = \left(\frac{mc^2}{4kT}\right)^{1/2} x_0. \tag{7.70}$$

The spectrum is correctly described by Eq. (7.66) for $x \ll x_{coh}$, but for $x \gtrsim x_{coh}$ we must consider inverse Compton effects, (see Fig. 7.4). In this region of the spectrum, if $x_{coh} \ll 1$, inverse Compton will go to saturation, and §7.6 shows that a Wien intensity will be produced [cf. Eq. (1.54)]:

$$I_\nu^W = \frac{2h\nu^3}{c^2} n = \frac{2h\nu^3}{c^2} e^{-\alpha} e^{-h\nu/kT}, \tag{7.71}$$

where the factor $e^{-\alpha}$ is related to the rate at which photons are produced. (Recall that the photon number is conserved in the scattering process.) The total flux in a spectrum of the form of Eq. (7.71) is

$$F^W(\text{erg s}^{-1}\,\text{cm}^{-2}) = \pi \int I_\nu^W \, d\nu = \frac{12\pi e^{-\alpha} k^4 T^4}{c^2 h^3}, \tag{7.72}$$

while the mean photon has an energy $\overline{h\nu} = 3kT$.

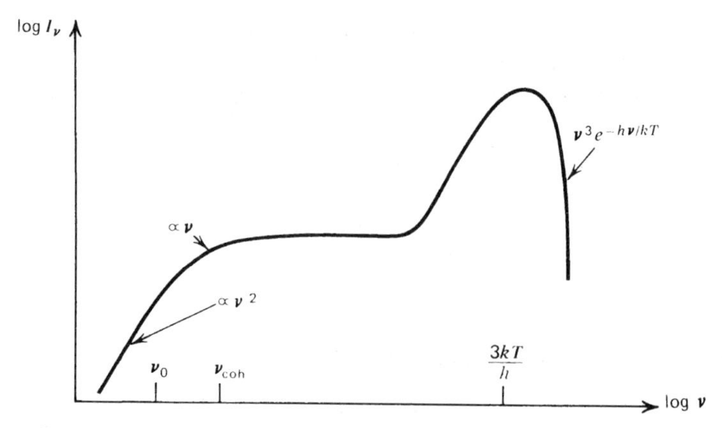

Figure 7.4 Spectrum from a thermal, nonrelativistic medium characterized by free-free emission and absorption and by saturated inverse Compton scattering. At low frequencies the spectrum is blackbody then becomes modified blackbody and, at high frequencies, becomes a Wien spectrum.

The rate at which energy is generated in the Comptonized spectrum can be calculated approximately by shifting all of the bremsstrahlung photons to energies kT:

$$\frac{dW^W}{dt\,dV}(\text{erg s}^{-1}\,\text{cm}^{-3}) \sim kT\int\left(\frac{\epsilon_\nu^{ff}}{h\nu}\right)d\nu$$

$$\sim \epsilon^{ff}\int_{\nu_{coh}}^{\infty} \bar{g}_{ff}(\nu, T)e^{-h\nu/kT}\frac{d\nu}{\nu}. \qquad (7.73)$$

Here ϵ_ν^{ff}(erg s^{-1} cm^{-3} Hz^{-1}) is the bremsstrahlung (free-free) energy generation rate given by Eq. (5.14), and ϵ^{ff}(erg s^{-1} cm^{-3}) given by Eq. (5.15) is the total energy per unit time per unit volume. This integral may be approximated by evaluating $\bar{g}$ at the lower limit, ν_{coh}, and letting $e^{-h\nu/kT}$ be a step function that is unity for $h\nu < kT$ and then zero for $h\nu > kT$. The result is, using the analytical approximation to $\bar{g}$ given in Fig. 5.2,

$$\frac{dW^W}{dt\,dV} \sim A(\rho, T)\epsilon^{ff}, \qquad (7.74a)$$

$$A(\rho, T) \equiv \frac{3}{4}\left[\ln(2.25/x_{coh})\right]^2. \qquad (7.74b)$$

Here $A(\rho, T)$ is the factor by which inverse Compton amplifies the bremsstrahlung power. Equation (7.74), including the more exact overall numerical factor, was first derived by Kompaneets (1957).

To calculate the emergent flux from Eq. (7.74), and hence the normalization of Eqs. (7.71) and (7.72), we must multiply Eq. (7.74) by a characteristic depth. If $x_t \ll 1$, then the medium is effectively thin for most photons and $F^W \sim RA\epsilon^{ff}$, where R is the size of the medium. If $x_t \gg 1$ then, since photons at energies $x > x_{coh}$ amplify quickly to $x \sim 1$, R is replaced by $\bar{R}$, where $\tau_*(\bar{R}, x=1) \sim 1$. The emergent intensity is shown in Fig. 7.4.

Unsaturated Comptonization with Soft Photon Input

Finally, we must consider situations in which $y \gg 1$, but in which $x_{coh} \sim 1$; that is, media for which the inverse Compton process is important but does not saturate to the Wien spectrum for most photons. In this case an analysis of the Kompaneets equation is required.

Let us consider a steady-state solution to this equation, under certain idealizations. For steady-state solutions in a finite medium it is necessary to consider both the input and the escape of photons. Denote the photon source by $Q(x)$. The photon escape is a spatial diffusion process. However, for photons which have scattered many times, it is a fair approximation to assume that the probability for a photon to escape per Compton scattering time is equal to the inverse of the mean number of scatterings, $\mathrm{Max}(\tau_{es}, \tau_{es}^2)$. With this approximation, one may consider a modified, steady-state Kompaneet's equation of the form

$$0 = \left(\frac{kT}{mc^2}\right) \frac{1}{x^2} \frac{\partial}{\partial x}\left[x^4(n' + n)\right] + Q(x) - \frac{n}{\mathrm{Max}(\tau_{es}, \tau_{es}^2)}, \tag{7.75}$$

where the n^2 term, usually small in astrophysical applications, has been dropped.

Assume now that $Q(x)$ is nonzero only for $x \leqslant x_s$, where $x_s \ll 1$; that is, we have an input of "soft" photons, rather than the bremsstrahlung input considered previously. For $x \gg 1$, the term in brackets shows that an approximate solution is

$$n \propto e^{-x}; \tag{7.76a}$$

that is, the spectrum falls roughly exponentially at photon energies much above the electron temperature, as would be expected for a thermal spectrum. On the other hand, for $x_s \ll x \ll 1$, the n term in brackets may be neglected in comparison with the n' term, and one obtains the approximate power-law solution:

$$n \propto x^m, \tag{7.76b}$$

$$m(m+3) - \frac{4}{y} = 0, \tag{7.76c}$$

$$m = -\frac{3}{2} \pm \sqrt{\frac{9}{4} + \frac{4}{y}}\,, \tag{7.76d}$$

where the Compton y parameter is given in Eq. (7.41a). The + root in Eq. (7.76d) is appropriate if $y \gg 1$ (leading to the low-frequency limit of the Wien law in the limit $y \to \infty, I_\nu \propto x^3 n \propto x^3$); for $y \ll 1$, the minus root is appropriate. For $y \sim 1$, one must take a linear combination of the two solutions, and no power law exists.

Figure 7.5 illustrates the spectrum resulting from unsaturated Comptonization. Note that measurement of only the shape of an unsaturated Compton spectrum with soft photon source determines both the electron temperature and the scattering optical depth of the source. The emergent intensity in the power-law regime satisfies

$$I_\nu \sim I_{\nu_s} \left(\frac{\nu}{\nu_s} \right)^{3+m}. \tag{7.77}$$

The spectrum is clearly sensitive to y. The input energy is significantly amplified for $m \geqslant -4$, that is, $y \geqslant 1$. This result is quite analogous to that for the relativistic case considered previously in §7.5. Unsaturated Compton spectra are treated in some detail in Shapiro, Lightman, and Eardley (1976) and Katz (1976).

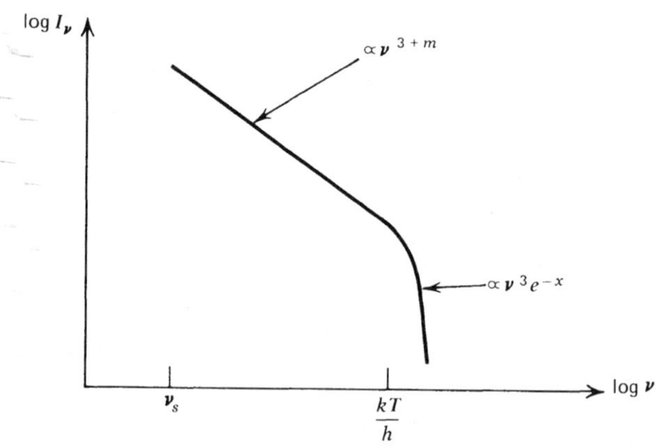

***Figure 7.5** Spectrum produced by unsaturated Comptonization of low energy photons by thermal electrons.*

PROBLEMS

7.1—A cloud of nonrelativistic electrons is maintained at temperature T. The cloud is thick to electron scattering, $\tau_{es} \gg 1$, but very thin to absorption, $\tau_*(h\nu = kT) \ll 1$. A copious supply of "soft" photons, each of characteristic energy $\epsilon_i \ll kT$, is injected into the cloud. As a result of inverse Compton scattering, these initially soft photons emerge from the cloud with characteristic energies $\epsilon_f \gg \epsilon_i$. It is found that ϵ_f increases rapidly with increasing τ_{es} as the latter is varied, until τ_{es} reaches a critical value τ_{crit}, above which the Comptonization process "saturates."

a. Find an approximate expression for ϵ_f as a function of ϵ_i, τ_{es}, T, and fundamental constants.

b. Find an approximate expression for τ_{crit}.

c. Find a single parameter of the fixed medium that determines whether inverse Compton is a significant effect.

7.2—Consider the observed X-ray source of Problem 5.2. From the deduced characteristics of the source, determine a lower limit to the central mass M such that inverse Compton effects in the emission mechanism are negligible.

7.3—Show that the photon energy in the electron rest frame is small compared to mc^2 for the following cases:

a. Electrons with $\gamma \sim 10^4$ scattering synchrotron photons produced in a magnetic field $B \sim 0.1$ G (typical of compact radio sources).

b. Electrons with $\gamma \sim 10^4$ scattering the 3 K photons of the cosmic microwave background.

7.4—Derive Eqs. (7.53) to (7.55) for the Kompaneets equation.

REFERENCES

Abramowitz, M., and Stegun, I. A., *op. cit.*
Blumenthal, G. R., and Gould, R. J., 1970, *Rev. Mod. Phys.*, **42**, 237.
Felten, J. E., and Rees, M. J., 1972, *Astron. Astrophys.*, **17**, 226.
Heitler, W., 1954, *The Quantum Theory of Radiation*, (Oxford, London).
Illarionov, A. F., and Sunyaev, R. A., 1972, *Sov. Astron. A. J.*, **16**, 45.
Jones, F. C., 1968, *Phys. Rev.*, **167**, 1159.
Katz, J. I., 1976, *Astrophys. J.*, **206**, 910.
Kompaneets, A. S., 1957, *Sov. Phys. JETP*, **4**, 730.
Pozdnyakov, L. A., Sobol, I. M., and Sunyaev, R. A., 1976, *Sov. Astron. Lett.*, **2**, 55.
Shapiro, S. L., Lightman, A. P., and Eardley, D. M., 1976, *Astrophys. J.*, **204**, 187.

8

PLASMA EFFECTS

So far we have assumed our propagation medium to be a vacuum. In astrophysical applications free charges often play a substantial role in determining the propagation properties of the medium. Loosely speaking, a globally neutral, ionized gas is called a *plasma*. In this chapter we give an elementary treatment of some basic plasma effects.

8.1 DISPERSION IN COLD, ISOTROPIC PLASMA

The Plasma Frequency

Maxwell's equations for a vacuum can still be used for a plasma if the charge and current densities ρ and $\mathbf{j}$ due to the plasma are explicitly included. If we assume a space and time variation of all quantities of the form $\exp i(\mathbf{k}\cdot\mathbf{r}-\omega t)$ these equations are [cf. Eqs. (2.19) with microscopic fields and free and bound charges included]

$$\begin{aligned} i\mathbf{k}\cdot\mathbf{E}&=4\pi\rho, \qquad i\mathbf{k}\cdot\mathbf{B}=0,\\ i\mathbf{k}\times\mathbf{E}&=i\frac{\omega}{c}\mathbf{B}, \quad i\mathbf{k}\times\mathbf{B}=\frac{4\pi}{c}\mathbf{j}-i\frac{\omega}{c}\mathbf{E}. \end{aligned} \tag{8.1}$$

Let us assume that our plasma consists of electrons with density n. The ions are neglected here, because they are very much less mobile than the electrons and contribute negligibly to the current. (They are important for certain wave motions other than radiation, however, and they do keep the plasma neutral globally.) We also assume that there is no external magnetic field; thus the plasma is *isotropic*. Each electron responds to the electric field according to Newton's law (for an electron charge $q=-e$)

$$m\dot{\mathbf{v}}=-e\mathbf{E}. \tag{8.2}$$

The magnetic force, being of order v/c, has been neglected. In terms of oscillating quantities $\mathbf{v}$ becomes

$$\mathbf{v}=\frac{e\mathbf{E}}{i\omega m}. \tag{8.3}$$

Since the current density is given by $\mathbf{j}=-ne\mathbf{v}$, we have

$$\mathbf{j}=\sigma\mathbf{E}, \tag{8.4}$$

where the *conductivity*, σ, satisfies

$$\sigma=\frac{ine^2}{\omega m}. \tag{8.5}$$

By means of the charge conservation equation we find:

$$-i\omega\rho+i\mathbf{k}\cdot\mathbf{j}=0,$$

so that

$$\rho=\omega^{-1}\mathbf{k}\cdot\mathbf{j}=\sigma\omega^{-1}\mathbf{k}\cdot\mathbf{E}. \tag{8.6}$$

Using these expressions for $\mathbf{j}$ and ρ and introducing the *dielectric constant* ϵ, defined by

$$\epsilon\equiv 1-\frac{4\pi\sigma}{i\omega}, \tag{8.7}$$

we find that Maxwell's equations become

$$\begin{aligned} &i\mathbf{k}\cdot\epsilon\mathbf{E}=0, && i\mathbf{k}\cdot\mathbf{B}=0,\\ &i\mathbf{k}\times\mathbf{E}=i\frac{\omega}{c}\mathbf{B}, && i\mathbf{k}\times\mathbf{B}=-i\frac{\omega}{c}\epsilon\mathbf{E}. \end{aligned} \tag{8.8}$$

These equations are now "source-free" and can be solved in precisely the same way as before. We find again that **k**, **E**, and **B** form a mutually orthogonal right-hand vector triad, but now the relation between k and ω becomes

$$c^2k^2 = \epsilon\omega^2. \tag{8.9}$$

Substituting in Eq. (8.5) for σ, we obtain an alternate expression for the dielectric constant

$$\epsilon = 1 - \left(\frac{\omega_p}{\omega}\right)^2, \tag{8.10}$$

where we have introduced the *plasma frequency* ω_p, defined by

$$\omega_p^2 = \frac{4\pi ne^2}{m}. \tag{8.11}$$

Numerically, we obtain

$$\omega_p = 5.63 \times 10^4 n^{1/2}\, s^{-1}, \tag{8.12}$$

where n is given in cm^{-3}. The dispersion relation connecting k and ω can now be written:

$$k = c^{-1}\sqrt{\omega^2 - \omega_p^2}\,, \tag{8.13a}$$

$$\omega^2 = \omega_p^2 + k^2c^2. \tag{8.13b}$$

We see immediately from these equations that for $\omega < \omega_p$, the wave number is *imaginary*

$$k = \frac{i}{c}\sqrt{\omega_p^2 - \omega^2}\,. \tag{8.14}$$

In this case the amplitude of the wave decreases exponentially on a scale of the order of $2\pi c/\omega_p$. Thus ω_p defines a *plasma cutoff frequency* below which there is no electromagnetic propagation. For example, the earth's ionosphere prevents extraterrestrial radiation at frequencies less than about 1 MHz from being observed at the earth's surface (corresponding to $n_{\text{average}} \sim 10^4\ cm^{-3}$).

Note from the purely imaginary nature of σ, Eq. (8.5), that **j** and **E** are 90° out of phase with each other ($i = e^{i\pi/2}$). Thus there is no time-averaged mechanical work done on the particles by the field in an isotropic plasma, and no dissipation.

The existence of the plasma cutoff yields an important method of probing the ionosphere. Let a pulse of radiation in a narrow range about ω be directed straight upward from the earth's surface. When there is a layer at which n is large enough to make $\omega_p > \omega$, the pulse will be *totally reflected* from the layer. The time delay of the pulse provides information on the height of the layer. By making such measurements at many different frequencies, the electron density can be determined as a function of height.

Group and Phase Velocity and the Index of Refraction

When $\omega > \omega_p$, there is propagation of electromagnetic radiation with *phase velocity*

$$v_{ph} \equiv \frac{\omega}{k} = \frac{c}{n_r}, \tag{8.15}$$

where n_r is the *index of refraction*

$$n_r \equiv \sqrt{\epsilon} = \sqrt{1 - \frac{\omega_p^2}{\omega^2}} \, . \tag{8.16}$$

The phase velocity always exceeds the speed of light. The *group velocity*

$$v_g \equiv \frac{\partial \omega}{\partial k} = c\sqrt{1 - \frac{\omega_p^2}{\omega^2}} \, , \tag{8.17}$$

on the other hand, is always less than c. The wave energy travels at the group velocity, as does any modulation of the wave (information coding). See Jackson (1975) for a standard discussion of v_{ph} and v_g and Problem 8.2 for an alternative treatment.

In a medium with variable electron density, and hence variable index of refraction, radiation travels along curved paths rather than in straight lines. Radio propagation in the ionosphere and solar corona is affected by such curved paths. The curved trajectories in inhomogeneous media may be obtained straightforwardly from application of Snell's law for ray bending (see e.g., Rossi, 1957) and are given by

$$\frac{d(n\hat{\mathbf{k}})}{dl} = \nabla n, \tag{8.18}$$

where n, $\hat{\mathbf{k}}$, and l are the index of refraction, ray direction, and ray path length, respectively. It can be shown (Problem 8.1) that it is the quantity I_ν/n_r^2 that is constant along the ray, rather than I_ν. This is a generalization of Liouville's theorem, Eq. (1.12).

An important application of the formula for group velocity is to pulsars. Each individual pulse from the pulsar has a spectrum covering a wide band of frequency. Therefore, the pulse will be *dispersed* by its interaction with the interstellar plasma, since each small range of frequencies travels at a slightly different group velocity and will reach earth at a slightly different time.

Suppose the pulsar is a distance d away. Then the time required for a pulse to reach earth at frequency ω is

$$t_p = \int_0^d \frac{ds}{v_g},$$

where s measures the line-of-sight distance from the pulsar to earth. The plasma frequencies in interstellar space are usually quite low ($\sim 10^3$ Hz), so we can assume $\omega \gg \omega_p$ and expand

$$v_g^{-1} = \frac{1}{c}\left(1 - \frac{\omega_p^2}{\omega^2}\right)^{-1/2} \approx \frac{1}{c}\left(1 + \frac{1}{2}\frac{\omega_p^2}{\omega^2}\right).$$

Thus we obtain

$$t_p \approx \frac{d}{c} + (2c\omega^2)^{-1}\int_0^d \omega_p^2\, ds. \tag{8.19}$$

The first term is the transit time for a vacuum; the second term is the plasma correction. What is usually measured is the rate of change of arrival time with respect to frequency, $dt_p/d\omega$. With the formula for ω_p^2 this can be written

$$\frac{dt_p}{d\omega} = -\frac{4\pi e^2}{cm\omega^3}\mathfrak{D}, \tag{8.20a}$$

where

$$\mathfrak{D} \equiv \int_0^d n\, ds \tag{8.20b}$$

is the *dispersion measure* of the ray. By assuming a typical value for the electron density in interstellar space ($n \sim 0.03\ \text{cm}^{-3}$) an estimate of the pulsar's distance can be obtained.

8.2 PROPAGATION ALONG A MAGNETIC FIELD; FARADAY ROTATION

We now want to extend somewhat the above discussion of plasma propagation effects by considering the effect of an external, fixed magnetic field $\mathbf{B}_0$. The properties of the waves will then depend on the direction of propagation relative to the direction of $\mathbf{B}_0$. For this reason the plasma is called *anisotropic*. We also make the cold plasma approximation here, and treat only the special case of propagation along the magnetic field.

Because of the magnetic field, a new frequency enters the problem, namely, the *cyclotron frequency*

$$\omega_B = \frac{eB_0}{mc}, \tag{8.21}$$

which is the frequency of gyration for an electron about the field lines. Numerically we obtain, for B_0 in gauss,

$$\omega_B = 1.67 \times 10^7 B_0\, s^{-1}, \tag{8.22a}$$

$$\hbar\omega_B = 1.16 \times 10^{-8} B_0\, eV. \tag{8.22b}$$

The dielectric constant is no longer a scalar; it becomes a tensor and has different effective values for waves of different directions. The medium now also discriminates between different polarizations. Only waves with special polarizations have the simple exponential forms we have been assuming, $\mathbf{E} \exp i(\mathbf{k}\cdot\mathbf{r} - \omega t)$ where $\mathbf{E}$ is *constant*.

If the fixed magnetic field $\mathbf{B}_0$ is much stronger than the field strengths of the propagating wave, then the equation of motion of an electron in the plasma is approximately

$$m\frac{d\mathbf{v}}{dt} = -e\mathbf{E} - \frac{e}{c}\mathbf{v} \times \mathbf{B}_0. \tag{8.23}$$

Assume that the propagating wave is circularly polarized and sinusoidal:

$$\mathbf{E}(t) = Ee^{-i\omega t}(\boldsymbol{\epsilon}_1 \mp i\boldsymbol{\epsilon}_2), \tag{8.24}$$

where the $-$ corresponds to right circular polarization and the $+$ corresponds to left circular polarization. Assume further, for simplicity, that the wave propagates along the fixed field $\mathbf{B}_0$:

$$\mathbf{B}_0 = B_0 \epsilon_3. \tag{8.25}$$

Substituting Eqs. (8.24) and (8.25) into (8.23), one finds that the steady-state velocity $\mathbf{v}(t)$ has the form

$$\mathbf{v}(t) = \frac{-ie}{m(\omega \pm \omega_B)} \mathbf{E}(t), \tag{8.26}$$

where ω_B is given in Eq. (8.21).

Comparison of Eq. (8.26) with Eqs. (8.3)–(8.5) and (8.7) then gives an expression for the dielectric constant

$$\epsilon_{R,L} = 1 - \frac{\omega_p^2}{\omega(\omega \pm \omega_B)}, \tag{8.27}$$

where the R, L corresponds to the $+$ and $-$ signs, respectively. These waves travel with different velocities. Therefore, a plane polarized wave, which is a linear superposition of a right-hand and a left-hand polarized wave, will not keep a constant plane of polarization, but this plane will *rotate* as it propagates. This effect is called *Faraday rotation.*

The phase angle ϕ through which the electric vector of a circularly polarized wave moves in traveling a distance d is simply $\mathbf{k} \cdot \mathbf{d}$. More generally, if the wave number is not constant along the path, the phase angle is

$$\phi_{R,L} = \int_0^d k_{R,L}\, ds, \tag{8.28a}$$

where

$$k_{R,L} = \frac{\omega}{c} \sqrt{\epsilon_{R,L}}\,. \tag{8.28b}$$

A plane-polarized wave is rotated through an angle $\Delta\theta$, equal to one-half the difference between ϕ_R and ϕ_L, as can be seen from Fig. 8.1. We assume that $\omega \gg \omega_p$ and $\omega \gg \omega_B$ so that

$$k_{R,L} \approx \frac{\omega}{c}\left[1 - \frac{\omega_p^2}{2\omega^2}\left(1 \mp \frac{\omega_B}{\omega}\right)\right]. \tag{8.29}$$

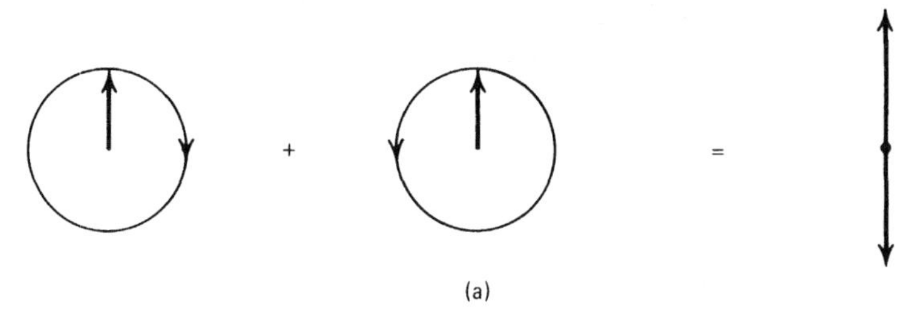

Figure 8.1a ***Decomposition of linear polarization into components of right and left circular polarization.***

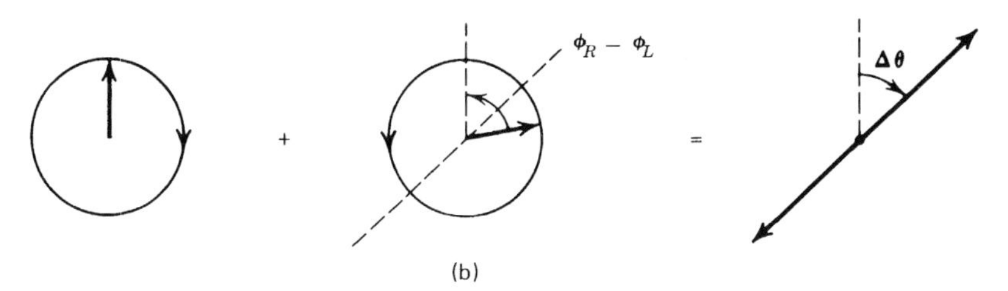

Figure 8.1b ***Faraday rotation of the plane of polarization.***

Thus we have the result

$$\begin{aligned}\Delta\theta &= \tfrac{1}{2}\int_0^d (k_R - k_L)\,ds \\ &= \tfrac{1}{2}\int_0^d (c\omega^2)^{-1}\omega_p^2\omega_B\,ds,\end{aligned} \tag{8.30}$$

or, substituting for ω_p^2 and ω_B, we obtain the formula for Faraday rotation:

$$\Delta\theta = \frac{2\pi e^3}{m^2c^2\omega^2}\int_0^d nB_{\parallel}\,ds. \tag{8.31}$$

As derived here, this formula holds only if the direction of **B** is always along the line of sight. However, it can be shown that this formula holds in general if we use $B_{\parallel}$, the component of **B** along the line of sight.

Since $\Delta\theta$ varies with frequency (as ω^{-2}) for the same line of sight, we can determine the value of the integral $\int nB_{\parallel}\,ds$ by making measurements at several frequencies. This can be used to deduce information about the interstellar magnetic field. However, if this field changes direction often along the line of sight (as we believe it does), then this method gives only a lower limit to actual field magnitudes.

8.3 PLASMA EFFECTS IN HIGH-ENERGY EMISSION PROCESSES

When fast particles radiate by means of a high-energy emission mechanism —like synchrotron, inverse Compton, or bremsstrahlung emission—this radiation is subject to all the plasma propagation effects mentioned previously. In particular, we can expect little observable radiation below the cutoff frequency ω_p, whereas above ω_p the phenomena of pulse dispersion and path curvature may occur. When magnetic fields are present, Faraday rotation will degrade the degree of polarization of synchrotron sources.

In addition, however, there are some specific effects on the high-energy emission processes themselves that can change the entire character of the emitted radiation. We shall describe two such effects, *Cherenkov radiation* and the *Razin effect*. Both of these require us to consider the induced motions and subsequent emission from the particles comprising the medium through which the fast particles are moving. Since we are only interested in the collective response of the medium, it is permissible to treat the medium in terms of a macroscopic dielectric constant ϵ. For certain parts of the following discussion we make the assumption that the dielectric constant is independent of frequency and wave number. This is not strictly true, as we have seen, but it allows us to obtain the principal results quickly. For more detailed derivations, without use of this assumption, see Ginzburg and Syrovatskii (1965) and Razin (1960). For our assumption, Maxwell's equations can be written as

$$\nabla \cdot \mathbf{E} = \frac{1}{\epsilon} 4\pi\rho, \qquad \nabla \cdot \mathbf{B} = 0,$$
$$\nabla \times \mathbf{E} = \frac{1}{c}\frac{\partial \mathbf{B}}{dt}, \qquad \nabla \times \mathbf{B} = \frac{4\pi}{c}\mathbf{j} + \frac{\epsilon}{c}\frac{\partial \mathbf{E}}{\partial t}. \tag{8.32}$$

It can easily be shown that these equations *formally* result from Maxwell's equation in vacuum by the substitutions

$$\mathbf{E} \to \sqrt{\epsilon}\,\mathbf{E}, \qquad c \to c/\sqrt{\epsilon}\,,$$
$$\mathbf{B} \to \mathbf{B}, \qquad \phi \to \sqrt{\epsilon}\,\phi, \tag{8.33}$$
$$e \to e/\sqrt{\epsilon}\,, \qquad \mathbf{A} \to \mathbf{A}.$$

These equations may be solved in the same manner as before for the retarded and Liénard–Wiechert potentials, using Eqs. (3.7a), (3.7b), and (3.10), and then making the substitutions indicated in Eqs. (8.33).

Cherenkov Radiation

A charge moving uniformly in a vacuum cannot radiate, as such radiation would violate the results of relativity theory. The same conclusion holds for a charge moving uniformly through a dielectric medium, providing the velocity of the charge is less than the phase velocity of light in the medium. This can be proved directly from the modified Liénard–Wiechert potentials. These potentials differ from the vacuum case only in the scale of some of the parameters, according to the substitutions in Eqs. (8.33); thus these changes do not affect the conclusion that the fields fall off as $1/R^2$ and do not carry energy over large distances.

If the medium has an index of refraction greater than unity, $n_r > 1$, the velocity of the charge can exceed the phase velocity. In this case the potentials differ qualitatively from those of the vacuum. From Eqs. (8.33) the factor $\kappa = 1 - \beta\cos\theta$ in Eqs. (3.7) becomes

$$\kappa = 1 - \beta n_r \cos\theta, \tag{8.34}$$

and this can *vanish* for an angle θ such that $\cos\theta = (n_r\beta)^{-1}$. The potentials become infinite at certain places, and this invalidates the usual arguments concerning the $1/R^2$ behavior of the fields. In consequence, the particle can now radiate.

Another qualitatively different effect appears when $v > c/n_r$, namely, that the potentials at a point may be determined by *two* retarded positions of the particle, rather than just one. This can be seen from Fig. 8.2. The points 1, 2, 3, and 4 denote successive positions of the particle, and the spheres represent "information spheres" generated at these positions, which move outward with the velocity c/n_r.

Looking at the case $v > c/n_r$ we note that space is divided into two distinct regions by a cone, the *Cherenkov cone*, such that points outside the cone feel no potentials as yet; inside the cone each point is intersected by two spheres, and thus each point feels the potentials due to two retarded positions of the particle.

The resulting radiation, called *Cherenkov radiation*, is confined within the cone and moves outward in a direction normal to the cone with the velocity c/n_r. Notice the similarity of this pattern with a shock pattern generated by a supersonic airplane; both are due to motion of a body at a velocity greater than that of wave propagation in the medium. The relation $\cos\theta = (\beta n_r)^{-1}$ can be understood from Fig. 8.3. Since $\cos\theta < 1$ and $v/c < 1$ it follows that

$$\frac{c}{n_r} < v < c \tag{8.35}$$

for Cherenkov radiation.

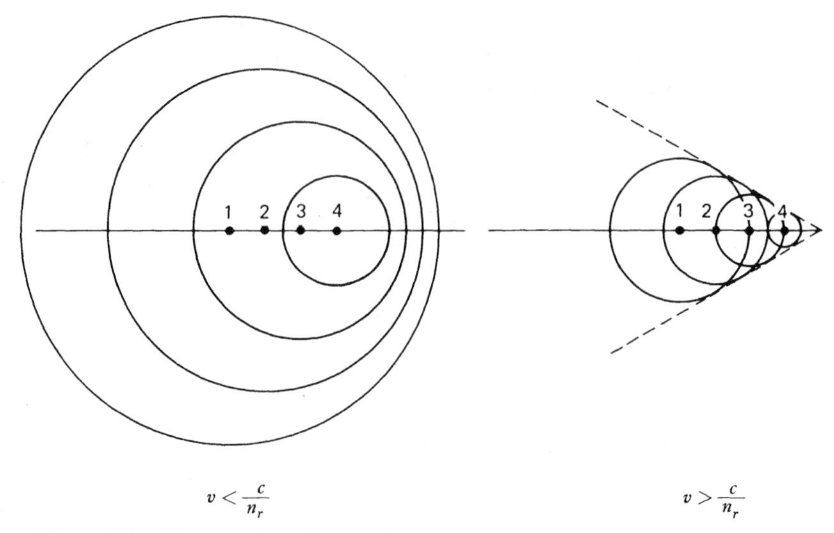

Figure 8.2 ***Propagation of wave fronts generated by a particle moving with velocity v through a refractive medium.***

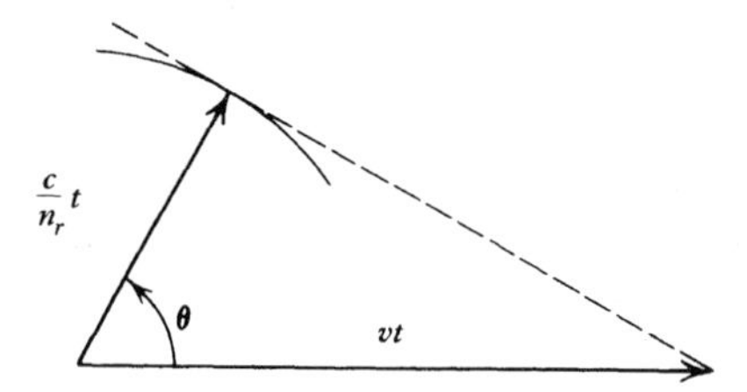

Figure 8.3 ***Geometry of Cherenkov cone.***

The precise direction of the radiation can be used as an energy measurement for fast particles in the laboratory or observatory. Cherenkov radiation due to high energy cosmic rays has been observed in the earth's atmosphere. Since the radiation is quite intense for fast particles, it acts as an effective mechanism for energy loss.

Razin Effect

When $n_r < 1$, as it is in a cold plasma, Cherenkov radiation cannot occur. In this case there is an effect that has important implications for synchrotron emission. The "beaming" effect associated with emission from a fast particle can be attributed to the factor $\kappa = 1 - \beta\cos\theta$ appearing in the denominators of the Liénard–Wiechert potentials. Making the above substitutions, this factor is now given by Eq. (8.34).

For $n_r < 1$, as in a plasma, the beaming effect is suppressed, for now there is no velocity and angle combination for which κ is small. This can be seen as follows: The critical angle defining the beaming effect has been shown to be given by $\theta_b \sim 1/\gamma = \sqrt{1-\beta^2}$ in a vacuum. Therefore, in a medium we have

$$\theta_b \sim \sqrt{1 - n_r^2\beta^2}\,, \tag{8.36}$$

using the substitutions of Eqs. (8.33). There are two cases of this formula that can be identified, depending on which factor, n_r or β, dominates in keeping θ_b from being small. If n_r is sufficiently close to unity, then θ_b is determined by β, as in the vacuum case. On the other hand, if n_r differs substantially from unity, then we have

$$\theta_b \sim \sqrt{1 - n_r^2} = \frac{\omega_p}{\omega}. \tag{8.37}$$

From this it can be seen that the medium will dominate beaming at low frequencies. At higher frequencies θ_b decreases until it becomes of order of the vacuum value $1/\gamma$, and thereafter the vacuum results apply. Therefore, the medium is unimportant when

$$\omega \gg \gamma\omega_p,$$

and the medium is important when

$$\omega \ll \gamma\omega_p.$$

This suppression of the beaming effect at low frequencies has a profound effect on synchrotron emission, as can be appreciated from the dominant role beaming has in the physical explanation of this process. Below the frequency $\gamma\omega_p$ the synchrotron spectrum will be cut off because of the suppression of beaming. This is called the *Razin effect*. It is obvious that this effect dominates the ordinary plasma cutoff, which occurs at the much lower frequency ω_p.

PROBLEMS

8.1—In a medium with dielectric constant n_r, show that I_ν/n_r^2 is constant along a ray.

8.2—Consider a traveling wave packet of amplitude

$$\psi(r,t)=\int_{-\infty}^{\infty} A(k)e^{i[kr-\omega(k)t]}\,dk,$$

where $\omega(k)$ is a real function of k. Define the centroid of the wave packet, $\langle r(t)\rangle$ by

$$\langle r(t)\rangle \equiv \frac{\int r|\psi(r,t)|^2\,dr}{\int |\psi(r,t)|^2\,dr}.$$

Show that the wave centroid travels with the velocity $\langle \partial\omega/\partial k\rangle$,

$$\frac{d}{dt}\langle r(t)\rangle = \langle \partial\omega/\partial k\rangle,$$

where

$$\langle \partial\omega/\partial k\rangle \equiv \frac{\int \partial\omega/\partial k|A(k)|^2\,dk}{\int |A(k)|^2\,dk}.$$

8.3—The signal from a pulsed, polarized source is measured to have an arrival time delay that varies with frequency as $dt_p/d\omega = 1.1\times10^{-5}$ s^2, and a Faraday rotation that varies with frequency as $d\Delta\theta/d\omega = 1.9\times10^{-4}$ s. The measurements are made around the frequency $\omega = 10^8$ s^{-1}, and the source is at unknown distance from the earth. Find the mean magnetic field, $\langle B_\parallel\rangle$, in the interstellar space between the earth and the source:

$$\langle B_\parallel\rangle \equiv \frac{\int nB_\parallel\,ds}{\int n\,ds}.$$

REFERENCES

Ginzburg, V. L. and Syrovatskii, S. I., 1965, *Ann. Rev. Astron. Astrophys.*, **3**, 297.
Jackson, J. D., 1975, *Classical Electrodynamics*, (Wiley, New York).
Pacolczyk, A. G., 1970, *Radio Astrophysics*, (Freeman, San Francisco).
Razin, V. A., 1960, *Izvestiya Vys. Ucheb. Zaved. Radiofiz.*, 3, 921.
Rossi, B., 1957, *Optics*, (Addison-Wesley, Reading, Mass.).

9

ATOMIC STRUCTURE

The classical theory of radiation is unable to treat physical processes in which the interaction between matter and radiation takes place by means of single (or a few) photons. We have already dealt with some elementary aspects of this interaction when we discussed the Planck law and the Einstein coefficients. However to really solve problems we need to find explicit expressions for the A and B coefficients or equivalents. This must involve detailed investigation of the structure of the matter that interacts with the radiation, its energy levels, and other physical properties. In this chapter we treat the structure of atoms, and in the next chapter we consider the radiative transitions of these atoms.

9.1 A REVIEW OF THE SCHRODINGER EQUATION

We begin with the *time-dependent Schrodinger equation* for a system with Hamiltonian H:

$$i\hbar\frac{\partial\Psi}{\partial t}=H\Psi. \tag{9.1}$$

Often we are interested in the stationary solutions found by separating the time and space parts of the wave function Ψ, which is possible if H is independent of time:

$$\Psi(\mathbf{r},t)=\psi(\mathbf{r})e^{iEt/\hbar}. \tag{9.2}$$

It follows that ψ satisfies the *time-independent Schrodinger equation*

$$H\psi=E\psi. \tag{9.3}$$

Here E is the energy and ψ is the wave function of the corresponding energy state. In the case of electrons surrounding a nucleus of charge Ze, neglecting spin, relativistic effects and nuclear effects, the Hamiltonian is

$$H=-\frac{\hbar^2}{2m}\sum_j \nabla_j^2-Ze^2\sum_j\frac{1}{r_j}+\sum_{i>j}\frac{e^2}{r_{ij}}. \tag{9.4}$$

Here the first term in H is the sum over electron kinetic energy, the second term is the Coulomb interaction energy between nucleus and electrons, and the third term is the Coulomb energy of the electrons interacting with themselves. We then obtain the equation

$$\left(-\frac{\hbar^2}{2m}\sum_j \nabla_j^2-E-Ze^2\sum_j\frac{1}{r_j}+\sum_{i>j}\frac{e^2}{r_{ij}}\right)\psi=0. \tag{9.5}$$

This determines an approximation to the atomic states. This equation can be put into dimensionless form by using the electron mass and charge as units of mass and charge, and using the first Bohr radius,

$$a_0\equiv\frac{\hbar^2}{me^2}=0.529\times10^{-8}\text{ cm} \tag{9.6}$$

as the unit of length. With this unit of length, the energy E is measured in units

$$\frac{e^2}{a_0}=4.36\times10^{-11}\text{ erg}=27.2\text{ eV}. \tag{9.7}$$

(This unit of energy equals two Rydbergs.) Characteristic sizes and binding energies of atoms will be of the order of the above values. In dimensionless form, the Schrodinger equation becomes

$$\left(\frac{1}{2}\sum_j \nabla_j^2+E+Z\sum_j\frac{1}{r_j}-\sum_{i>j}\frac{1}{r_{ij}}\right)\psi=0. \tag{9.8}$$

9.2 ONE ELECTRON IN A CENTRAL FIELD

Even in complete atoms with N electrons it is useful to consider single-electron states. We assume that each electron moves in the potential of the nucleus plus the averaged potential due to the other $N-1$ electrons. This is called the *self-consistent field approximation*. When, in addition, this averaged potential is assumed to be spherically symmetric, it is called the *central field approximation* and represents one of the most powerful concepts in atomic theory. It provides a useful classification of atomic states and also a starting point for treating correlations as perturbations.

In the central field approximation each electron feels a different potential, which may be regarded as a shielded nuclear charge. When the electron is far from the nucleus and outside the cloud of other electrons, the potential is

$$V(r) \to \frac{Z-N+1}{r}, \qquad r \to \infty.$$

When the electron is close to the nucleus, so that all the other electrons are further away, we have

$$V(r) \to -\frac{Z}{r} + C, \qquad r \to 0.$$

Wave Functions

In classical mechanics a central potential implies the constancy of orbital angular momentum. The same is true in quantum mechanics. If H depends only on the magnitude of r, we can make the separation

$$\psi(r,\theta,\phi) = r^{-1}R(r)\,Y(\theta,\phi). \tag{9.9}$$

The functions $Y(\theta,\phi)$ are the *spherical harmonics*, defined by

$$Y = Y_{lm}(\theta,\phi) = \left[\frac{(l-|m|)!}{(l+|m|)!}\,\frac{2l+1}{4\pi}\right]^{1/2}(-1)^{(m+|m|)/2}P_l^{|m|}(\cos\theta)e^{im\phi}, \tag{9.10}$$

where P_l^m is the associated Legendre function, and l and m are integers. The functions Y_{lm} are eigenfunctions of the orbital angular momentum operator $\mathbf{L}=\mathbf{r}\times\mathbf{p}$. That is,

$$\mathbf{L}^2 Y_{lm} = l(l+1)\, Y_{lm}, \tag{9.11a}$$

$$L_z\, Y_{lm} = m Y_{lm}, \tag{9.11b}$$

where angular momentum is in units of $\hbar$. The values of l are $l=0,1,2,3,4,\ldots,$ called s states, p states, d states, f states, g states, and so on, respectively. The value m ranges from $-l$ to $+l$ in integer steps. The functions Y_{lm} are orthonormal:

$$\int d\Omega\, Y^*_{lm}(\theta,\phi)\, Y_{l'm'}(\theta,\phi) = \delta_{l,l'}\delta_{m,m'}. \tag{9.12}$$

Note that the angular eigenfunctions, unlike the radial functions below, are independent of the form of the potential, $V(r)$, as long as it is spherically symmetric.

The radial part of the wave function satisfies the equation

$$\frac{1}{2}\frac{d^2R_{nl}}{dr^2} + \left[E - V(r) - \frac{l(l+1)}{2r^2}\right]R_{nl} = 0. \tag{9.13}$$

We see that R depends on l but not on m. The index n labels the energy states. Generally for a given value of l, the states in increasing order of energy are labeled:

$$n = l+1,\ l+2,\ l+3,\ldots.$$

The radial functions have the normalization

$$\int_0^\infty R_{nl}(r) R_{n'l}(r)\, dr = \delta_{n,n'}. \tag{9.14}$$

(We have not put a complex conjugation here, since the Rs can always be chosen as real). In addition to the above discrete eigenfunctions, there is also a continuous set of eigenfunctions, corresponding to unbound states.

The solutions for the pure Coulomb case, when $V(r) = -Z/r$, are

$$R_{nl}(r) = -\left\{\frac{Z(n-l-1)!}{n^2[(n+l)!]^3}\right\}^{1/2} e^{-\rho/2}\rho^{l+1} L^{2l+1}_{n+l}(\rho), \tag{9.15a}$$

$$E_n = -Z^2/2n^2, \tag{9.15b}$$

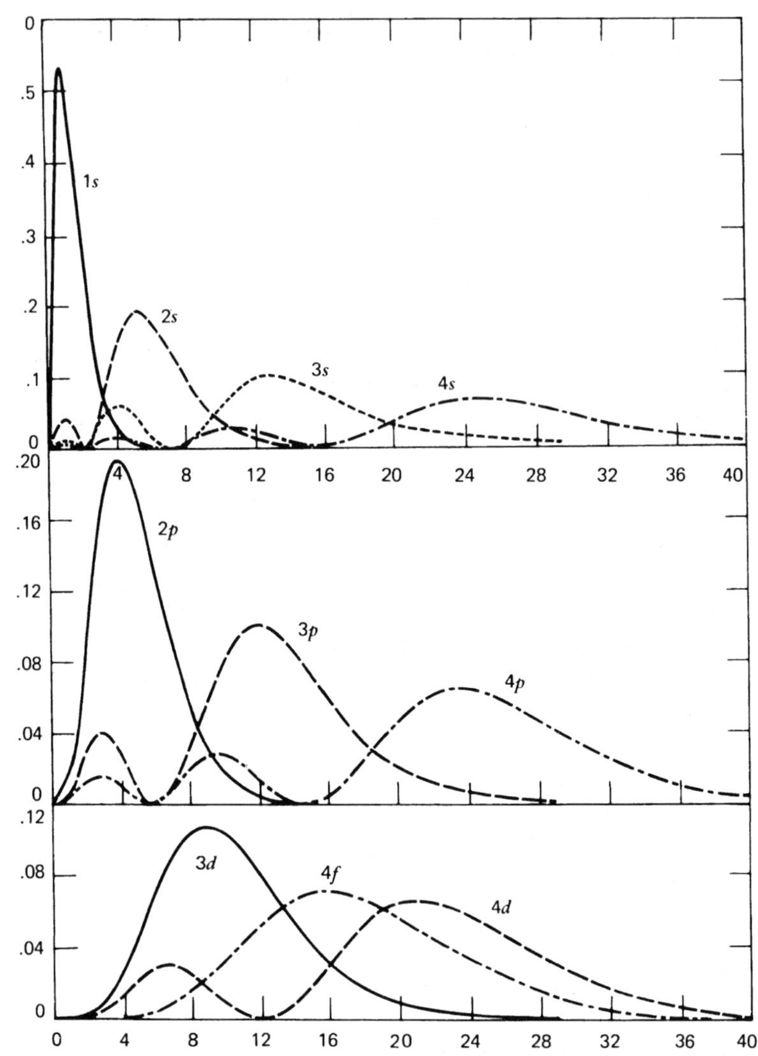

***Figure 9.1** Radial probability distribution for an electron in several of the lowest levels of hydrogen. The abscissa is the radius in atomic units. (Taken from Condon, E. and Shortley, G. 1963, The Theory of Atomic Spectra, Cambridge, Cambridge University Press.)*

where $\rho = 2Zr/n$. The functions L_{n+l}^{2l+1} are the associated Laguerre polynomials. The first three radial functions are:

$$R_{10} = 2Z^{3/2} r e^{-Zr}, \tag{9.16a}$$

$$R_{20} = \left(\frac{Z}{2}\right)^{3/2} (2 - Zr) r e^{-Zr/2}, \tag{9.16b}$$

$$R_{21} = \left(\frac{Z}{2}\right)^{3/2} \frac{Zr^2}{\sqrt{3}} e^{-Zr/2}. \tag{9.16c}$$

The quantity R_{nl}^2 is the probability that the electron is between r and $r + dr$. Figure 9.1 shows the probability distribution for the lowest states of hydrogen.

Spin

The electron possesses an intrinsic angular momentum s, with $|s| = \frac{1}{2}$. There are thus two states, $m_s = \pm\frac{1}{2}$, for the spin. To incorporate spin into the theory in a completely satisfactory way one should use the relativistic Dirac equation. However, for nonrelativistic cases it is usually sufficient to treat the spin in terms of wave functions with two components. The wave functions corresponding to the values $m_s = \pm\frac{1}{2}$ are defined as

$$|\tfrac{1}{2}\rangle \equiv \alpha = \begin{pmatrix} 1 \\ 0 \end{pmatrix} \qquad |-\tfrac{1}{2}\rangle \equiv \beta = \begin{pmatrix} 0 \\ 1 \end{pmatrix}. \tag{9.17}$$

A single particle state must now include specification of m_s as well as n, l, and m.

9.3 MANY-ELECTRON SYSTEMS

Statistics: The Pauli Principle

We now have a set of single-particle states specified by n, l, m, and m_s. (These are called *orbitals*). From these we want to construct states of the whole system. As a first step let us form products of the sort

$$u_a(1) u_b(2) \cdots u_k(N),$$

where each subscript $a, b \cdots k$ represents the set of values (n, l, m, m_s) and the numbers $1, 2, \ldots, N$ represent the space and spin coordinates of the 1st,

2nd,...,Nth particle. The functions u are the orbitals with spatial part ψ_{nlm}, multiplied by a spin part α or β.

Such products are satisfactory from one point of view: they form a complete set in terms of which any state of the system of N electrons can be represented. They fail, however, to satisfy a basic principle of quantum mechanics, namely, that all electrons are *identical* and that it should not be possible to say that particle 1 is in orbital a, particle 2 is in orbital b, and so on. We may avoid this by forming linear combinations of the above products, including every permutation P of the particles among the orbitals. Since there are $N!$ permutations, the weight we choose must have magnitude $(N!)^{-\frac{1}{2}}$. Its phase is determined by the *Pauli exclusion principle*, which states that no two electrons can occupy the same orbital. Thus we choose the phase as

$$\epsilon_p = \pm 1$$

according as the permutation is an even or odd permutation of some standard ordering. Thus if two electrons are put into the same orbital, the linear combination will vanish, so that no physical (normalizable) state exists. Therefore, the basis states for the whole system are

$$(N!)^{-\frac{1}{2}} \sum_P \epsilon_p P u_a(1) u_b(2) \cdots u_k(N). \tag{9.18}$$

This may be conveniently written as the *Slater determinant*

$$\frac{1}{\sqrt{N!}} \begin{vmatrix} u_a(1) & u_a(2) & \cdots & u_a(N) \\ u_b(1) & u_b(2) & \cdots & u_b(N) \\ \vdots & & & \\ u_k(1) & u_k(2) & \cdots & u_k(N) \end{vmatrix}. \tag{9.19}$$

In this form it is clear that when two electrons occupy the same orbital, two rows of this determinant are equal and it therefore vanishes.

Particles with the above symmetry for their wave functions are called *Fermi–Dirac* particles or simply *fermions*. There is complete antisymmetry of the wave function under interchange of two particles, $\psi(1,2,\ldots N) = \epsilon_p P\psi(1,2,\ldots N)$, as can be seen by interchanging two columns in the above determinant.

Hartree–Fock Approximation: Configurations

An important method for choosing the orbitals used to construct atomic states is based on a variational principle for the expectation value of the energy. The *exact* energy states of the system are determined by the variational condition

$$\delta\langle H\rangle=\delta\int\psi^* H\psi\, d(1)d(2)\cdots d(N)=0,$$

where δ is an arbitrary variation of the normalized trial wave function ψ. We now can determine approximate energy states by using a restricted variation in which ψ is a properly antisymmetrized product of orbitals (a Slater determinant) and considering only variations with respect to a choice of these orbitals. When the details of this variation are carried out one obtains the *Hartree–Fock equations* for each orbital. These are Schrodinger equations with two types of potentials: (1) a term representing the electrostatic potential of the nucleus and of the averaged charge density of all other electrons and (2) a term having no classical analogue, called the *exchange potential*. This exchange term has its origin in the Pauli principle and may be regarded as an expression of an effective repulsion of electrons with the same spin (see Problem 9.1).

There is no real "potential" in the N-electron problem corresponding to this exchange repulsion, only the antisymmetry of the wave functions, which prevents two electrons with the same spin from occupying the same volume element. It is only when one formulates the N-electron problem in terms of single-particle states that the repulsion manifests itself by means of an effective potential in the equations. The essentially nonclassical nature of the exchange potential is clear, since it takes a "nonlocal" form, which cannot easily be interpreted classically.

If the Hartree–Fock potentials are averaged over all angles, one obtains a central potential, which is used to compute the orbitals. It is found that these orbitals give a fair description of the gross structure of atomic systems, including the main features of the periodic table.

The *configuration* of an atomic system is defined by specifying the nl values of all the electron orbitals: nl^x means x electrons in the orbital defined by n and l. There are $2(2l+1)$ electron states available to each l value because m has $2l+1$ values for each l and there are two possible spins. A fairly complete table of ground-level configurations is given in Table 9.1. We see here the regular filling of shells up to the case of $Ar(Z=18)$. Then there is a nonuniformity in that $K(Z=19)$ fills the $4s$ orbital rather than the $3d$ orbital. This is because the effective potential

Table 9.1

Neutral atoms

Atom		K	L		M			N			O	Ground
		1s	2s	2p	3s	3p	3d	4s	4p	4d	5s	level
H	1	1										$^2S_{1/2}$
He	2	2										$^1S_0$
Li	3	2	1									$^2S_{1/2}$
Be	4	2	2									$^1S_0$
B	5	2	2	1								$^2P^0_{1/2}$
C	6	2	2	2								$^3P_0$
N	7	2	2	3								$^4S^0_{1\frac{1}{2}}$
O	8	2	2	4								$^3P_1$
F	9	2	2	5								$^2P^0_{1\frac{1}{2}}$
Ne	10	2	2	6								$^1S_0$
Na	11	2	2	6	1							$^2S_{1/2}$
Mg	12				2							$^1S_0$
Al	13				2	1						$^2P^0_{1/2}$
Si	14		10		2	2						$^3P_0$
P	15				2	3						$^4S^0_{1\frac{1}{2}}$
S	16		Ne core		2	4						$^3P_2$
Cl	17				2	5						$^2P^0_{1\frac{1}{2}}$
Ar	18				2	6						$^1S_0$
K	19	2	2	6	2	6		1				$^2S_{1/2}$
Ca	20							2				$^1S_0$
Sc	21						1	2				$^2D_{1\frac{1}{2}}$
Ti	22						2	2				$^3F_2$
V	23			18			3	2				$^4F_{1\frac{1}{2}}$
Cr	24						5	1				$^7S_3$
Mn	25			A core			5	2				$^6S_{2\frac{1}{2}}$
Fe	26						6	2				$^5D_4$
Co	27						7	2				$^4F_{4\frac{1}{2}}$
Ni	28						8	2				$^3F_4$
Cu	29	2	2	6	2	6	10	1				$^2S_{1/2}$
Zn	30							2				$^1S_0$
Ga	31							2	1			$^2P^0_{1/2}$
Ge	32				28			2	2			$^3P_0$
As	33							2	3			$^4S^0_{1\frac{1}{2}}$
Se	34							2	4			$^3P_2$
Br	35							2	5			$^2P^0_{1\frac{1}{2}}$
Kr	36							2	6			$^1S_0$
Rb	37	2	2	6	2	6	10	2	6		1	$^2S_{1/2}$
Sr	38										2	$^1S_0$
Y	39									1	2	$^2D_{1\frac{1}{2}}$
Zr	40									2	2	$^3F_2$
Nb	41					36				4	1	$^6D_{1/2}$
Mo	42									5	1	$^7S_3$
Tc	43					Kr core				5	2	$^6S_{2\frac{1}{2}}$
Ru	44									7	1	$^5F_5$
Rh	45									8	1	$^4F_{4\frac{1}{2}}$
Pd	46									10		$^1S_0$

Atom		K L M N	N	O				P			Q	Ground
			4f	5s	5p	5d	5f	6s	6p	6d	7s	level
Ag	47			1								$^2S_{1/2}$
Cd	48			2								$^1S_0$
In	49			2	1							$^2P^0_{1/2}$
Sn	50			2	2							$^3P_0$
Sb	51			2	3							$^4S^0_{1\frac{1}{2}}$
Te	52			2	4							$^3P_2$
I	53			2	5							$^2P^0_{1\frac{1}{2}}$
Xe	54			2	6							$^1S_0$
Cs	55			2	6			1				$^2S_{1/2}$
Ba	56			8				2				$^1S_0$
La	57					1		2				$^2D_{1\frac{1}{2}}$
Ce	58		1	2	6	1		2				$^1G^0_4$
Pr	59		3					2				$^4I^0_{4\frac{1}{2}}$
Nd	60		4					2				$^5I_4$
Pm	61		5					2				$^6H^0_{2\frac{1}{2}}$
Sm	62	(K 1s, L 2s 2p, M 3s 3p 3d, N 4s 4p 4d)	6					2				$^7F_0$
Eu	63		7					2				$^8S^0_{3\frac{1}{2}}$
Gd	64		7	8		1		2				$^9D_2$
Tb	65		9					2				$^6H^0_{7\frac{1}{2}}$
Dy	66		10					2				$^5I_8$
Ho	67		11					2				$^4I^0_{7\frac{1}{2}}$
Er	68		12					2				$^3H_6$
Tm	69		13					2				$^2F^0_{3\frac{1}{2}}$
Yb	70		14					2				$^1S_0$
Lu	71		14			1		2				$^2D_{1\frac{1}{2}}$
Hf	72		14	2	6	2		2				$^3F_2$
Ta	73					3		2				$^4F_{1\frac{1}{2}}$
W	74					4		2				$^5D_0$
Re	75		46+22			5		2				$^6S_{2\frac{1}{2}}$
Os	76					6		2				$^5D_4$
Ir	77					7		2				$^4F_{1\frac{1}{2}}$
Pt	78					9		1				$^3D_3$
Au	79		14	2	6	10		1				$^2S_{1/2}$
Hg	80							2				$^1S_0$
Tl	81							2	1			$^2P^0_{1/2}$
Pb	82	46 Pd core	46+32					2	2			$^3P_0$
Bi	83							2	3			$^4S^0_{1\frac{1}{2}}$
Po	84							2	4			$^3P_2$
At	85							2	5			$^2P^0_{1\frac{1}{2}}$
Rn	86							2	6			$^1S_0$
Fr	87		14	2	6	10		2	6		1	$^2S_{1/2}$
Ra	88										2	$^1S_0$
Ac	89		46+32							1	2	$^2D_{1\frac{1}{2}}$
Th	90									2	2	$^3F_2$
Pa	91						2			1	2	$^4K_{5\frac{1}{2}}$
U	92						3			1	2	$^5L^0_6$

due to the electron cloud gives more binding to electrons that penetrate closer to the nucleus and thus feel the higher Coulomb field; such electrons are just the low-l electrons.

Closed shells generally are not much influenced by changes in the outer, partially filled shells, so that often one will only specify the configuration of the outer shell, such as: $Al-3s^2 3p$. Radiative transitions, at least at optical frequencies, usually affect only outer electrons.

By using the Pauli principle in this way, one can understand qualitatively the building up of the periodic table of elements.

The Electrostatic Interaction; *LS* Coupling and Terms

The specification of the electron configuration, the n, l values of all electrons, leaves a great deal of unspecified information, since we are not given the values of m_l and m_s. Note that in the central field approximation all of these states are degenerate, since the central field Hamiltonian is spherically symmetric and does not depend on spin. To proceed further we write the exact Hamiltonian as

$$H=\sum_i \frac{P_i^2}{2m}-Z\sum \frac{1}{r_i}+\sum V_i(r_i)+H_1\equiv H_0+H_1. \tag{9.20}$$

We have added and subtracted the central field potentials due to the smeared-out electrons. We regard this as a perturbation problem in which H_0 is the zeroth-order potential, whose states are just the configurations we have been discussing. The perturbation part H_1 is

$$H_1=\sum_{i>j}\frac{1}{r_{ij}}-\sum_i V_i(r_i)+H_{so}\equiv H_{es}+H_{so}+\cdots, \tag{9.21}$$

where H_{so} is the spin-orbit interaction to be discussed later, and where there are additional terms that are to be regarded as negligible. The first two terms represent the residual electrostatic interaction between the electrons after the averaged central field has been subtracted. This is what we simply call the *electrostatic interaction*, H_{es}.

For the present we are concerned with the splitting of the configurations by the electrostatic interaction. We note first of all that the individual orbital angular momenta will not remain constant under this interaction, although their total $\mathbf{L}=\Sigma_i \mathbf{l}_i$ will be constant. Also the sum of the spin angular momenta, $\mathbf{S}=\Sigma \mathbf{s}_i$, will be constant.

According to degenerate perturbation theory the first-order energy corrections must be found by evaluating the diagonal matrix elements between the particular linear combinations of the unperturbed states that diagonalize the perturbation. Another way of characterizing these linear combinations is that they are eigenstates of operators that commute with the perturbation. We note that two such operators are **L** and **S** so that the whole perturbation problem is simplified (and in many cases completely solved) by forming those linear combinations of unperturbed states that represent states of total spin and total orbital angular momenta.

In this way we find the configurations split into *terms* with particular values of L and S (the magnetic numbers m_S and m_L do not enter by rotational symmetry arguments). These terms then split further by the action of the spin-orbit interaction. The fact that the electrostatic interaction is the dominant splitting interaction of a configuration for many atoms (especially of low Z) and that the remaining spin-orbit splitting is much smaller makes this perturbation scheme and its attendant characterization and labeling of states a very useful one. It is called *LS coupling* or *Russell–Saunders coupling*.

Let us discuss the origin of this electrostatic splitting from a physical point of view. The electrons repel each other, and therefore their mutual electrostatic energy is positive. The farther away the electrons get, the lower will be the contribution of the electrostatic energy to the total energy. This leads to an important set of rules governing the splitting of the configuration energies as a function of spin and orbital angular momentum. First we note that a large spin implies that the individual spins are aligned in the same direction. By the nature of the Pauli principle, we have that the electrons will be further apart on the average. Thus the rule: *terms with larger spin tend to lie lower in energy*. There is a similar effect regarding the orbital angular momentum L. A large L implies that the individual $\mathbf{l}_i$ are aligned so that the sense of orbiting around the atom is the same for most electrons. Such a pattern lends itself to the electrons keeping farther apart on the average than when they orbit in opposite directions. This effect is usually smaller than the preceding, thus the rule: *of those terms of a given configuration with a given spin those with largest L tend to lie lower in energy*. These two rules are known as *Hund's rules* and apply strictly only to the ground configuration.

9.4 PERTURBATIONS, LEVEL SPLITTINGS, AND TERM DIAGRAMS

Equivalent and Nonequivalent Electrons and Their Spectroscopic Terms

A problem of great importance is the evaluation of the possible spectroscopic terms that can arise from a given configuration of single particle states. This is a matter of listing the possible values of m_l and m_s for the electrons outside of the closed shells and then determining what values of S and L can be constructed from them, subject to limitations imposed by indistinguishability and the Pauli exclusion principle. The reason that only the electrons in the closed shells need be considered is the following: Closed shells are spherically symmetric ($L=0$) and have very little interaction with external electrons. This fact results from a property of the spherical harmonics: for given n and l, if all possible electron states are filled, the total electron density distribution is precisely spherically symmetric. For example, for $l=1$,

$$|Y_{10}|^2+|Y_{1-1}|^2+|Y_{11}|^2=\frac{3}{4\pi}\cos^2\theta+\frac{3}{8\pi}\sin^2\theta+\frac{3}{8\pi}\sin^2\theta=\frac{3}{4\pi}.$$

It is useful to distinguish the cases of *nonequivalent* electrons and *equivalent* electrons. Nonequivalent electrons are those differing in either n or l values, whereas equivalent electrons have the same n and l values. For two equivalent s electrons, for example, we write s^2; if they are nonequivalent, we write $s \cdot s$ or ss'.

The terms of nonequivalent electrons are fairly simple to find. For sample, the configuration $1s2s$ can only have $L=0$, since both electrons have $l=0$. The spin can be $S=0,1$, corresponding to the two ways of orienting the spins. Thus we have the two possible terms ^1S and ^3S, where the letter refers to the total L value and the superscript refers to the number of m_s values, namely, $(2S+1)$. The $S=0$ and $S=1$ total spin states are called *singlet* and *triplet* states, respectively, in accordance with the number of m_s values. If the electrons are equivalent, say $1s^2$, then the triplet term cannot occur, since this would imply both spins are the same, and all sets of quantum numbers would be identical. Thus the only term for the equivalent electrons is ^1S.

The distinction between the spectroscopic combination of equivalent and nonequivalent electrons can be seen in the following illustration. Consider the combination of two p electrons. If they have different values of n, so that they are nonequivalent, the possible L-S combinations are $S=0,1$, $L=0,1,2$, leading to the spectroscopic terms ^1S, ^1P, ^1D, ^3S, ^3P, ^3D and $1+3+5+3+9+15=36$ distinguishable states, corresponding to the 6×6 product of the one-electron states. Now, suppose the two p electrons have the same n values and are thus equivalent. Then all the 36 states are not available: some are ruled out by the Pauli exclusion principle, and some are ruled out because they are not distinguishable from others. To count the distinguishable permitted states, we construct Table 9.2, giving possible combinations of m_{l_1}, m_{l_2}, m_{s_1}, m_{s_2}, marking OUT for Pauli excluded states and labeling only distinguishable states. We find there are 15 distinguishable states allowed.

Table 9.2

m_{l1}	m_{l2}	m_{s1}	m_{s2}	Label	m_{l1}	m_{l2}	m_{s1}	m_{s2}	Label
+1	+1	+½	+½	OUT	0	−1	+	+	11
		+	−	1			+	−	12
		−	+	1			−	+	13
		−	−	OUT			−	−	14
+1	0	+	+	2	−1	+1	+	+	6
		+	−	3			+	−	8
		−	+	4			−	+	7
		−	−	5			−	−	9
+1	−1	+	+	6	−1	0	+	+	11
		+	−	7			+	−	13
		−	+	8			−	+	12
		−	−	9			−	−	14
0	+1	+	+	2	−1	−1	+	+	OUT
		+	−	4			+	−	15
		−	+	3			−	+	15
		−	−	5			−	−	OUT
0	0	+	+	OUT					
		+	−	10					
		−	+	10					
		−	−	OUT					

Which spectroscopic terms do these combinations correspond to? We simply use the fact that

$$m_L = m_{l_1} + m_{l_2}, \tag{9.22a}$$

$$m_S = m_{s_1} + m_{s_2}. \tag{9.22b}$$

Since the combination $m_L = \pm 2$, $m_S = \pm 1$ does not occur, the ^3D state can be ruled out. On the other hand, state 2 requires a ^3P configuration. State 1 requires a $^1D_2$ configuration. These two configurations take up $3\times 3 + 1\times 5 = 14$ of the 15 distinguishable states. The only remaining configuration can be ^1S, with one associated state. Thus the allowed terms for two equivalent p electrons are

$$^1S,\ ^3P,\ ^1D.$$

When more than two equivalent electrons are involved, the counting is straightforward, but more tedious.

Table 9.3

TERMS OF NON-EQUIVALENT ELECTRONS

Electron Configuration	Terms
$s\ s$	^1S, ^3S
$s\ p$	^1P, ^3P
$s\ d$	^1D, ^3D
$p\ p$	^1S, ^1P, ^1D, ^3S, ^3P, ^3D
$p\ d$	^1P, ^1D, ^1F, ^3P, ^3D, ^3F
$d\ d$	^1S, ^1P, ^1D, ^1F, ^1G, ^3S, ^3P, ^3D, ^3F, ^3G
$s\ s\ s$	^2S, ^2S, ^4S
$s\ s\ p$	^2P, ^2P, ^4P
$s\ s\ d$	^2D, ^2D, ^4D
$s\ p\ p$	^2S, ^2P, ^2D, ^2S, ^2P, ^2D, ^4S, ^4P, ^4D
$s\ p\ d$	^2P, ^2D, ^2F, ^2P, ^2D, ^2F, ^4P, ^4D, ^4F
$p\ p\ p$	$^2S(2)$, $^2P(6)$, $^2D(4)$, $^2F(2)$, $^4S(1)$, $^4P(3)$, $^4D(2)$, $^4F(1)$
$p\ p\ d$	$^2S(2)$, $^2P(4)$, $^2D(6)$, $^2F(4)$, $^2G(2)$, $^4S(1)$, $^4P(2)$, $^4D(3)$, $^4F(2)$, $^4G(1)$
$p\ d\ f$	$^2S(2)$, $^2P(4)$, $^2D(6)$, $^2F(6)$, $^2G(6)$, $^2H(4)$, $^2I(2)$ $^4S(1)$, $^4P(2)$, $^4D(3)$, $^4F(3)$, $^4G(3)$, $^4H(2)$, $^4I(1)$

Similar arguments can be used to obtain the terms for other configurations, although the details become quite tedious for complicated cases. In these cases tables such as Table 9.3 may be consulted.

A useful rule concerning the terms of equivalent electrons is that the terms for a shell more than half filled are the same as for the complementary number of electrons needed to fill the shell. Since 6 electrons are required to fill the p shell, the terms corresponding to p and p^5 are the same; also p^2 and p^4. This rule is simply proved by noting that the total spin and orbital angular momentum of a closed shell are both zero, as mentioned previously. In enumerating the various values of m_L and m_S it makes no difference if we use m_l and m_s of the missing electrons, since only the magnitudes of the sums Σm_{l_i} and Σm_{s_i} are relevant. This is sometimes stated as the equivalence of electrons and holes in a shell.

Parity

Besides the quantum numbers L and S there is another important quantum number called the *parity* of the configuration. This is simply ± 1 or (even, odd) according to the even or oddness of the sum Σl_i extended over all the electrons of the configuration. Since the sum of the l_i for a closed shell is even, we may restrict the sum to incomplete shells. Physically the parity corresponds to the symmetry or antisymmetry of the wave function when all spatial coordinates are reflected: $x \to -x$, $y \to -y$, $z \to -z$. For even parity $\psi \to \psi$, and for odd parity $\psi \to -\psi$. Since this property of the wave function is maintained under the usual interactions with which we deal, if a wave function has a certain parity at one time, it will keep that parity for all times. It should be noted that although the individual orbital angular momenta l_i do not in general have meaning, the evenness or oddness of their sum does. Note also that the sum Σl_i does not, in general, equal the total L of the configuration.

The parity of a configuration is usually given as a superscript "O" on the terms arising from this configuration when the parity is odd; when the parity is even no superscript appears. Thus a *s-p* configuration leads to terms $^1P^O$ and $^3P^O$, whereas *s-d* leads to terms ^1D and ^3D. (Sometimes the parity is not indicated at all, so that the absence of a superscript does not always mean even parity).

Spin-Orbit Coupling

The next step in the resolution of the degenerate levels of a configuration is through the *spin-orbit* coupling. In *L-S* coupling this is assumed to be much smaller than the electrostatic interaction. The effect is to split each term into a set of *levels*, each of which is labeled by the one remaining quantum number, the total angular momentum J. The magnetic quantum number M_J, or simply M, does not participate in the splitting, unless there are external fields to break the rotational symmetry of the internal interactions.

The basic spin-orbit interaction may be illustrated by an individual electron moving in a central electrostatic force field. In the rest frame of the electron this electric field will be perceived as having a magnetic field component

$$\mathbf{B} = -\frac{1}{c}\mathbf{v}\times\mathbf{E} = \frac{\mathbf{l}}{mecr}\frac{dU}{dr}. \tag{9.23}$$

Here $\mathbf{v}$ is the electron's velocity, $\mathbf{l} = m\mathbf{v}\times\mathbf{r}$ is its orbital angular momentum, and $U(r)$ is the equivalent electrostatic potential. This magnetic field interacts with the electron's magnetic moment, which is

$$\boldsymbol{\mu} = -\frac{e}{mc}\mathbf{s}. \tag{9.24}$$

This is twice the value one obtains by considering the electron to be a classical charge and mass distribution of the same shape, and it requires the Dirac equation of relativistic quantum mechanics for its derivation. (See, e.g., Bjorken and Drell, 1964.)

From the above, we might expect the interaction energy to be $U_{\text{int}} = -\boldsymbol{\mu}\cdot\mathbf{B}$. However, an exact derivation from the Dirac equation yields a value of one-half this. The discrepancy can be traced to the use of the instantaneous rest frame of the electron, which is constantly changing as the electron orbits. The effect of this acceleration can be described by *Thomas precession* (see Leighton, 1959), which is one-half the naively expected rate, but in the opposite direction, leading to the final result

$$H_{so} = \frac{1}{2m^2c^2}\mathbf{s}\cdot\mathbf{l}\frac{1}{r}\frac{dU}{dr}. \tag{9.25}$$

This is often written, for the sum of the interactions of all electrons,

$$H_{so} = \sum_i \xi_i(r_i)\mathbf{s}_i\cdot\mathbf{l}_i, \tag{9.26a}$$

where

$$\xi_i(r_i) = \frac{1}{2m^2c^2r}\frac{dU(r_i)}{dr}. \tag{9.26b}$$

When we find matrix elements of this H_{so} between states of S and L, the individual spin and orbital angular momenta become averaged over in such a way that an equivalent interaction for our purposes is simply

$$H_{so} = \xi\mathbf{S}\cdot\mathbf{L}, \tag{9.27a}$$

where

$$\mathbf{S} = \sum \mathbf{s}_i, \qquad \mathbf{L} = \sum \mathbf{l}_i \tag{9.27b}$$

and ξ is an appropriate average of the ξ_i. (For details see Bethe and Jackiw, 1968.)

With this simplified spin-orbit term we are in a position to find the splittings of a given term as a function of the total angular momentum quantum number J. To do this we note that

$$\mathbf{J}^2 = (\mathbf{L}+\mathbf{S})\cdot(\mathbf{L}+\mathbf{S}) = \mathbf{L}^2 + \mathbf{S}^2 + 2\mathbf{L}\cdot\mathbf{S}, \tag{9.28}$$

so that

$$H_{so} = \tfrac{1}{2}\xi(\mathbf{J}^2 - \mathbf{L}^2 - \mathbf{S}^2). \tag{9.29}$$

Note also that $\mathbf{J}^2$, $\mathbf{L}^2$ and $\mathbf{S}^2$ are mutually commuting operators, since $\mathbf{L}$ commutes with $\mathbf{L}^2$ and $\mathbf{S}$ commutes with $\mathbf{S}^2$. Therefore, when we take diagonal elements of this quantity between states of given L, S, and J we obtain

$$\langle H_{so}\rangle = \tfrac{1}{2}C[J(J+1) - L(L+1) - S(S+1)], \tag{9.30}$$

where C is a constant related to the average of $\xi(r)$ over the spatial part of the wave function.

For fixed L and S, that is, for a given term, the energy shift is proportional to $J(J+1)$, so that the consecutive splittings are given by

$$\begin{aligned} E_{J+1} - E_J &= \tfrac{1}{2}C[(J+1)(J+2) - J(J+1)] \\ &= C(J+1). \end{aligned} \tag{9.31}$$

Therefore, we have the *Lande interval rule*: *the spacing between two consecutive levels of a term is proportional to the larger of the two J values involved.* This rule is very useful in determining the J values of levels empirically.

The J value of a level is given as a subscript on the term symbol: ${}^3P_2$, ${}^2S_{\frac{1}{2}}$. Often the allowed values of J are given on the term symbol, separated by commas, for example, ${}^2P_{1/2,3/2;}$ ${}^3D_{1,2,3}$. The number of J values in any term is equal to the smaller of $(2L+1)$ and $(2S+1)$.

The ordering of the energies within the levels of a term are with increasing J if the shell is less than half-full, that is, the constant C above is positive. Such a term is called *normal*. On the other hand for shells more than half full the ordering is with decreasing J. Such terms are called *inverted*. An illustration of this is the two cases of the ground levels of carbon and oxygen. Each has the same terms, as the configurations p^2 and p^4, respectively. The ground term is a ${}^3P$ in both cases, but the ground level is ${}^3P_0$ for C, and a ${}^3P_2$ for O.

The progressive splitting of a configuration into terms and levels is illustrated by Fig. 9.2.

The degeneracy of each of the levels is $(2J+1)$, corresponding to the values of the magnetic quantum numbers $M_J = -J, \ldots -1, 0, 1, \ldots J$. These levels remain degenerate, unless external fields are applied, for example, a magnetic field (Zeeman effect) or an electric field (Stark effect). It is easily verified from Fig. 9.2a that for the case of the $4p4d$ configuration the total number of states represented by the final fine structure levels is 60. This is the same as the number of states represented by the configuration: $2\cdot(2l+1)\cdot 2(2l'+1) = 60$, where $l=1$, $l'=2$.

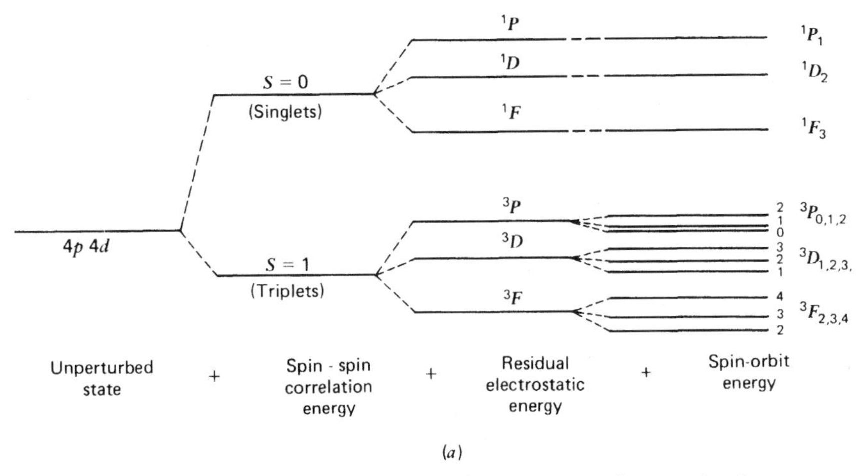

(a)

***Figure 9.2a** Schematic diagram illustrating the terms of energy levels generated by a 4p4d configuration in L-S coupling.*

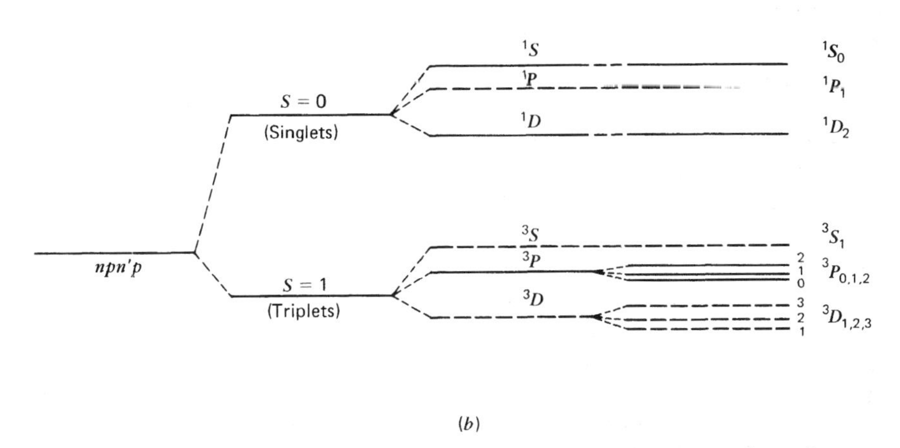

(b)

***Figure 9.2b** Same as a, but for two p electrons. Dashed levels are absent from the multiplet if the electrons are equivalent (n = n'). (Taken from Leighton, R., 1959, Principles of Modern Physics, McGraw-Hill, New York.)*

The usual mode of presentation of this information is in a *term diagram*, which separates the terms of different S values, and within each group separates according to L values. The energies are represented by lines drawn on the proper vertical scale.

Zeeman Effect

As is shown in Chapter 3, electrical particles of charge e, mass m, oscillating at frequency ω_0 radiate dipole radiation of frequency ω_0. As is easily shown (Leighton, 1959), a classical analysis indicates that in the presence of a magnetic field of strength B, the radiation is split into three separate frequencies, $\omega_+ \equiv \omega_0 + eB/2mc$, ω_0, and $\omega_- \equiv \omega_0 - eB/2mc$. This splitting, due to the Lorentz force on the electron, also has well-defined polarization properties: If the radiation is viewed at right angles to $\mathbf{B}$, all three components are visible, with the component ω_0 plane polarized and $\omega_\pm$ circularly polarized. If the radiation is viewed along the magnetic field, the undeviated component ω_0 is no longer visible. These classical line patterns are termed *normal* Zeeman lines.

Unfortunately, the observed Zeeman splittings are generally *anomalous*; that is, they disagree with the classical prediction because of quantum mechanical effects. As in the spin-orbit coupling discussed previously, the interaction energy between the electrons of total magnetic moment $\boldsymbol{\mu}$ and the external magnetic field is

$$U_B = -\boldsymbol{\mu}\cdot\mathbf{B}. \tag{9.32}$$

The total magnetic moment is the sum over all electrons of the orbital and spin magnetic moments of the individual electrons

$$\boldsymbol{\mu} = -\sum\left[\tfrac{1}{2}\left(\frac{e}{mc}\right)\mathbf{l}_i + \left(\frac{e}{mc}\right)\mathbf{s}_i\right]. \tag{9.33}$$

The different proportionality factors multiplying $\mathbf{l}_i$ and $\mathbf{s}_i$ result from the quantum mechanical nature of intrinsic spin, [cf. Eq. (9.24)]. Now, since the energy of Zeeman splitting is generally much smaller than that of the fine-structure levels, we may treat the former as a perturbation, with L and S remaining good quantum numbers. Thus Eq. (9.33) becomes, using Eqs. (9.27b) and (9.28)

$$\begin{aligned}\boldsymbol{\mu} &= -\tfrac{1}{2}\left(\frac{e}{mc}\right)(\mathbf{L}+2\mathbf{S})\\ &= -\tfrac{1}{2}\left(\frac{e}{mc}\right)(\mathbf{J}+\mathbf{S}).\end{aligned} \tag{9.34}$$

The torque of the external magnetic field causes the magnetic moment $\boldsymbol{\mu}$ to precess around **B**. However, this precession frequency is much smaller than the precession frequency of **S** around **J** (because of the much more energetic **L**·**S** coupling). Thus the component of $\boldsymbol{\mu}$ along **J** can be considered fixed, with the component along **S** precessing around. The time-averaged component of $\boldsymbol{\mu}$ along **B** (assumed to lie along the z axis) can be approximated by the component along **J** multiplied by the component of **J** along **B**:

$$U_B \approx \tfrac{1}{2}\left(\frac{e}{mc}\right)\left[(\mathbf{J}+\mathbf{S})\cdot\mathbf{J}\right]\frac{(\mathbf{J}\cdot\mathbf{B})}{|\mathbf{J}|^2}$$

$$= \tfrac{1}{2}\left(\frac{e\hbar B}{mc}\right) g M_J, \tag{9.35a}$$

where

$$M_J = J_Z$$

and

$$g(J,L,S) \equiv 1 + \frac{J(J+1)+S(S+L)-L(L+1)}{2J(J+1)}. \tag{9.35b}$$

Here we have used Eq. (9.28) to evaluate Eq. (9.35). The quantity g is called the Lande g factor; if the proportionality factors multiplying $\mathbf{l}_i$ and $\mathbf{s}_i$ in Eq. (9.33) were equal, g would be independent of J, L and S.

The frequency of a transition from level 1 to level 2 is

$$\omega_{12} = \tfrac{1}{2}\left(\frac{eB}{mc}\right)\left(g_1 M_{J_1} - g_2 M_{J_2}\right). \tag{9.36}$$

If $\Delta M_J = 0, \pm 1$ (see Chapter 10) and if $g_1 = g_2$, the splittings would agree with the classical theory. However, in general J, L and S change in the transition in such a way that g also changes, leading to a variety of different splittings.

Role of the Nucleus; Hyperfine Structure

Up to this point we have made several simplifying assumptions concerning the nucleus: (1) infinite mass; (2) point particle; (3) interaction with electrons only through the Coulomb field of its total charge Ze. The violations of these assumptions produce small effects on the atomic electron states called *hyperfine structure*. The small effects on the states are not so important in themselves as are the splittings of the states into several substates, since this is much easier to observe. The splittings may be divided into two groups which have rather different origins:

I—*Isotope Effect:* An atomic nucleus of charge Ze can have a number of different masses, depending on the total number of neutrons it possesses. The various species of nuclei with the same atomic number Z are called *isotopes*. Each isotope will have a slightly different set of atomic energy levels, because of finite (noninfinite) mass and finite (nonzero) size effects, which differ for each isotope. In any naturally occurring material there will be a distribution over the various isotopes in proportions that depend on the origin, age, and history of the material. The spectra produced by such an isotopic mixture show splittings of lines, each component of which comes from a different isotope.

One may regard isotope splittings as due purely to the production of spectra by differing atomic species, where the differences are extremely small. The splittings as such do not occur in a single atom, and it would be meaningless to speak of an atomic transition between the split states, as this would require a nuclear transformation.

II—*Nuclear Spin:* Like electrons, other subatomic particles possess spin and associated magnetic moments. The nucleus therefore also has a total spin angular momentum $\mathbf{I}$, with eigenvalues $I(I+1)$ for its square and M_I for its z component. We may express the magnetic moment $\boldsymbol{\mu}_N$ by means of a nuclear g factor:

$$\boldsymbol{\mu}_N = g\frac{e}{2Mc}\mathbf{I}. \tag{9.37}$$

For the proton, for example, where $M \sim 1840\ m_e$, we have $g = 5.5855$. (Recall that for the electron $g = 2.00232$.) Since g factors are normally of order unity we see that nuclear magnetic moments are about 10^{3-4} smaller than that of the electron.

The nuclear magnetic moment interacts with the magnetic moments of the atomic electrons, and each previously described atomic state is further split by this interaction. In analogy with the L-S coupling scheme we now introduce the total angular momentum vector $\mathbf{F} = \mathbf{J} + \mathbf{I}$ and label the hyperfine states by the quantum number F. For example, when $I = 2$ for a $^3D_3$ state, we have five splittings, corresponding to $F = 1$ to 5.

In contrast to the isotope effect, the nuclear spin effects produce splittings within a single atom, and the states so produced may be reached by an appropriate atomic electron transition, such that the orientation of $\mathbf{J} = \mathbf{L} + \mathbf{S}$ changes relative to **I**.

An example of extreme importance in astrophysics is the ground level of neutral atomic hydrogen, which is a $^2S_{\frac{1}{2}}$ level. The proton spin is $\frac{1}{2}$ so that two hyperfine states occur, the ground state with $F=0$ and an excited state with $F=1$. The energy difference between these states corresponds to a frequency of 1420 MHz, or a wavelength of 21 cm. Radiative transitions from $F=1$ to $F=0$ are extremely rare for a given atom, that is, about once every 10^7 years, but with the enormous abundance of neutral hydrogen this nonetheless gives rise to an observable 21-cm line.

9.5 THERMAL DISTRIBUTION OF ENERGY LEVELS AND IONIZATION

Thermal Equilibrium: Boltzmann Population of Levels

The relative populations of the various atomic levels is a difficult question in general, since it depends on the detailed processes by which any level becomes populated or depopulated. An exception is the case of *thermal equilibrium*, where the populations are completely determined by the temperature T. Then in any collection of atoms of a specific type the number in any given level is proportional to $ge^{-\beta E}$, where $\beta \equiv 1/kT$, $k=$ Boltzmann's constant, and $g=$ statistical weight (degeneracy) of the level. In L-S coupling the g factors are simply $g=(2J+1)$. It is customary to measure energies using the ground level as a zero point; let us call these energies E_i for the ith level. If N_i is the population (number per unit volume) of the ith level and N is the total population of the atom we have the *Boltzmann law*:

$$N_i = \frac{N}{U} g_i e^{-\beta E_i}. \tag{9.38}$$

Here U is the constant of proportionality; it is called the *partition function*. We may find U by demanding that

$$N = \sum N_i,$$

where the sum is over all levels. This yields

$$U = \sum g_i e^{-\beta E_i}. \tag{9.39}$$

At sufficiently low temperatures only the first term in the sum is significant, and we obtain

$$U = g_0,$$

where g_0 is the degeneracy of the ground level.

At finite temperatures we run into a mathematical difficulty: the sum $\sum g_i e^{-\beta E_i}$ diverges. This occurs because $g = 2J+1$ approaches infinity while $e^{-\beta E_i}$ approaches a constant as the ionization continuum is approached. Physically this is resolved by recognizing that in an actual gas the atoms are not at infinite distances, so that the idealized model of an atom extending to infinity is not valid. The high principal quantum number n values that cause the divergence are just those states that are affected by the presence of the neighboring atoms. These high-n electrons can be easily ripped off by perturbations from the neighbors, so that an atom reaches its effective ionization potential at some large but finite value of the principal quantum number, n_{max}. This lowering of the ionization potential can be taken into account approximately by cutting off the summation over levels at the value $n = n_{max}$. One limit on n_{max} may be deduced from the condition that the Bohr orbit corresponding to $n = n_{max}$ be of order of the interatomic distances

$$n_{max}^2 a_0 Z^{-1} \sim N^{-1/3},$$

$$n_{max} \sim \left(\frac{Z}{a_0}\right)^{1/2} N^{-1/6}. \tag{9.40}$$

For hydrogen at $N = 10^{12}$ cm^{-3}, for example, we would have $n_{max} \sim 10^2$. Actually, there are other effects operating here as well (e.g., Debye shielding) which depend on temperature as well, so that the computation of n_{max} is quite involved. A really basic understanding of the cutoff has probably not yet been achieved. Fortunately, for many cases of interest the precise value of the cutoff is not too critical. In the range of temperatures up to 10^4K, U is in most cases equal to g_0, the exceptions being low-ionization potential elements like the alkali metals.

The Saha Equation

So far we have considered the distribution among the levels of a single atom in thermal equilibrium. Now we want to determine the distribution of an atomic species among its various stages of ionization. The resulting equation is called the *Saha equation*. We now derive this equation for the case of a neutral atom and its first stage of ionization.

We start with the generalization of the Boltzmann law:

$$\frac{dN_0^+(v)}{N_0}=\frac{g}{g_0}\exp\left[-\frac{\left(\chi_I+\frac{1}{2}m_e v^2\right)}{kT}\right], \tag{9.41}$$

where χ_I is the ionization potential. Here $dN_0^+(v)$ is the differential number of ions in the ground level with the free electron in velocity range $(v, v+dv)$, and N_0 is the number of atoms in the ground level. The statistical weight of the atom in its ground state is g_0. The statistical weight g is the product of the statistical weight of the ion in its ground state g_0^+ and the differential electron statistical weight g_e:

$$g=g_0^+\cdot g_e. \tag{9.42}$$

The statistical weight g_e is given by

$$g_e=\frac{2\,dx_1dx_2dx_3dp_1dp_2dp_3}{h^3}, \tag{9.43}$$

where the factor 2 comes about from the two spin states. The volume element satisfies $dx_1dx_2dx_3=1/N_e$, where N_e = electron density, since we are applying Boltzmann's law to a region containing one electron. Since the electrons have an isotropic velocity distribution, we have

$$dp_1dp_2dp_3=4\pi m_e^3v^2dv.$$

Thus Eq. (9.41) becomes

$$\frac{dN_0^+(v)}{N_0}=\frac{8\pi m_e^3}{h^3}\frac{g_0^+}{N_e g_0}\exp\left[-\frac{\left(\chi_I+\frac{1}{2}m_e v^2\right)}{kT}\right]v^2dv. \tag{9.44}$$

To find the total N_0^+, irrespective of the electron's velocity, we integrate over all v:

$$\frac{N_0^+N_e}{N_0}=\frac{8\pi m_e^3}{h^3}\frac{g_0^+}{g_0}e^{-\chi_I/kT}\left(\frac{2kT}{m_e}\right)^{3/2}\int_0^\infty e^{-x^2}x^2\,dx,$$

where the substitution $x\equiv(m_e/2kT)^{1/2}v$ has been made. The integral has the value $\pi^{1/2}/4$. Thus we obtain

$$\frac{N_0^+N_e}{N_0}=\left(\frac{2\pi m_e kT}{h^2}\right)^{3/2}\frac{2g_0^+}{g_0}e^{-\chi_I/kT}. \tag{9.45}$$

To find the number of atoms or ions in any state, not just the ground state, we use the Boltzmann laws [cf. Eqs. (9.38)],

$$\frac{N_0}{N}=\frac{g_0}{U(T)}, \qquad \frac{N_0^+}{N^+}=\frac{g_0^+}{U^+(T)}. \tag{9.46}$$

We then obtain *Saha's equation*:

$$\frac{N^+N_e}{N}=\frac{2U^+(T)}{U(T)}\left(\frac{2\pi m_e kT}{h^2}\right)^{3/2}e^{-\chi_I/kT}. \tag{9.47}$$

Here N and N^+ are the total number densities of neutral atoms and first ionized atoms, respectively, and U and U^+ are the corresponding partition functions.

A similar derivation shows that there is a Saha equation connecting any two successive stages of ionization:

$$\frac{N_{j+1}N_e}{N_j}=\frac{2U_{j+1}(T)}{U_j(T)}\left(\frac{2\pi m_e kT}{h^2}\right)^{3/2}e^{-\chi_{j,j+1}/kT}, \tag{9.48}$$

where the subscripts here refer to stages of ionization. These equations are often stated in terms of pressures rather than number densities. The ideal gas law is

$$P=NkT,$$

so that

$$\frac{P_{j+1}P_e}{P_j}=\frac{2U_{j+1}(T)}{U_j(T)}\left(\frac{2\pi m_e}{h^2}\right)^{3/2}(kT)^{5/2}e^{-\chi_{j,j+1}/kT}. \tag{9.49}$$

To calculate the ionizational equilibrium of a mixture of various elements, some further equations must be used. First there must be an equation giving the conservation of nuclei

$$\sum N_j^{(i)}=N^{(i)}, \tag{9.50a}$$

where $N_j^{(i)}$ is the number density of species i in the jth stage of ionization, and $N^{(i)}$ is the total number density over all stages of ionization (the number density of nuclei of that species). Also, there is an equation for conservation of charge (number of electrons):

$$N_e=\sum_i\sum_j Z_j N_j^{(i)}. \tag{9.50b}$$

Here Z_j is the charge (in units of e) of the jth stage of ionization.

The actual solution to these equations must proceed numerically, in most cases by an iterative procedure. For many cases of physical interest, most of a given species is found in a few (one to three) ionization stages for any one set of conditions (see Problem 9.4). This reduces the numerical problems considerably, so that a solution can usually be obtained after a few iterations.

The ionization equilibrium of pure hydrogen can be worked out analytically (neglecting the H^- ion as unimportant) (see Problem 9.5), but this is an exception. Also, one must be quite careful in such situations to take into consideration species that have a low ionization potential, even if such species are not abundant. This is because the electron density may be completely determined by ionization of these trace constituents. Because of this, a "pure" hydrogen case rarely occurs in nature.

It is common in astrophysics to denote neutral and ionized hydrogen by HI and HII, respectively. In general, an element Q which is in its nth ionization state is denoted by Q followed by the Roman numeral for $n+1$.

PROBLEMS

9.1—Consider two electronic orbitals u_a and u_b occupied by two electrons, 1 and 2. Neglect the electrostatic repulsion of the two electrons.

a. Show that the mean square distance, $\langle R^2 \rangle$, between the two electrons is

$$\langle R^2 \rangle = (\mathbf{r}^2)_a + (\mathbf{r}^2)_b - 2|\mathbf{r}_a||\mathbf{r}_b| + 2|\mathbf{r}_{ab}|^2$$

where

$$\mathbf{r}_a \equiv \int u_a^* \mathbf{r} u_a d^3r,$$

$$(\mathbf{r}^2)_a \equiv \int u_a^* \mathbf{r}^2 u_a d^3r,$$

$$\mathbf{r}_{ab} \equiv \int u_a^* \mathbf{r} u_b d^3r.$$

(The integration here also imply a summation over spins.)

b. For states a and b defined by n, l, m, m_s show that $\mathbf{r}_a = \mathbf{r}_b = 0$, so that

$$\langle R^2 \rangle = (\mathbf{r}^2)_a + (\mathbf{r}^2)_b + 2|\mathbf{r}_{ab}|^2.$$

c. For electrons having different spins show that

$$\mathbf{r}_{ab}=0$$

so that for such electrons

$$\langle R^2\rangle=(\mathbf{r}^2)_a+(\mathbf{r}^2)_b,$$

which is the same as for the classical uncorrelated motion of two particles.

d. Thus show that electrons having the same spins are on the average further apart than electrons having different spins. This is an example of an *electron correlation effect*.

9.2—Give the spectroscopic terms arising from the following configurations, using L-S coupling. Include parity and J values. Give your arguments in detail for deriving these results.

a. $2s^2$

b. $2p3s$

c. $3p4p$

Find the terms corresponding to the following configuration.

d. $2p^4 3p$

9.3—For each of the *configurations* in the problem above evaluate its degeneracy from the l values involved [you may omit (d) here]. Next evaluate the degeneracy of each of the terms from the L and S values. Finally, evaluate the degeneracy of each of the levels from the J values. Show that these degeneracies are consistent, in that the degeneracy of any configuration is equal to the sum of the degeneracies of the terms it generates, and that the degeneracy of any term is equal to the sum of the degeneracies of the levels it generates.

9.4—The thermal de Broglie wavelength of electrons at temperature T is defined by $\lambda=h/(2\pi mkT)^{1/2}$. The degree of degeneracy of the electrons can be measured by the number of electrons in a cube λ on a side:

$$\xi\equiv N_e\lambda^3=4.1\times10^{-16}N_eT^{-3/2}.$$

For many cases of physical interest the electrons are very nondegenerate, the quantity $\gamma \equiv \ln \xi^{-1}$ being of order 10 to 30. We want to investigate the consequences for the Boltzmann and Saha equations of γ being large and only weakly dependent on temperature. For the present purposes assume that the partition functions are independent of temperature and of order unity.

a. Show that the value of temperature at which the stage of ionization passes from j to $j+1$ is given approximately by

$$kT \sim \frac{\chi}{\gamma}$$

where χ is the ionization potential between stages j and $j+1$. Therefore, this temperature is much smaller than the ionization potential expressed in temperature units.

b. The rapidity with which the ionization stage changes is measured by the temperature range ΔT over which the ratio of populations N_j/N_{j+1} changes substantially. Show that

$$\frac{\Delta T}{T} \sim \left[\frac{d \log(N_{j+1}/N_j)}{d \log T} \right]^{-1} \sim \gamma^{-1}$$

Therefore, ΔT is much smaller than T itself, and the change occurs rapidly.

c. Using the Boltzmann equation and result (a) above, show that when γ is large, an atom or ion stays mostly in its ground state before being ionized.

9.5—A cold neutral hydrogen gas of density ρ resides inside a metal container. The container walls are then heated to temperature T. Find the equilibrium value of the ratio δ of ionized to neutral hydrogen as a function of ρ and T.

a. Find a single, dimensionless parameter $\Delta(\rho, T)$ that determines δ (cf. 9.4 above).

b. Derive an explicit algebraic expression for $\delta(\Delta)$. You may assume that the partition function is constant and equal to the ground state statistical weight.

REFERENCES

Bethe, H. A., and Jackiw, R., 1968. *Intermediate Quantum Mechanics* 2nd ed., (Benjamin, Reading, Mass.).

Bjorken, J. D., and Drell, S. D. 1964, *Relativistic Quantum Mechanics*, (McGraw-Hill, New York).

Kuhn, H., 1962, *Atomic Spectra* (Academic, New York).

Leighton, R. 1959, *Principles of Modern Physics*, (McGraw-Hill, New York).

Mihalas, D., 1978, *Stellar Atmospheres* 2nd ed., (Freeman, San Francisco).

10

RADIATIVE TRANSITIONS

10.1 SEMI-CLASSICAL THEORY OF RADIATIVE TRANSITIONS

So far we have looked only at those properties of atomic systems—such as ionization potentials and statistical mechanics—that depend solely on the energies of the various states. Now we want to investigate the nature of the light produced in transitions between these states. There are two major objectives here: first, to give so-called *selection rules* for radiative transitions and second, to determine the *strengths* of the radiation. The first of these is in some sense a special case of the second, but we shall regard it separately. The rules we give will be mostly applicable to L-S coupling and, additionally, to electric dipole transitions, although we do discuss some generalizations.

We use the so-called *semi-classical* theory of radiation, in which the atom is treated quantum mechanically, but the radiation field is treated classically. It is found that this theory correctly predicts the induced radiation processes, that is, those processes described by Einstein B coefficients, but that it fails to predict the spontaneous process, described by the Einstein A coefficient. This is not a great difficulty, because the Einstein coefficients are related, and any one can be used to derive the other two. The physical argument used to justify the semi-classical approach is the following: the classical limit of radiation is the one in which the number of photons per photon state is large. Thus the induced processes, which are proportional to the number of photons, dominate the spontaneous process, which is independent of the number of photons. Because of the linearity of the induced processes in the number of photons, these processes may be extrapolated to small photon numbers, i.e. the quantum regime. The spontaneous rate can then be found by the Einstein relations.

The Electromagnetic Hamiltonian

The relativistic generalization of the Hamiltonian for a particle in an external electromagnetic field is

$$H=\left[(c\mathbf{p}-e\mathbf{A})^2+m^2c^4\right]^{1/2}+e\phi. \tag{10.1}$$

If we expand this in the nonrelativistic limit, ignoring the (constant) rest mass, we obtain

$$\begin{aligned} H&=\frac{1}{2m}\left(\mathbf{p}-\frac{e\mathbf{A}}{c}\right)^2+e\phi \\ &=\frac{p^2}{2m}-\frac{e}{mc}\mathbf{A}\cdot\mathbf{p}+\frac{e^2A^2}{2mc^2}+e\phi. \end{aligned} \tag{10.2}$$

In Eq. (10-2) we have used the "Coulomb gauge," (see §2.5 for a discussion of Gauge transformations),

$$\nabla\cdot\mathbf{A}=\phi=0, \tag{10.3}$$

so that the momentum operator $\mathbf{p}$ commutes with $\mathbf{A}$ in their scalar product:

$$\mathbf{p}\cdot\mathbf{A}=\mathbf{A}\cdot\mathbf{p}.$$

We may estimate the ratio of the two terms in $\mathbf{A}$:

$$\eta\equiv\frac{epA/mc}{e^2A^2/2mc^2}=\frac{2ev/c}{\alpha^2a_0A},$$

where α is the *fine-structure* constant

$$\alpha\equiv\frac{e^2}{\hbar c}\simeq\frac{1}{137}. \tag{10.4}$$

Since $v/c \sim \alpha$ [cf. Eqs. (9.6) and (9.7)] for atoms and $A \sim \lambda E$, where E is the electric field and λ is the wavelength, we have

$$\eta^2 \sim \frac{4\hbar\omega}{2\pi\alpha a_0^2 \lambda E^2}.$$

Since $\lambda \sim a_0/\alpha$ and $n_{ph} \sim E^2/\hbar\omega$ is the photon density, we have

$$\eta^2 \sim \left(n_{ph} a_0^3\right)^{-1} \gg 1 \tag{10.5}$$

as the condition that the linear term in A dominates the quadratic one. In other words, the number of photons inside the atom at one time is small. In fact, the term quadratic in A contributes to two-photon processes, which we ignore here under the assumption that the number of photons is sufficiently small. Note that the photon density at which this assumption fails is $n_{ph} \sim 10^{25}$ cm^{-3}, whereas at the sun's surface we have only $n_{ph} \sim 10^{12}$ cm^{-3}. Ordinarily, the neglect of the A^2 term is justified.

We now want to apply this to an atomic system of electrons. To do this we regard the sum of terms of the sort $(-e/mc)\mathbf{p}\cdot\mathbf{A}$ as a perturbation to the atomic Hamiltonian, and we use time-dependent perturbation theory to calculate the transition probabilities between the atomic states. (We continue to work in the Coulomb gauge, so that $\phi = 0$ and $\nabla\cdot\mathbf{A} = 0$.)

The Transition Probability

First, we split the Hamiltonian of Eq. (10.2) into a time-independent and a time-dependent piece:

$$H = H^0 + H^1. \tag{10.6}$$

Here H^0 is the atomic Hamiltonian, assumed independent of time, and H^1 is the perturbation due to the external electromagnetic field. The atomic eigenvalues E_k and eigenfunctions ϕ_k of H^0 are given by

$$H^0\phi_k = E_k\phi_k. \tag{10.7}$$

Therefore, the zeroth-order time dependent wave functions are $\phi_k \exp(-iE_k t/\hbar)$. We may expand the actual wave function in this complete set

$$\psi(t) = \sum a_k(t)\phi_k \exp(-iE_k t/\hbar). \tag{10.8}$$

It is now straightforward to show from the Schrodinger equation (e.g.,

Merzbacher, 1961) that the probability per unit time for a transition from state i to state f, w_{fi}, is given by

$$w_{fi}=\frac{4\pi^2}{\hbar^2 T}|H^1_{fi}(\omega_{fi})|^2 \tag{10.9}$$

where

$$H^1_{fi}(\omega)\equiv(2\pi)^{-1}\int_0^T H^1_{fi}(t')e^{i\omega t'}\,dt' \tag{10.10a}$$

$$H^1_{fi}(t)\equiv\int\phi_f^* H^1\phi_i\,d^3x, \tag{10.10b}$$

$$\omega_{fi}\equiv\frac{E_f-E_i}{\hbar}. \tag{10.10c}$$

Here the perturbation is assumed to be active only during the time interval 0 to T.

For a number of atomic electrons we have the perturbation from an external field

$$H^1=\frac{-e}{mc}\sum\mathbf{A}\cdot\mathbf{p}_j=\frac{ie\hbar}{mc}\mathbf{A}\cdot\sum\boldsymbol{\nabla}_j, \tag{10.11}$$

since $\mathbf{p}_j\rightarrow-i\hbar\boldsymbol{\nabla}_j$. We assume that $\mathbf{A}(\mathbf{r},t)$ has the form

$$\mathbf{A}(\mathbf{r},t)=\mathbf{A}(t)e^{i\mathbf{k}\cdot\mathbf{r}},$$

where $\mathbf{A}(t)$ vanishes outside the interval $0<t<T$. T is assumed to be large enough that a well-defined frequency of the wave exists. Then we obtain

$$H^1_{fi}(\omega_{fi})=\mathbf{A}(\omega_{fi})\cdot\frac{ie\hbar}{mc}\left\langle f\middle|e^{i\mathbf{k}\cdot\mathbf{r}}\sum\boldsymbol{\nabla}_j\middle|i\right\rangle, \tag{10.12}$$

where

$$\left\langle f\middle|e^{i\mathbf{k}\cdot\mathbf{r}}\sum\boldsymbol{\nabla}_j\middle|i\right\rangle\equiv\int\phi_f^* e^{i\mathbf{k}\cdot\mathbf{r}}\sum\boldsymbol{\nabla}_j\phi_i\,d^3x,$$

which does not depend on time. Here d^3x denotes integration over the coordinates of all particles. $\mathbf{A}(\omega)$ is defined in the same manner as $H^1_{fi}(\omega)$, [cf. Eq. (10.10a)]. The transition rate is then

$$w_{fi}=\frac{4\pi^2 e^2}{m^2c^2 T}|A(\omega_{fi})|^2\left|\left\langle f\middle|e^{i\mathbf{k}\cdot\mathbf{r}}\mathbf{1}\cdot\sum\boldsymbol{\nabla}_j\middle|i\right\rangle\right|^2, \tag{10.13}$$

where $\mathbf{l}$ is a unit vector specifying the polarization of the wave: $\mathbf{A}=A\mathbf{l}$. We want to express $A(\omega_{fi})$ in terms of the intensity of the electromagnetic wave traveling in direction $\mathbf{n}$. This intensity is [cf. §2.3]

$$I=\langle \mathbf{S}\cdot\mathbf{n}\rangle=\frac{c}{4\pi T}\int_{-\infty}^{\infty}E^2(t)\,dt=\frac{c}{T}\int_0^{\infty}|E(\omega)|^2\,d\omega. \tag{10.14}$$

Also, for the monochromatic intensity we have [cf. Eq. (2.34)]

$$\frac{dW}{dA\,d\omega\,dt}\equiv\mathcal{J}(\omega)=\frac{c|E(\omega)|^2}{T}.$$

But since $\mathbf{E}=-c^{-1}\partial\mathbf{A}/\partial t$ we have $\mathbf{E}(\omega)=-i\omega c^{-1}\mathbf{A}(\omega)$ so that

$$\mathcal{J}(\omega)=\frac{\omega^2}{cT}|\mathbf{A}(\omega)|^2. \tag{10.15}$$

Thus we obtain

$$w_{fi}=\frac{4\pi^2e^2}{m^2c}\frac{\mathcal{J}(\omega_{fi})}{\omega_{fi}^2}\left|\left\langle f\left|e^{i\mathbf{k}\cdot\mathbf{r}}\mathbf{l}\cdot\sum\nabla_j\right|i\right\rangle\right|^2. \tag{10.16a}$$

This formula applies equally to absorption or to induced emission. The two processes can be simply related. The probability rate for the inverse process is the same, except w_{fi} is replaced by w_{if}, and the integral is replaced by $\langle i|e^{i\mathbf{k}\cdot\mathbf{r}}\mathbf{l}\cdot\Sigma\nabla_j|f\rangle$. If we interchange labels f and i integrate by parts, noting $\mathbf{l}\cdot\mathbf{k}=0$ for a plane wave, we have

$$w_{if}=\frac{4\pi^2e^2}{m^2c}\frac{\mathcal{J}(\omega_{fi})}{\omega_{fi}^2}\left|\left\langle f\left|e^{-i\mathbf{k}\cdot\mathbf{r}}\mathbf{l}\cdot\sum\nabla_j\right|i\right\rangle^*\right|^2, \tag{10.16b}$$

which is the same as (10.16a). Thus we have

$$w_{fi}=w_{if}, \tag{10.17}$$

the "principle of detailed balance."

10.2 THE DIPOLE APPROXIMATION

The transition probabilities contain terms of the form

$$\int\phi_f^*e^{i\mathbf{k}\cdot\mathbf{r}}\mathbf{l}\cdot\nabla_j\phi_i\,d^3x. \tag{10.18}$$

We now wish to justify an expansion of the exponential

$$e^{i\mathbf{k}\cdot\mathbf{r}} = 1 + i\mathbf{k}\cdot\mathbf{r} + \tfrac{1}{2}(i\mathbf{k}\cdot\mathbf{r})^2 + \cdots$$

This is appropriate, since

$$\mathbf{k}\cdot\mathbf{r} \sim ka_0 \sim \frac{a_0 \Delta E}{\hbar c} \sim \frac{Z\alpha}{2} \ll 1,$$

at least for moderate Z. The lowest order of this approximation, in which $e^{i\mathbf{k}\cdot\mathbf{r}}$ is set equal to unity, gives rise to the *dipole approximation*. When the results of this approximation yield a zero result for certain transition rates, however, one needs to go to the higher terms in the expansion to derive the actual rates. These higher order terms give rise to *electric quadrupole*, *octupole*, and so on and *magnetic dipole*, *quadrupole*, and so on. Since the quantity $Z\alpha$ is also the order of magnitude estimate for v/c of the electrons in an atom [cf. Eq. (9.15b)], therefore, an equivalent condition for the applicability of the dipole approximation is

$$\frac{v}{c} \ll 1.$$

The expansion in $\mathbf{k}\cdot\mathbf{r}$ may be regarded as an expansion in v/c. Note that as higher order terms in v/c are retained, one must also add correction terms of these orders to the nonrelativistic form of the Schrodinger equation, Eq. (10.2). The reason that electric quadrupole and magnetic dipole radiation have roughly the same order of magnitude is that the magnetic force is already down by a factor v/c from the electric force.

By setting $e^{i\mathbf{k}\cdot\mathbf{r}} = 1$, the integral (10.18) becomes

$$\int \phi_f^* (\mathbf{l}\cdot\nabla_j)\phi_i \, d^3x = i\hbar^{-1}(\mathbf{l}\cdot\mathbf{p}_j)_{fi} \tag{10.19}$$

where $(\ )_{fi}$ denotes the matrix elements between states f and i. A useful alternative expression may be found by using the commutation relations

$$\mathbf{r}_j\mathbf{p}_j^2 - \mathbf{p}_j^2\mathbf{r}_j = 2i\hbar\mathbf{p}_j.$$

It follows that $\mathbf{r}_j$ commutes with the Hamiltonian

$$H^0 = \frac{1}{2m}\sum \mathbf{p}_j^2 + V(\mathbf{r}_1, \mathbf{r}_2, \ldots, \mathbf{r}_N) \tag{10.20}$$

in the following way:

$$\hbar^{-2}(\mathbf{r}_j H_0 - H_0 \mathbf{r}_j) = i\hbar^{-1}\mathbf{p}_j.$$

Using this to replace $i\hbar^{-1}\mathbf{p}_j$ in the matrix element yields

$$i\hbar^{-1}(\mathbf{l}\cdot\mathbf{p}_j)_{fi} = m\hbar^{-2}\int \phi_f^* \mathbf{l}\cdot(\mathbf{r}_j H_0 - H_0 \mathbf{r}_j)\phi_i \, d^3x$$

$$= m\hbar^{-2}(E_i - E_f)\int \phi_f^* \mathbf{l}\cdot\mathbf{r}\phi_i \, d^3x, \tag{10.21}$$

where we have used the fact that H_0 acting on its eigenfunctions yields the corresponding eigenvalues. Thus the transition rate is

$$w_{fi} = \frac{4\pi^2}{\hbar^2 c}|(\mathbf{l}\cdot\mathbf{d})_{fi}|^2 \mathcal{J}(\omega_{fi}), \tag{10.22}$$

where

$$\mathbf{d} \equiv e\sum_j \mathbf{r}_j \tag{10.23}$$

is the *electric dipole operator*.

Often we are only concerned with unpolarized radiation from atoms with random orientations. We then average the above formula over all angles, which gives

$$\langle|(\mathbf{l}\cdot\mathbf{d})_{fi}|^2\rangle = \frac{1}{3}|d_{fi}|^2,$$

since

$$\langle\cos^2\theta\rangle = \frac{1}{3}.$$

Here we interpret the quantity $|d_{fi}|^2$ to mean the combination

$$\mathbf{d}_{fi}^* \cdot \mathbf{d}_{fi} = |(d_x)_{fi}|^2 + |(d_y)_{fi}|^2 + |(d_z)_{fi}|^2. \tag{10.24}$$

Thus the average transition rate is

$$\langle w_{fi}\rangle = \frac{4\pi^2}{3c\hbar^2}|d_{fi}|^2 \mathcal{J}(\omega_{fi}). \tag{10.25}$$

10.3 EINSTEIN COEFFICIENTS AND OSCILLATOR STRENGTHS

We can relate this to our previous discussion in terms of the Einstein B coefficients (§1.6). Letting u and l refer to the upper and lower-states, respectively, we have

$$\langle w_{lu} \rangle = B_{lu} J_{\nu_{ul}}. \tag{10.26a}$$

Note that $J_{\nu_{ul}} = (4\pi)^{-1} \mathcal{I}(\nu_{ul})$, since the intensity considered here is undirectional. Also, we have the relation that $\mathcal{I}(\nu_{ul}) = 2\pi \mathcal{I}(\omega_{ul})$ so that

$$\langle w_{lu} \rangle = \frac{1}{2} B_{lu} \mathcal{I}(\omega_{ul}). \tag{10.26b}$$

Comparing this with the above expression gives

$$B_{lu} = \frac{8\pi^2 |d_{lu}|^2}{3c\hbar^2} = \frac{32\pi^4 |d_{lu}|^2}{3ch^2}. \tag{10.27}$$

From the Einstein relations (Eqs. 1.72) we have for nondegenerate levels

$$B_{lu} = B_{ul}$$

$$A_{ul} = \frac{4\omega_{fi}^3}{3c^3\hbar} |d_{ul}|^2 = \frac{64\pi^4 \nu_{ul}^3 |d_{ul}|^2}{3c^3 h}. \tag{10.28a}$$

If the levels are *degenerate*, the transition rate is found by averaging over the initial states and summing over the final states. Thus the Einstein A coefficient is given by

$$A_{ul} = \frac{64\pi^4 \nu_{ul}^3}{3hc^3} \frac{1}{g_u} \sum |d_{ul}|^2, \tag{10.28b}$$

where the sum is over all substates of the upper and lower levels. In this case the Einstein relations have their usual statistical weight factors.

It is convenient to define the *absorption oscillator strength* f_{lu} by the relationship

$$B_{lu} = \frac{4\pi^2 e^2}{h\nu_{ul} mc} f_{lu}, \tag{10.29a}$$

$$f_{lu} = \frac{2m}{3\hbar^2 g_l e^2} (E_u - E_l) \sum |d_{lu}|^2. \tag{10.29b}$$

The reason for naming it such is that the B coefficient associated with a classical oscillator can be defined in terms of the total energy extracted from a beam of radiation [cf. Eqs. (3.65), (1.66), and (1.74)]

$$\int_0^\infty \sigma(\nu)\,d\nu = \frac{\pi e^2}{mc} = B_{lu}^{\text{classical}} \frac{h\nu_{lu}}{4\pi}, \tag{10.30}$$

so that

$$B_{lu}^{\text{classical}} = \frac{4\pi^2 e^2}{h\nu_{lu} mc}. \tag{10.31}$$

The oscillator strength (or *f value*) is just that factor which corrects this classical result. One can in this way picture the quantum mechanical process as being due to a number (usually fractional) f_{lu} of equivalent classical electron oscillators of the same frequency ν. Normally f_{lu} is of order unity, so that it is a particularly useful quantity to characterize the strengths of transitions.

It is also convenient to define an *emission oscillator strength* by the formula

$$B_{ul} = \frac{4\pi^2 e^2}{h\nu_{lu} mc} f_{ul}. \tag{10.32}$$

Since $g_l B_{lu} = g_u B_{ul}$ and $\nu_{ul} = -\nu_{lu}$, we have the general relation

$$g_l f_{lu} = -g_u f_{ul}. \tag{10.33}$$

Thus emission oscillator strengths are *negative*. We may write the A coefficient in terms of the emission and absorption oscillator strengths:

$$g_u A_{ul} = -\frac{8\pi^2 e^2 \nu_{ul}^2}{mc^3} g_u f_{ul} = \frac{8\pi^2 e^2 \nu_{ul}^2}{mc^3} g_l f_{lu}. \tag{10.34}$$

One modification of the oscillator strength concept is necessary when the upper state happens to lie in a continuum. In this case it is meaningless to talk about the probability of a transition to a single state, but rather we need to define the *probability per unit energy (or frequency) range*. With this in mind we define the derivatives of f such that $(df/d\epsilon)d\epsilon$ is the strength for a transition from state i to a set of continuum states in an energy range $d\epsilon$. The frequency of the emitted photon is given by $h\nu = \epsilon + \chi$ where χ is the ionization potential from the state i. This is illustrated in Fig. 10.1.

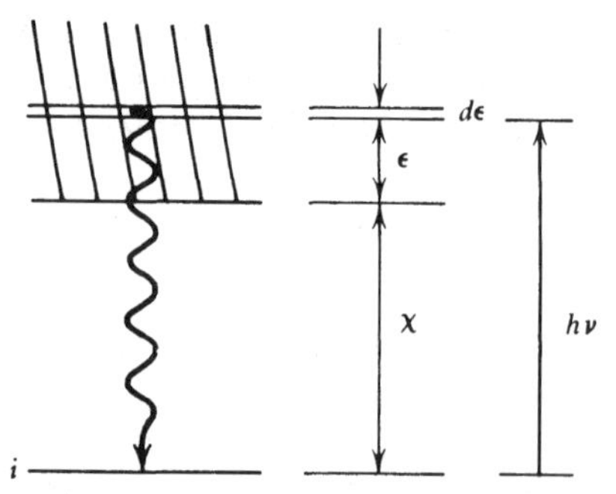

***Figure 10.1** Transition between level i and the continuum (shaded). Here χ is the ionization potential.*

The *continuum oscillator strength* f_c is the total oscillator strength to all continuum states:

$$f_c = \int_0^\infty \frac{df}{d\epsilon}\,d\epsilon = \int_{\nu_0}^\infty \frac{df}{d\nu}\,d\nu \tag{10.35}$$

where $h\nu_0 = \chi$.

The oscillator strengths must be found by direct calculation or by experiment. The theoretical determination of f values (or A values) is difficult, but with the advent of large computers, much can now be done to obtain accurate, reliable results. The basic difficulty is that most approximate wave functions for complex atoms, such as Hartree–Fock, tend to be most accurate at small radii where the associated contribution to the total energy is most important. However, the transition probabilities depend more critically on the wave functions at large radii. We can see this simply by noting that the energies depend on averages of *inverse distances*,

$$\int \psi^* r^{-1} \psi\, d^3\mathbf{r},$$

while the dipole operator depends on averages of *distance*,

$$\int \psi^* r \psi\, d^3\mathbf{r}.$$

For this reason one needs better wave functions than those ordinarily available.

There are a number of general result relating oscillator strengths, known as *sum rules*. They are of great value in determining approximate values or bounds for f values that cannot easily be measured or calculated, and also in obtaining absolute f values from relative f values. The simplest and most general sum rule is the *Thomas–Reiche–Kuhn sum rule*:

$$\sum_{n'} f_{nn'} = N, \tag{10.36}$$

where N is the total number of electrons in the atom, and the summation is over all states of the atom. For each initial state this rule gives a relation involving transitions to all other states. Equation (10.36) follows from the expressions for f_{if} Eqs. (10.23) and (10.29b), and the easily proved identity

$$\frac{2m}{3\hbar^2} \sum_{k} (E_k - E_s)|\mathbf{r}_{sk}|^2 = 1.$$

In many cases, such as when there is a closed shell and a smaller number q of electrons outside the closed shells that are involved in a more limited set of transitions, we also have

$$\sum_{n'} f_{nn'} = q, \tag{10.37}$$

where the sum is now only over those states which involve transitions of these outer electrons.

The sum can be split into two sums, depending on whether n' is a state above or below n:

$$\sum_{\substack{n' \\ E_{n'} > E_n}} f_{nn'} + \sum_{\substack{n' \\ E_{n'} < E_n}} f_{nn'} = q. \tag{10.38}$$

The first sum gives the contribution due to absorption from the state n, and the second sum gives the contribution due to emission from state n to all lower states. Since in the second sum these emission oscillator strengths are negative, we have

$$\sum_{\substack{n' \\ E_{n'} > E_n}} f_{nn'} \geqslant q, \tag{10.39}$$

the equality holding only for the ground state or for an excited state that cannot radiate by a dipole transition (metastable state).

Other types of sum rules also exist under more restrictive assumptions about the nature of the atomic states (e.g., single configuration, L-S coupling, j-j coupling, single electron).

10.4 SELECTION RULES

In general, there will always be some probability for radiative transition between two states, but in some cases this probability can be exceedingly small. This occurs when the states involved fall approximately into a classification scheme (like L-S coupling) for which the transition probability would be strictly zero if that scheme held rigorously. For example, a transition probability may be strictly zero in the dipole approximation but nonzero for higher order multipole radiation or two-photon emission.

The precise statements of when a transition probability vanishes under some specified set of assumptions are called *selection rules*. We are primarily concerned with dipole selection rules, so that the crucial question involves when the dipole matrix element $\mathbf{d}_{fi}$ vanishes. The most general result is *Laporte's rule*: *there are no transitions between states of the same parity*. This is easily proved by recalling the definition

$$\mathbf{d}_{fi} \equiv e \int \phi_f^* \sum_j \mathbf{r}_j \phi_i \, d^3x.$$

If we reflect all coordinates we note $\Sigma \mathbf{r}_j \to -\Sigma \mathbf{r}_j$ while $\phi_f^* \phi_i$ is unchanged if f and i have the same parity. Thus the integral is equal to its negative, and vanishes.

For states with a specific configuration assignment the parity is $(-1)^{\Sigma l_i}$ where the l_i are the angular momentum quantum numbers of the individual orbitals. Thus we deduce that the configuration must change by at least one orbital, from Laporte's rule. There are no dipole transitions between states of the same configuration.

A sharpened selection rule applies to the transitions between configurations: The configuration must change by precisely one orbital. This is proved by noting that a given configuration may be expressed as a superposition of determinental wave functions, which in turn are superpositions of products of one-particle orbitals. The dipole operator is a sum of the $\mathbf{r}_j$ over all electrons, so that ultimately one can write the matrix element $\mathbf{d}_{fi}$ as a sum of matrix elements of a single $\mathbf{r}_j$ between product wave functions corresponding to the two configurations:

$$\int u_a^* u_b^* \cdots u_k^*(\mathbf{r}_j) u_{a'} u_{b'} \cdots u_{k'} \, d^3x.$$

The particular one-particle wave functions having the coordinates $\mathbf{r}_j$ will integrate out to some result (in general nonzero), but all the other integrals will be simply the orthonormality integrals of the functions u; therefore, in order not to give a zero result, all the corresponding functions must be the same, except for the one involving $\mathbf{r}_j$. The only way to ensure that all the terms in the grand summation will not vanish is to make all orbitals the same except for one. This selection rule is known as the *one-electron jump rule*. It can be violated by states that are superpositions of several configurations (configuration interaction), but it will be obeyed for *L-S* coupling, which assumes no such configuration interaction.

As indicated above, under the assumption of configuration assignments we may evaluate the dipole matrix element by evaluating a simplified matrix element that connects states of the jumping electron orbitals only. Recall these orbitals have a very simple representation (Eq. 9.9) times a spin function. Since **r** and **s** commute, we see that the spin cannot change, so that $m_s = m_{s'}$; thus we may deal with the space parts alone. The dipole matrix element between two such orbitals will involve the integral

$$\int r R_{nl} R_{n'l'} \, dr,$$

which is called the *radial integral*. It will also involve an integral over spherical harmonics. An examination of this latter integral, using the fact that the dipole operator is a vector, leads to the selection rules (see Problem 10.6)

$$\Delta l = \pm 1, \tag{10.40a}$$

$$\Delta m = 0, \pm 1. \tag{10.40b}$$

In a multielectron atom these rules apply to the jumping electron. These rules completely determine the spectra of one-electron atoms, such as HI and HeII, and also the alkali metals.

There are also selection rules for many electron atoms that involve the total quantities L, S, and J. One general result (which applies even to higher multipole radiation) is that the transition $J = 0$ to $J = 0$ is *forbidden*, because the photon carries off one unit of angular momentum. In *L-S* coupling we find that we must have

$$\Delta S = 0 \tag{10.41a}$$

$$\Delta L = 0, \pm 1, \tag{10.41b}$$

$$\Delta J = 0, \pm 1. \qquad (\text{except } J = 0 \text{ to } J = 0) \tag{10.41c}$$

The rule $\Delta S = 0$ follows from the fact that the dipole operator does not involve spin. We note that $\Delta L = 0$ is allowed here but that $\Delta l = 0$ is not. This is because there is no direct relation of L to the parity; for example, for two equivalent p electrons we have the state ^3P which has odd L but even parity, and ^1S which has even L and even parity.

For higher multipole radiation the selection rules for J remain unchanged ($\Delta J = 0, \pm 1$, except $J = 0$ to $J = 0$), but the parity rule becomes: for magnetic dipole and electric quadrupole radiation, parity is unchanged.

For magnetic dipole transitions *the configuration does not change*. This allows for many of the forbidden lines in the ground configurations of C, N, O, for example, and for the important 21-cm lines.

10.5 TRANSITION RATES

One case in which a fairly complete discussion of transition rates can be given purely theoretically is the pure Coulomb case of hydrogen (and for other hydrogen-like ions, such as HeII and LiIII). The frequency of a photon absorbed or emitted in a transition between two discrete levels with principal quantum numbers n' and n' is given by

$$h\nu = Ry(n^{-2} - n'^{-2}), \tag{10.42a}$$

where

$$Ry \equiv \frac{e^2}{2a_0} = 13.6 \text{ eV}. \tag{10.42b}$$

When the upper level is in the continuum, so that there is a free electron with energy $\epsilon = \frac{1}{2}mv^2$, we have

$$h\nu = Ry/n^2 + \epsilon. \tag{10.43}$$

To liberate a free electron one needs a photon of at least the *threshold energy*, $h\nu_n = \chi_n =$ ionization potential from the initial state n.

Bound-bound Transitions for Hydrogen

To calculate the dipole oscillator strength we must evaluate the dipole operator matrix element. This will involve integrals over the radial wave functions $R_{nl}(r)$ of the form

$$\int R_{nl} R_{n'(l-1)} r\, dr. \tag{10.44}$$

By the selection rule (10.40a) we know that $\Delta l = 1$. Since these radial functions are analytically known [Laguerre polynomials; see Eq. (9.15a)], the integrals of Eq. (10.44) can be performed, but are complicated (Gordon, 1929.) When the integral is performed, it can then be summed over all l appropriate to a given n and n'. The Lyman-α transition ($n=1$ $n'=2$) in hydrogen is treated explicitly in Problem 10.3 and yields the f value

$$gf = \frac{2^{14}}{3^9} = 0.8324. \tag{10.45}$$

For other members of the Lyman series ($n'=1$), the result is (Menzel and Pekeris, 1935)

$$g_1 f_{1n} = \frac{2^9 n^5 (n-1)^{2n-4}}{3(n+1)^{2n+4}}. \tag{10.46}$$

In general, the expression for f can be reduced to a closed form. Note that for high values of n the oscillator strengths decrease rapidly

$$g_1 f_{1n} = \frac{2^9}{3n^3} \frac{\left(1-\frac{1}{n}\right)^{2n-4}}{\left(1+\frac{1}{n}\right)^{2n-4}}$$

$$\sim \frac{2^9}{3n^3} \frac{e^{-2}}{e^2} \sim 3.1 \frac{1}{n^3}. \tag{10.47}$$

Further values of oscillator strengths for the bound-bound transitions can be obtained from Table 10.1.

Table 10.1

n / n'	1	2	3	n / n'	1	2	3
2	4.162×10^{-1}			19	2.295×10^{-4}	5.167×10^{-4}	8.364×10^{-4}
3	7.910×10^{-2}	6.408×10^{-1}		20	1.966×10^{-4}	4.418×10^{-4}	7.117×10^{-4}
4	2.899×10^{-2}	1.193×10^{-1}	8.420×10^{-1}	21	1.698×10^{-4}	3.803×10^{-4}	6.111×10^{-4}
5	1.394×10^{-2}	4.467×10^{-2}	1.506×10^{-1}	22	1.476×10^{-4}	3.302×10^{-4}	5.286×10^{-4}
6	7.800×10^{-3}	2.209×10^{-2}	5.585×10^{-2}	23	1.276×10^{-4}	2.885×10^{-4}	4.608×10^{-4}
7	4.814×10^{-3}	1.271×10^{-2}	2.768×10^{-2}	24	1.137×10^{-4}	2.534×10^{-4}	4.040×10^{-4}
8	3.184×10^{-3}	8.037×10^{-3}	1.604×10^{-2}	25	1.005×10^{-4}	2.240×10^{-4}	3.558×10^{-4}
9	2.216×10^{-3}	5.429×10^{-3}	1.023×10^{-2}	26	8.931×10^{-5}	1.987×10^{-4}	3.155×10^{-4}
10	1.605×10^{-3}	3.851×10^{-3}	6.981×10^{-3}	27	7.963×10^{-5}	1.772×10^{-4}	2.809×10^{-4}
11	1.201×10^{-3}	2.836×10^{-3}	4.996×10^{-3}	28	7.138×10^{-5}	1.587×10^{-4}	2.513×10^{-4}
12	9.215×10^{-4}	2.150×10^{-3}	3.711×10^{-3}	29	6.431×10^{-5}	1.427×10^{-4}	2.243×10^{-4}
13	7.226×10^{-4}	1.672×10^{-3}	2.839×10^{-3}	30	5.809×10^{-5}	1.288×10^{-4}	2.034×10^{-4}
14	5.774×10^{-4}	1.326×10^{-3}	2.223×10^{-3}	31	5.260×10^{-5}	1.167×10^{-4}	1.840×10^{-4}
15	4.687×10^{-4}	1.070×10^{-3}	1.776×10^{-3}	32	4.784×10^{-5}	1.060×10^{-4}	1.670×10^{-4}
16	3.855×10^{-4}	8.770×10^{-4}	1.443×10^{-3}	33	4.359×10^{-5}	9.654×10^{-5}	1.521×10^{-4}
17	3.211×10^{-4}	7.273×10^{-4}	1.189×10^{-3}	34	3.982×10^{-5}	8.829×10^{-5}	1.389×10^{-4}
18	2.703×10^{-4}	6.098×10^{-4}	9.914×10^{-4}	35	3.656×10^{-5}	8.084×10^{-5}	1.272×10^{-4}

Bound-free Transitions (Continuous Absorption) for Hydrogen

When the upper state lies in the continuum, there can be absorption in a continuous range of frequencies. Since the absorption results in an electron being liberated from the atom, this process is also called *photoionization*. We express our results in terms of the cross section for the transition. The differential transition rate, dw, for a transition from bound state i to a continuum state f, with electron in momentum range dp and solid angle range $d\Omega$, is

$$dw=\frac{4\pi^2e^2}{m^2c}\frac{\mathcal{J}(\omega)}{\omega^2}|\langle f|e^{i\mathbf{k}\cdot\mathbf{r}}\,\boldsymbol{\nabla}|i\rangle|^2\left[\frac{dn}{dp\,d\Omega}dp\,d\Omega\right]. \tag{10.48}$$

Here the term in brackets is the number of free electron states available, that is, the "density of states" $dn/dp\,d\Omega$ multiplied by the differential range $dp\,d\Omega$, and the remaining factor is identical to our expression for the transition rate for bound-bound transitions, Eq. (10.16). By energy conservation, we have that the frequency interval $d\omega$ of incident photons is related to the momentum interval dp of nonrelativistic electrons by

$$\hbar\,d\omega=\frac{p\,dp}{m}. \tag{10.49}$$

We also have, by definition, that the number of photons per unit area per unit time per unit frequency in the incident beam satisfies

$$\frac{dN}{dA\,dt\,d\omega}=\frac{\mathcal{J}(\omega)}{\hbar\omega}. \tag{10.50}$$

If the final electron is localized to a volume V, then the density of states for a given final spin state is [cf. Eq. (9.43)]

$$\frac{dn}{dp\,d\Omega}=\frac{p^2V}{h^3}. \tag{10.51}$$

Combining Eqs. (10.48) to (10.51), we obtain for the differential bound-free cross section,

$$\frac{d\sigma_{bf}}{d\Omega}=\frac{\alpha vV}{2\pi\omega}|\langle f|e^{i\mathbf{k}\cdot\mathbf{r}}\,\boldsymbol{\nabla}|i\rangle|^2, \tag{10.52}$$

where $v=p/m$ is the final electron velocity.

For the simple case of a bound-free transition from the ground state of hydrogen, ionized by a photon of frequency ω, this differential cross section is evaluated explicitly for $\hbar\omega \gg Ry$ in Problem 10.4. The total cross section, $\sigma_{bf} = \int (d\sigma/d\Omega)\,d\Omega$, is

$$\sigma_{bf}(\hbar\omega \gg Ry) \approx \frac{(2\alpha)^{9/2}\pi Z^5 c^{7/2}}{3a_0^{3/2}\omega^{7/2}}. \tag{10.53}$$

For the more general case of a bound-free transition from state n and l, a detailed calculation (Karzas and Latter, 1961) gives

$$\sigma_{bf} = \frac{512\pi^7 m e^{10} Z^4}{3\sqrt{3}\, ch^6 n^5}\,\frac{g(\omega, n, l, Z)}{\omega^3}, \tag{10.54}$$

where g is the bound-free Gaunt factor. If χ_n is the ionization potential for the initial level, σ_{bf} is zero for $\omega < \omega_n$ where

$$\omega_n \equiv \frac{\chi_n}{\hbar} = \frac{\alpha^2 mc^2 Z^2}{2\hbar n^2}, \tag{10.55}$$

rises abruptly to Eq. (10.54) at threshold $\omega = \omega_n$ and then decreases roughly as ω^{-3}. Near threshold, the Gaunt factor g is unity, to within 20%.

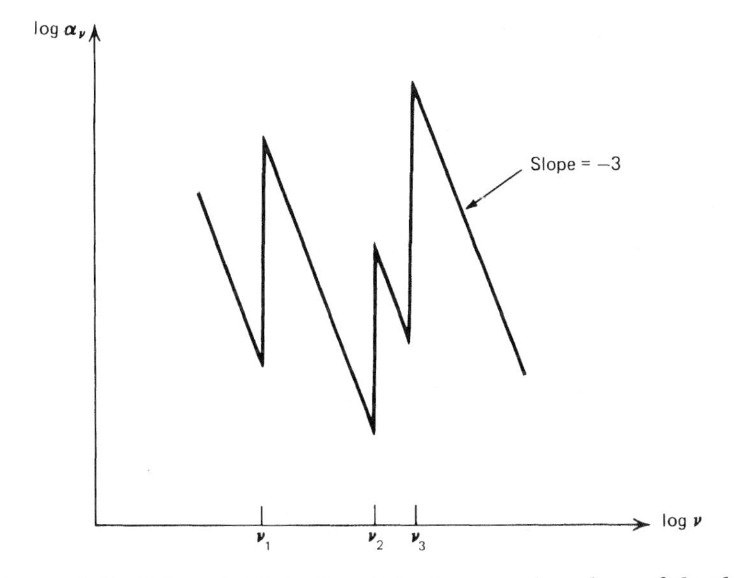

***Figure 10.2** Schematic illustration of the frequency dependence of the absorption coefficient. The sharp rises, absorption edges, occur at the frequency of ionization of a particular level.*

For convenience, Eq. (10.54) can also be written in the form

$$\sigma_{bf} = \left(\frac{64\pi n g}{3\sqrt{3}\, Z^2} \right) \alpha a_0^2 \left(\frac{\omega_n}{\omega} \right)^3. \tag{10.56}$$

We can also write our results in terms of the absorption coefficient, $\alpha_\nu = N_n \sigma$, where N_n is the atomic density at the absorbing level. The total absorption coefficient equals the sum of terms of this form and is illustrated schematically in Fig. 10.2. The *absorption edges* correspond to the onset of absorption from different levels. The relative strength of these edges depends on the number of atoms in each level. For example, if the material is in thermodynamic equilibrium, these numbers are given by the Boltzmann law.

Radiative Recombination; Milne Relations

The process inverse to photoionization is *radiative recombination*, in which an electron is captured by an ion into a bound state n with emission of a photon. There are connections between rates for photoionization and recombination, analogous to the Einstein relations. These are called the *Milne relations* and are examples of general *detailed balancing* relations. If we want to apply these directly to a single capture event we first have to consider the distribution function for the electrons, that is, how many electrons are moving in each speed range. However, it is also quite useful, and usually sufficient, to deal with a *thermal distribution* of electrons. The detailed balance relations can then be obtained by the simple requirement that the radiation field in equilibrium is the Planck function $B_\nu(T)$. Since the coefficients refer to atomic properties, they then can be used for any distributions of electrons and radiation.

Let $\sigma_{fb}(v)$ be the cross section for recombination for electrons of velocity v. Then the number of recombinations per unit time per unit volume due to thermal electrons in speed range dv is

$$N_+ N_e \sigma_{fb} f(v) v\, dv, \tag{10.57}$$

where N_e is the electron density, N_+ is the ion density, and $f(v)$ is the Maxwellian velocity distribution. The number of photoionizations per time per volume for a blackbody radiation field ($I_\nu = B_\nu$) in frequency range $d\nu$ is, (cf. Problem 1.2), with N_n the neutral atom density,

$$\frac{4\pi}{h\nu} N_n \sigma_{bf} (1 - e^{-h\nu/kT}) B_\nu\, d\nu, \tag{10.58}$$

where the factor $(1-e^{-h\nu/kT})$ now gives the net photoionization rate when "stimulated recombinations" are subtracted out. Then equating (10.57) and (10.58) and using the Planck function and Eq. (10.49), we obtain

$$\frac{\sigma_{bf}}{\sigma_{fb}}=\frac{N_+N_e}{N_n}e^{h\nu/kT}\frac{f(v)c^2h}{8\pi m\nu^2}. \tag{10.59}$$

But, we also know

$$f(v)=4\pi\left(\frac{m}{2\pi kT}\right)^{3/2}v^2\exp\left(\frac{-mv^2}{2kT}\right), \tag{10.60}$$

and from Saha's equation [cf. Eq. (9.47)]

$$\frac{N_+N_e}{N_n}=\left(\frac{2\pi mkT}{h^2}\right)^{3/2}\frac{g_eg_+}{g_n}e^{-\chi/kT}.$$

Using the result

$$h\nu=\tfrac{1}{2}mv^2+\chi, \tag{10.61}$$

we obtain the Milne relation:

$$\frac{\sigma_{bf}}{\sigma_{fb}}=\frac{m^2c^2v^2}{\nu^2h^2}\frac{g_eg_+}{2g_n}. \tag{10.62}$$

Since we have already found σ_{bf}, we can compute σ_{fb}.

In this way recombination coefficients can be computed for given velocity distribution, say Maxwellian. We have the following results for the thermal recombination coefficient onto the nth level of hydrogen: (Gaunt factor = 1)

$$\langle v\sigma_{fb}\rangle=\int vf(v)\sigma_{fb}\,dv \tag{10.63}$$

where $f(v)$ is the speed distribution of the thermal electrons, Eq. (10.60). Substitution of Eqs. (10.56), (10.61), and (10.62) into (10.63) then yields (Cillié 1932)

$$\langle v\sigma_{fb}\rangle=3.262\times10^{-6}M(n,T), \tag{10.64a}$$

where

$$M(n,T)=\frac{e^{\chi_n/kT}}{n^3T^{3/2}}E_1\left(\frac{\chi_n}{kT}\right), \tag{10.64b}$$

and where

$$E_1(x)\equiv\int_x^\infty \frac{e^{-t}}{t}\,dt. \tag{10.64c}$$

In evaluating Eqs. (10.64) we have used $g_e=2$, $g_+=1$, $g_n=2n^2$ as the values for the statistical weight factors.

Also of interest is the recombination coefficient summed over all bound states n. A convenient approximation is (Seaton, 1959)

$$\sum_n \langle v\sigma_{fb}\rangle = 5.197\times10^{-14}\lambda^{1/2}\left(0.4288+\tfrac{1}{2}\ln\lambda+0.469\lambda^{-1/3}\right) \tag{10.65a}$$

where

$$\lambda\equiv1.579\times10^5/T. \tag{10.65b}$$

Recombination can proceed in other ways besides radiative recombination. *Three-body recombination* is usually quite slow at astrophysical densities, since it requires a close encounter of three bodies simultaneously. However, *dielectronic recombination* (see, e.g., Massey and Gilbody 1974) can be very important for some ions.

The Role of Coupling Schemes in the Determination of f Values

When particular coupling schemes are appropriate, it is possible to relate the f values for different transitions by means of formulas (or tables). For L-S coupling (Russell–Saunders coupling) we can interrelate the f values of all lines between two given terms; this set of lines is called a *multiplet*. The relative strengths of the lines within a multiplet depend only on the term types of the two terms involved. For example, if we have an upper ^2P term and a lower ^2S term, the transition $^2S_{1/2}-{}^2P_{3/2}$ is twice as strong as the transition $^2S_{1/2}-{}^2P_{1/2}$. The factor of 2 is due to there being two times the number of states in $J=3/2$ as in $J=1/2$. This is the situation with the Lyman $-\alpha$ ($Ly\alpha$) transition in HI. If we know the *total* strength of the multiplet, we can then find the strengths of the individual line components.

Thus since the total gf is 0.8324 [cf (10.45)], we have (gf) $1/2-3/2=$ 0.5549, (gf) $1/2-1/2=0.2775$. Tables to deduce the relative strength of lines within a multiplet can be found in Allen (1974) and Aller (1963).

Another use of the L-S coupling scheme is to deduce the relative strengths of *multiplets* between two *configurations*. This kind of calculation is affected more by deviations from L-S coupling than the preceding, so that it is not as reliable. The set of multiplets arising out of transitions between two configurations is called a *transition array*, and the relative strengths of multiplets within a transition array is discussed in the above references.

Other coupling schemes give their own rules for relating f values, but we do not discuss these here. In cases where a particular coupling scheme is not applicable, or its applicability is dubious, we must obtain f values for the desired transitions either directly by experiment or by a more sophisticated theoretical calculation.

10.6 LINE BROADENING MECHANISMS

Atomic levels are not infinitely sharp, nor are the lines connecting them. This was already recognized in our discussion of the Einstein coefficients, where we introduced the line profile function $\phi(\nu)$ to account for the nonzero width of the line. Many physical effects determine the line shape, and we can only deal with a few here (see, e.g., Griem 1974; Mihalas 1978).

Doppler Broadening

Perhaps the simplest mechanism for line broadening is the Doppler effect. An atom is in thermal motion, so that the frequency of emission or absorption in its own frame corresponds to a different frequency for an observer. Each atom has its own Doppler shift, so that the net effect is to spread the line out, but not to change its total strength.

The change in frequency associated with an atom with velocity component v_z along the line of sight (say, z axis) is, to lowest order in v/c, given by Eq. (4.12)

$$\nu-\nu_0=\frac{\nu_0 v_z}{c}. \tag{10.66}$$

Here ν_0 is the rest-frame frequency. The number of atoms having velocities in the range v_z to $v_z + dv_z$ is proportional to the Maxwellian distribution

$$\exp\left(-\frac{m_a v_z^2}{2kT}\right) dv_z$$

where m_a is the mass of an atom. From the above we have the relations

$$v_z = \frac{c(\nu - \nu_0)}{\nu_0}, \tag{10.67a}$$

$$dv_z = \frac{c\, d\nu}{\nu_0}. \tag{10.67b}$$

Therefore, the strength of the emission in the frequency range ν to $\nu + d\nu$ is proportional to

$$\exp\left[-\frac{m_a c^2(\nu - \nu_0)^2}{2\nu_0^2 kT}\right] d\nu,$$

and the profile function is

$$\phi(\nu) = \frac{1}{\Delta\nu_D\sqrt{\pi}} e^{-(\nu - \nu_0)^2/(\Delta\nu_D)^2}. \tag{10.68}$$

Here the *Doppler width* $\Delta\nu_D$ is defined by

$$\Delta\nu_D = \frac{\nu_0}{c}\sqrt{\frac{2kT}{m_a}}. \tag{10.69}$$

The constant $(\Delta\nu_D\sqrt{\pi})^{-1}$ in the formula for $\phi(\nu)$ is determined by the normalization condition $\int\phi(\nu)\, d\nu = 1$ under the (reasonable) assumption that $\Delta\nu_D \ll \nu_0$. The *line-center* cross section for each atom, neglecting stimulated emission, is therefore

$$\sigma_{\nu_0} = B_{12}\frac{h\nu_0}{4\pi}\phi(\nu_0) = \frac{1}{\Delta\nu_D\sqrt{\pi}}\frac{h\nu_0}{4\pi}B_{12}$$

$$= \frac{\pi e^2}{mc} f_{12} \frac{1}{\Delta\nu_D\sqrt{\pi}} \tag{10.70}$$

for the case of Doppler broadening. Numerically this is

$$\sigma_{\nu_0} = 1.16 \times 10^{-14} \lambda_0 \sqrt{A/T}\, f_{12}\ \mathrm{cm}^2, \tag{10.71}$$

where λ_0 is in $\mathring{A}$, T in K, and A is the atomic weight for the atom.

In addition to thermal motions there can also be turbulent velocities associated with macroscopic velocity fields. When the scale of the turbulence is small in comparison with a mean free path (called *microturbulence*) these motions are often accounted for by an effective Doppler width

$$\Delta\nu_D = \frac{\nu_0}{c}\left(\frac{2kT}{m_a} + \xi^2\right)^{1/2}, \tag{10.72}$$

where ξ is a root mean-square measure of the turbulent velocities. This assumes that the turbulent velocities also have a Gaussian distribution.

Natural Broadening

A certain width to the atomic level is implied by the uncertainty principle, namely, that the spread in energy ΔE and the duration Δt in the state must satisfy $\Delta E\,\Delta t \sim \hbar$. We note that the spontaneous decay of an atomic state n proceeds at a rate

$$\gamma = \sum_{n'} A_{nn'},$$

where the sum is over all states n' of lower energy. If radiation is present, we should add the induced rates to this. The coefficient of the wave function of state n, therefore, is of the form $e^{-\gamma t/2}$ and leads to a decay of the electric field by the same factor. (The energy then decays proportional to $e^{-\gamma t}$.) Therefore, we have an emitted spectrum determined by the decaying sinusoid type of electric field, as given in §2.3 and Fig. 2.3. Thus the profile is of the form

$$\phi(\nu) = \frac{\gamma/4\pi^2}{(\nu-\nu_0)^2 + (\gamma/4\pi)^2}. \tag{10.73}$$

This is called a *Lorentz* (or *natural*) *profile.*

Actually, the above result applies to cases in which only the upper state is broadened (e.g., transitions to the ground state). If both the upper and lower state are broadened, then the appropriate definition for γ is

$$\gamma = \gamma_u + \gamma_l, \tag{10.74}$$

where γ_u and γ_l are the widths of the upper and lower states involved in the transition. Thus, for example, we can have a weak but broad line if the lower state is broadened substantially.

Collisional Broadening

The Lorentz profile applies even more generally to certain types of collisional broadening mechanisms. For example, if the atom suffers collisions with other particles while it is emitting, the phase of the emitted radiation can be altered suddenly (see Fig. 10.3). If the phase changes completely randomly at the collision times, then information about the emitting frequencies is lost. If the collisions occur with frequency ν_{col}, that is, each atom experiences ν_{col} collisions per unit time on the average, then the profile is (see Problem 10.7).

$$\phi(\nu) = \frac{\Gamma/4\pi^2}{(\nu-\nu_0)^2 + (\Gamma/4\pi)^2}, \tag{10.75a}$$

where

$$\Gamma = \gamma + 2\nu_{\text{col}}. \tag{10.75b}$$

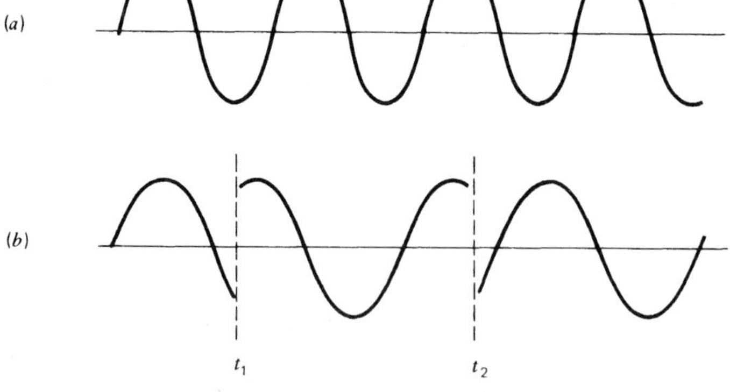

***Figure 10.3** Time-dependence of the electric field of emitted radiation which is (a) purely sinusoidal and (b) subject to random phase interruptions by atomic collisions.*

Combined Doppler and Lorentz Profiles

Quite often an atom shows both a Lorentz profile plus the Doppler effect. In these cases we can write the profile as an average of the Lorentz profile over the various velocity states of the atom:

$$\phi(\nu)=\frac{\Gamma}{4\pi^2}\int_{-\infty}^{\infty}\frac{(m/2\pi kT)^{1/2}\exp(-mv_z^2/2kT)}{(\nu-\nu_0-\nu_0 v_z/c)^2+(\Gamma/4\pi)^2}\,dv_z. \tag{10.76}$$

We can write this more compactly using the definition of the *Voigt function*

$$H(a,u)\equiv\frac{a}{\pi}\int_{-\infty}^{\infty}\frac{e^{-y^2}\,dy}{a^2+(u-y)^2}. \tag{10.77}$$

Then Eq. (10.76) can be written as

$$\phi(\nu)=(\Delta\nu_D)^{-1}\pi^{-1/2}H(a,u), \tag{10.78}$$

where

$$a\equiv\frac{\Gamma}{4\pi\,\Delta\nu_D}, \tag{10.79a}$$

$$u\equiv\frac{\nu-\nu_0}{\Delta\nu_D}. \tag{10.79b}$$

For small values of a, the center of the line is dominated by the Doppler profile, whereas the "wings" are dominated by the Lorentz profile. (See problem 10.5).

PROBLEMS

10.1—What radiative transitions are allowed between the fine structure levels of a ^3P term and those of a ^3S term? Draw a diagram showing the levels with spacings determined by the Lande interval rule. How many spectral lines will be produced, and how will they be spaced relative to one another? Consider the different possibilities of ^3P being normal or inverted and being the upper or lower term.

10.2—Which of the following transitions are allowed under L-S coupling selection rules for electric dipole radiation and which are not? Explain which rules, if any, are violated.

a. $3s\ ^2S_{1/2} \leftrightarrow 4s\ ^2S_{1/2}$

b. $2p\ ^2P_{1/2} \leftrightarrow 3d\ ^2D_{5/2}$

c. $3s3p\ ^3P_1 \leftrightarrow 3p^2\ ^1D_2$

d. $2p3p\ ^3D_1 \leftrightarrow 3p4d\ ^3F_2$

e. $2p^2\ ^3P_0 \leftrightarrow 2p3s\ ^3P_0$

f. $3s2p\ ^1P_1 \leftrightarrow 2p3p\ ^1P_1$

g. $2s3p\ ^3P_0 \leftrightarrow 3p4d\ ^3P_1$

h. $1s^2\ ^1S_0 \leftrightarrow 2s2p\ ^1P_1$

i. $2p3p\ ^3S_1 \leftrightarrow 2p4d\ ^3D_2$

j. $2p^3\ ^2D_{3/2} \leftrightarrow 2p^3\ ^2D_{1/2}$

10.3—Derive Eq. (10.45) for the Lyman-α oscillator strength.

10.4—Derive Eq. (10.53) for the bound-free cross section, using the nonrelativistic Born approximation.

10.5—Line radiation is emitted from an optically thin, thermal source. Assuming that the only broadening mechanisms are Doppler and natural broadening, show that the observed half-width of the line is independent of the temperature T for $T \ll T_c$ and increases as the square root of T for $T \gg T_c$, where T_c is some critical temperature. For the Lyman-α line of hydrogen estimate T_c in terms of fundamental constants, and give its numerical value.

10.6—Derive the simple dipole selection rule, Eq. (10.40).

10.7—Derive the profile function, Eq. (10.75), when phase-destroying collisions occur with frequency ν_{col}.

REFERENCES

Allen, C. W., 1974, *Astrophysical Quantities*, (Althone Press, London).
Aller, L. H., 1963, *The Atmospheres of the Sun and Stars*, (Ronald, New York).

Bethe, H. A., and Jackiw, R., op. cit.
Bethe, H. A., and Salpeter, E. E., 1957, *Quantum Mechanics of One- And Two-Electron Atoms*, (Springer-Verlag, Berlin).
Cillié, G., 1932, *Mon. Not. Roy. Astr. Soc.* **92**, 820.
Gordon, W., 1929, *Ann. Phys.*, **2**, 1031.
Griem, H. R., 1974, *Spectral Line Broadening by Plasmas*, (Academic, New York).
Karzas, W. J., and Latter, R., 1961, *Astrophys. J. Suppl.*, **6**, 167.
Massey, H. S. W., and Gilbody, H. B., 1974, *Electronic and Ionic Impact Phenomena, vol. IV*, (Clarendon, Oxford).
Menzel, D. H., and Pekeris, C. L., 1935, *Mon. Not. Roy. Astr. Soc.*, **96**, 77.
Merzbacher, E., 1961, *Quantum Mechanics*, (Wiley, New York).
Mihalas, D., 1978, *Stellar Atmospheres*, (Freeman, San Francisco).
Seaton, M., 1959, *Mon. Not. Roy. Astr. Soc.*, **119**, 81.

11

MOLECULAR STRUCTURE

When two or more atoms join together into a molecule there is considerable complexity, as compared to a single atom. Many of the simple symmetries of the atom, such as complete rotational symmetry about the nucleus, are lost, and this means fewer quantum numbers are available to help sort out the molecular states. On the other hand, there are a few consolations:

1. For *diatomic* molecules (to which we restrict ourselves exclusively) there is still rotational symmetry about a line.
2. Some of the most important transitions in molecules involve *rotation* and/or *vibration* of the nuclei with respect to each other; these transitions do not occur in atoms and are actually quite a bit simpler than any atomic transitions. The primary difficulties in understanding molecules are *electronic* states.

11.1 THE BORN–OPPENHEIMER APPROXIMATION: AN ORDER OF MAGNITUDE ESTIMATE OF ENERGY LEVELS

A great simplification in the understanding of molecules was made when it was realized that the motions of the electrons and nuclei could be treated separately. This comes about because of the great disparity between the masses of the electron and a typical nucleus, which have ratios m/M in the range $\sim 10^{-4} - 10^{-5}$. From the uncertainty relations we see that this implies that the electrons are much faster than the nuclei and characteristically have much higher energies. Let a be a typical molecular size. Then the momentum of an electron is of order $\hbar/a$ and will have energy states with typical spacings

$$E_{\text{elect}} \sim \frac{\hbar^2}{ma^2}. \tag{11.1}$$

For typical molecular sizes ($\sim 10^{-8}$ cm) this amounts to about a few eV.

The slowly moving nuclei only sense the electrons as a kind of smoothed-out cloud. Therefore, as the nuclei move the electrons have sufficient time to adjust adiabatically to the new nuclear positions. The nuclei then feel only an *equivalent potential* that depends on the internuclear distance and on the particular electronic state. This separation of nuclear and electronic motions is called the *Born–Oppenheimer approximation*.

For stable molecules the internuclear potential has a minimum at some point (see Fig. 11.1). Vibrations about the minimum can occur and can be estimated roughly by comparing to a harmonic oscillator. We can approximate the potential as $\frac{1}{2}M\omega^2\xi^2$, where ξ is the displacement of the

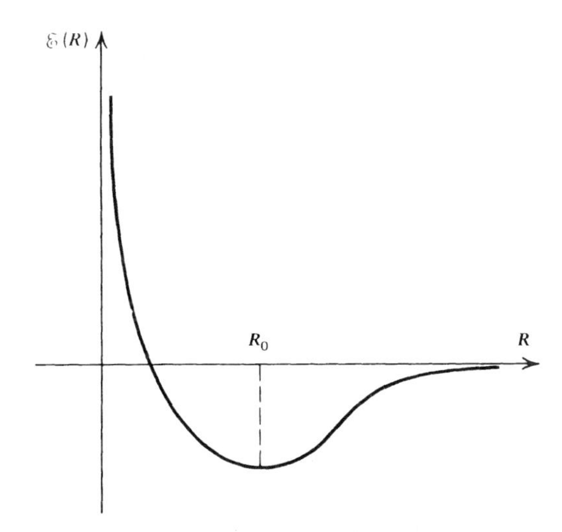

***Figure 11.1** Potential between two atoms in a molecule as a function of their separation R.*

nucleus from its equilibrium position and ω is the frequency of vibration. When ξ is of order a the electronic energies must change to something of order $\hbar^2/2ma^2$, so we set

$$\frac{1}{2}M\omega^2a^2 \sim \frac{\hbar^2}{2ma^2},$$

so that

$$E_{\text{vib}} \sim \hbar\omega \sim \left(\frac{m}{M}\right)^{1/2}\frac{\hbar^2}{ma^2} \sim \left(\frac{m}{M}\right)^{1/2}E_{\text{elect}}. \tag{11.2}$$

These energies are typically tenths or hundredths of an eV, lying in the *infrared.*

The nuclei can also rotate about each other. Let us estimate the energies involved in such motions. If the angular momentum of this motion is $l\hbar$ ($l=0,1,2,\ldots$), then the energy of rotation is

$$E_{\text{rot}} \sim \frac{\hbar^2 l(l+1)}{2I}, \tag{11.3}$$

where I is the moment of inertia of the molecule: $I \sim Ma^2$. Thus for small values of l (low-lying rotational states)

$$E_{\text{rot}} \sim \left(\frac{m}{M}\right)E_{\text{elect}}. \tag{11.4}$$

These energies are of order 10^{-3} eV, lying in the far infrared or *radio.*

The various energies of the molecule are approximately additive

$$E = E_{\text{elect}} + E_{\text{vib}} + E_{\text{rot}}, \tag{11.5}$$

and the contributions are in the approximate ratios

$$E_{\text{elect}} : E_{\text{vib}} : E_{\text{rot}} = 1 : \left(\frac{m}{M}\right)^{1/2} : \frac{m}{M}. \tag{11.6}$$

11.2 ELECTRONIC BINDING OF NUCLEI

We give below a couple of simple examples in which approximate solutions are found for the molecular potential as a function of separation distance of the nuclei. These solutions provide qualitative understanding of the nature of the potential minimum.

The H_2^+ Ion

The simplest molecule is H_2^+, formed when two protons are held together by one electron. The Hamiltonian for the ion is, in the same units as Eq. (9.8),

$$H = -\frac{\nabla^2}{2} - \frac{1}{|\mathbf{r}-\mathbf{R}_A|} - \frac{1}{|\mathbf{r}-\mathbf{R}_B|} + \frac{1}{|\mathbf{R}_A-\mathbf{R}_B|} \tag{11.7}$$

(see Fig. 11.2). We have neglected the kinetic energy of the nuclei and have assumed that their positions are fixed. This problem can be treated approximately by a variational method. We assume that the electron is in a state that is a superposition of two hydrogen atomic states, each centered on a different nucleus:

$$\psi(\mathbf{r}) = \alpha\psi_A(\mathbf{r}) + \beta\psi_B(\mathbf{r}), \tag{11.8}$$

where, for ψ_A and ψ_B both ground states [cf. Eq. (9.16)]

$$\psi_A(\mathbf{r}) = \pi^{-1/2} e^{-|\mathbf{r}-\mathbf{R}_A|}, \tag{11.9a}$$

$$\psi_B(\mathbf{r}) = \pi^{-1/2} e^{-|\mathbf{r}-\mathbf{R}_B|}. \tag{11.9b}$$

The potential is symmetric about the midpoint of the molecule $(\mathbf{R}_A + \mathbf{R}_B)/2$, so we can classify the states by their parities: thus either $\alpha = \beta$ or

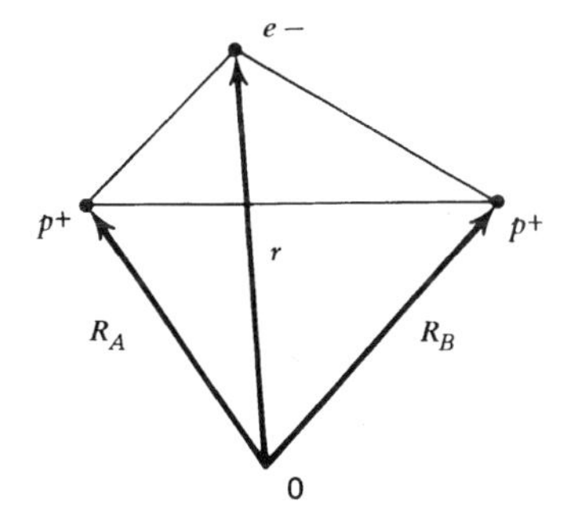

Figure 11.2 Schematic illustration of the location of particles in an H_2^+ ion.

$\alpha = -\beta$, and we can write

$$\psi_\pm(\mathbf{r}) = C_\pm[\psi_A(\mathbf{r}) \pm \psi_B(\mathbf{r})]. \tag{11.10}$$

The normalization constant $C_\pm$ must be found by integrating $|\psi_\pm(\mathbf{r})|^2$ over all space:

$$1 = \int |\psi_\pm(\mathbf{r})|^2 d^3\mathbf{r} = |C_\pm|^2 \int |\psi_A(\mathbf{r}) \pm \psi_B(\mathbf{r})|^2 d^3\mathbf{r}.$$

Thus we obtain for $C_\pm$

$$|C_\pm|^{-2} = 2 \pm 2S(R), \tag{11.11}$$

where $S(R)$ is the *overlap integral*

$$\begin{aligned} S(R) &\equiv \mathrm{Re} \int \psi_A^*(\mathbf{r})\psi_B(\mathbf{r})\, d^3r \\ &= \left(1 + R + \tfrac{1}{3}R^2\right)e^{-R}, \end{aligned} \tag{11.12}$$

and where

$$R \equiv |\mathbf{R}_A - \mathbf{R}_B|. \tag{11.13}$$

Equation (11.12) is derived in problem 11.2. A quite similar evaluation applies to the other integrals below (see Baym, 1969). Choosing $C_\pm$ to be real, we obtain

$$C_\pm = [2 \pm 2S(R)]^{-1/2}. \tag{11.14}$$

The expectation value of H is

$$\begin{aligned} \langle H_\pm \rangle \equiv \varepsilon_\pm(R) &= \int \psi_\pm^* H \psi_\pm\, d^3\mathbf{r} \\ &= \frac{\langle A|H|A\rangle + \langle B|H|B\rangle \pm 2\langle A|H|B\rangle}{2 \pm 2S(R)}, \end{aligned} \tag{11.15}$$

where

$$\begin{aligned} \langle A|H|A\rangle &\equiv \int \psi_A^*(\mathbf{r}) H \psi_A(\mathbf{r})\, d^3\mathbf{r} \\ &= \langle B|H|B\rangle = -\frac{1}{2} + (1 + R^{-1})e^{-2R}. \end{aligned} \tag{11.16}$$

Note the $1/R$ term, arising from the repulsive force of the nuclei at small distances and causing large positive energies at small R. The term $-\frac{1}{2}$ is the energy of the $1s$ state of atomic hydrogen (in atomic units). Also, we have the result

$$\begin{aligned}\langle A|H|B\rangle &\equiv \mathrm{Re}\int \psi_A^* H \psi_B \, d^3r \\ &= \left(-\tfrac{1}{2}+R^{-1}\right)S(R) - \mathrm{Re}\int \psi_A^* |\mathbf{r}-\mathbf{R}_B|^{-1} \psi_B \, d^3\mathbf{r}. \qquad (11.17)\end{aligned}$$

The integral here is the *exchange integral*

$$\int \psi_A^* |\mathbf{r}-\mathbf{R}_B|^{-1} \psi_B \, d^3\mathbf{r} = (1+R)e^{-R}. \qquad (11.18)$$

Plotting the sum of all these terms we have two curves of $\varepsilon_{\pm}(R)$, one for even, one for odd parity (see Fig. 11.3). We seek a minimum of these curves with respect to R. The odd parity state has no minimum, and therefore, there is no bound molecular state with odd parity. However, an even parity state does exist at internuclear separation 1.3 Å and at a relative binding of -1.76 eV. Experimentally, it is found that $R_0 = 1.03$ Å and $\Delta E = -2.8$ eV, which is some indication of the crudeness of our approximations.

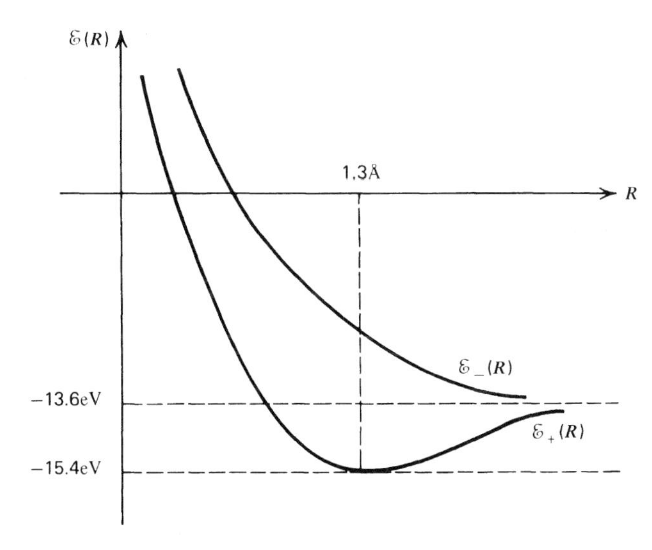

***Figure 11.3** Energy of an H_2^+ ion as a function of the proton separation R. ε_+ and ε_- denote the even and odd parity solutions.*

A single electron wave function is called a molecular *orbital*. In particular, a molecular orbital such as $\psi_{\pm}$, chosen to be a *l*inear *c*ombination of *a*tomic *o*rbitals, is called *LCAO*. Orbitals such as ψ_+ are called *bonding orbitals*, and orbitals such as ψ_- are called *antibonding orbitals*.

In this case we can understand the reason why ψ_+ is a bonding orbital and ψ_- is not. Since ψ_- has odd parity, it vanishes at the midpoint of the molecule; but even stronger, it vanishes everywhere on the midplane because of rotational symmetry around the internuclear line. Thus the electron has a low probability of being between the two nuclei where it can perform a bonding function. On the other hand, ψ_+ is larger along the midplane, which leads to a higher concentration of the electron there, and this in turn produces the bonding.

When R is large, we simply have a wave function that is a superposition of two separated hydrogen atoms in $1s$ states. Since the wave functions do not overlap, this is equivalent to saying that the electron can be bound to either proton with equal probability. The energy of this state is correctly given by the above wave functions, since we have constructed it out of exact $1s$ functions, having the exponential form $e^{-|\mathbf{r}-\mathbf{R}|}$.

In the opposite limit, when $R \to 0$, the two protons come together, and we have a He^+ atom. The electronic energy of this case is not well approximated by our wave function, because the exponential for He^+ should be $e^{-2|\mathbf{r}-\mathbf{R}|}$. For this reason the binding of the electrons to the protons is underestimated by our wave functions near $R=0$. This explains why we obtained a binding energy significantly less than the experimental value. Some account of this can be made by taking modified atomic orbitals that have an arbitrary scaling factor η, $\psi_A(r) \to \psi_A(\eta r)$ and by using η as a variational parameter. In our case this makes the wave functions correct at $R=0$ and improves the binding energy estimate.

The H_2 Molecule

The next simplest molecule is the neutral hydrogen molecule H_2. Because of the two electrons, we must take account of the Pauli principle. As a first approximation let us take two molecular orbitals for the H_2^+ molecule and form a wave function from these. Since we are concerned with finding the ground state, we expect that we want two *binding* orbitals of the type ψ_+. The space part of the wave function will then be symmetric; thus we must choose an antisymmetric spin part, that is, the singlet spin state. Thus choose

$$\psi_s(1,2) = \frac{1}{2[1+S(r)]}[\psi_A(\mathbf{r}_1) + \psi_B(\mathbf{r}_1)][\psi_A(r_2) + \psi_B(\mathbf{r}_2)]\chi_s.$$

There is a difficulty with this wave function that can be seen when we expand out the space parts:

$$\psi_s \propto [\psi_A(\mathbf{r}_1)\psi_A(\mathbf{r}_2)+\psi_B(\mathbf{r}_1)\psi_B(\mathbf{r}_2)]+[\psi_A(\mathbf{r}_1)\psi_B(\mathbf{r}_2)+\psi_A(\mathbf{r}_2)\psi_B(\mathbf{r}_1)].$$

Let us examine the meaning of these terms as $R \to \infty$. The first set of terms correspond to a proton plus a H^- ion, while the second set corresponds to two separated H atoms. We know that the binding of the H^- ion is very weak, so that we expect that it is the second set of terms that will lead to strong binding in H_2, and not the first set. Since we are doing a variational calculation of sorts, we are at liberty to use any information we have to bring to bear on the selection of trial functions. Thus we simply eliminate the first set of terms; this gives the *valence bond* or *London–Heiter* method (as opposed to the molecular orbital method):

$$\psi_s(1,2)=\frac{1}{\sqrt{2(1+S^2)}}[\psi_A(\mathbf{r}_1)\psi_B(\mathbf{r}_2)+\psi_A(\mathbf{r}_2)\psi_B(\mathbf{r}_1)]\chi_s. \qquad (11.19)$$

Note that the normalization is now dependent on the square of the overlap integral S. Note also that this state has even parity. A similar result

$$\psi_t(1,2)=\frac{1}{\sqrt{2(1-S^2)}}[\psi_A(\mathbf{r}_1)\psi_B(\mathbf{r}_2)-\psi_A(\mathbf{r}_2)\psi_B(\mathbf{r}_1)]\chi_t \qquad (11.20)$$

holds for the triplet states. This state has odd parity.

With these trial functions the internuclear potentials can be computed as before. The details are complicated, however, and are omitted. The results are quite similar in form to Fig. 11.3. The curve $\varepsilon_+(R)$ has a minimum at a value less than -27.2 eV, which is the value for two separated H atoms. Thus a H_2 molecule can exist in the singlet spin state.

Similar problems to those in the H_2^+ molecule occur here when we go to the limit $R \to 0$. The electronic states should approach the ground state of the He atom, but because our wave functions have been defined in terms of H-like functions, this limit is rather badly approximated. Similar rescaling can be used to improve the results. Extensive variational calculations have been done on H_2, and the results compare extremely well to experiment.

One seeming contradiction implied by the above results is that for atoms we argued that electrons with aligned spins (large total spin) led to the lowest Coulomb energies and thus to the tightest binding. Now we find that it is the low spin (singlet) state that binds, while the triplet state does not. This paradox is explained by the fact that for molecules it is the electron density between the nuclei that leads to binding and this effect outweighs the lower interelectron Coulomb energy in the high spin states.

Another point of interest involves the large R behavior of the internuclear potential. In a second-order perturbation expansion it is found that two H-atoms will attract each other with a R^{-6} *Van der Waals* potential. Thus the triplet curve eventually becomes attractive at large R. However, the depth of the resulting potential is insufficient to lead to binding in H_2, although it can lead to binding in other molecules.

11.3 PURE ROTATION SPECTRA

Energy Levels

In the ground state a diatomic molecule is very near to the bottom of the potential between two nuclei. (Because of zero point motions we cannot say that they are precisely at the bottom.) In this state the easiest way to excite it into higher energy states is to cause the molecule to rotate. This follows from the discussion of §11.1, where it was shown that the energy required to excite a vibrational mode or an electronic state was much greater than typical rotation energies. Therefore, it is possible to have transitions solely among the rotational states when the molecule is in its lowest vibrational and electronic states. Such transitions give rise to a *pure rotational spectrum*, which typically lies in the radio or far-IR regimes.

Since the moment of inertia of a diatomic molecule around the line connecting the nuclei is negligible, the appropriate axis of rotation to consider is perpendicular to this line, through the center of mass of the two nuclei. The moment of inertia about this axis is

$$I = \mu r_0^2, \tag{11.21}$$

where r_0 is the equilibrium internuclear distance, and μ is the reduced mass, defined below. If we denote by $\mathbf{K}$ the angular momentum operator for rotation, then the Hamiltonian is $H=(1/2I)\mathbf{K}^2$, which leads to the energy eigenvalues

$$E_K = \frac{\hbar^2}{2I} K(K+1). \tag{11.22}$$

Corrections to this essentially classical formula can be found by considering the radial wave equation for the nuclei of a diatomic molecule,

$$\frac{1}{r^2}\frac{d}{dr}\left(r^2\frac{d\psi}{dr}\right) + \frac{2\mu}{\hbar^2}\left[E - V_n(r) - \frac{\hbar^2 K(K+1)}{2\mu r^2}\right]\psi = 0, \tag{11.23}$$

where μ is the reduced mass of the diatomic molecule

$$\mu \equiv \frac{M_1 M_2}{M_1 + M_2}. \tag{11.24}$$

Here M_1 and M_2 are the masses of the two nuclei and r is their separation; $V_n(r)$ is the potential of the nuclei, in electronic state n; and K is the angular momentum quantum number of the molecule.

Vibrational and rotational energy levels of the molecule may be approximately calculated by expanding the "effective potential"

$$U_n \equiv V_n + \frac{\hbar^2 K(K+1)}{2\mu r^2} \equiv V_n + V_K \tag{11.25}$$

about its minimum in a Taylor series. Letting r_0 and r_K be the equilibrium radii of V_n and U_n respectively, that is, $\partial V/\partial r|_{r_0}=0$ and $\partial U_n/\partial r|_{r_K}=0$ and letting the " spring constant" k_{n0}, A_n and V_{n0} be defined by

$$V(r) \sim V_{n0} + \tfrac{1}{2} k_{n0}(r-r_0)^2 + A_n(r-r_0)^3 + \cdots, \tag{11.26}$$

one obtains for r_K

$$r_K = r_0 + \frac{\hbar^2 K(K+1)}{k_n \mu r_0^3} + O\left(\frac{V_k}{V_n}\right)^2. \tag{11.27}$$

An approximate expression for $U_n(r)$ is then

$$U_n(r) = V_{n0} + \frac{\hbar^2 K(K+1)}{2\mu r_K^2} + \frac{1}{2}\left[k_{n0} + \frac{3\hbar^2 K(K+1)}{\mu r_K^4}(1 + 2A_n r_K/k_n)\right](r-r_K)^2 + \cdots \tag{11.28}$$

Note that Eq. (11.22), derived classically, has the same form as the first two terms of $U_n(r)$, which define *rotational energy levels* E_{nk} satisfying [cf. Eq. (11.28)]

$$E_{nk} = V_{n0} + \frac{\hbar^2 K(K+1)}{2\mu r_k^2} \approx V_{n0} + \frac{\hbar^2 K(K+1)}{2\mu r_0^2}\left[1 - \frac{2\hbar^2 K(K+1)}{k_n \mu r_0^4}\right]. \tag{11.29}$$

The second term in brackets [cf. Eqs. (11.28) and (11.29)] corresponds to a stretching of the molecule in response to centrifugal forces, which increases the moment of inertia and therefore decreases the kinetic energy of rotation for fixed angular momentum.

Selection Rules and Emission Frequencies

Whether a transition between two K values can be accompanied by the emission or absorption of radiation is governed by *selection rules.* For dipole radiation there are two such rules:

1. $d \neq 0$ (11.30a)
2. $\Delta K = -1$ (emission) or $\Delta K = +1$ (absorption). (11.30b)

Here d is the *permanent dipole moment* of the molecule:

$$d \equiv Z_1 e r_1 + Z_2 e r_2 + d_e, \tag{11.31}$$

where d_e is the electronic contribution.

These selection rules can be understood physically. A rotating system will radiate classically only if its dipole moment changes. Clearly, if $d=0$, it cannot radiate classically, which explains the first rule. The second

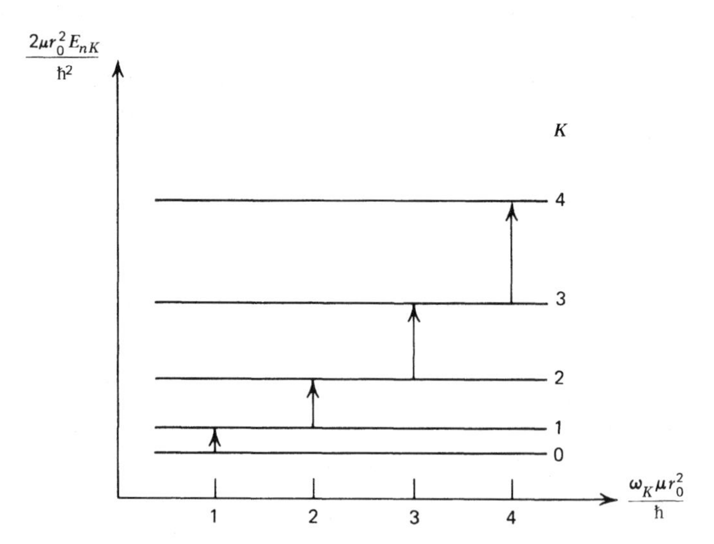

***Figure 11.4** Term diagram for energy levels and frequencies in pure rotational transitions.*

follows from angular momentum considerations, essentially identical to those leading to the selection rule of Eq. (10.40). See also Problem 10.6.

An immediate consequence of rule 1 is that a *homonuclear diatomic molecule cannot show pure rotation spectrum in the dipole approximation* [cf. Eq. (11.31)]. This rules out molecules such as H_2, O_2, C_2, although weak pure rotation spectra due to higher order radiation have been observed.

Rule 2 allows us to immediately write down the emission frequencies for rotational transitions [cf. Eqs. (11.29)]:

$$\omega_K \equiv \frac{E_{nK+1} - E_{nK}}{\hbar} = \frac{\hbar(K+1)}{\mu r_0^2}\left[1 - \frac{4\hbar^2(K+1)^2}{k_n \mu r_0^4}\right]. \tag{11.32}$$

This can also be depicted in a term diagram in Fig. 11.4. The frequencies are almost equidistant, but get slightly closer together with high K.

11.4 ROTATION-VIBRATION SPECTRA

Energy Levels and the Morse Potential

Because the energies required to excite vibrational modes are much larger than those required to excite rotation, it is unlikely to have a pure vibrational spectrum in analogy to the pure rotational spectrum. There is, instead, what is called a *rotation-vibration spectrum*, in which both the vibrational state and the rotational state can change together. We can, however, consider cases in which the electronic state remains the same.

The third term in $U_n(r)$ of Eq. (11.28) is the potential of a harmonic oscillator, leading to *vibrational energy* levels

$$E_{nv} = \hbar\omega_{nK}\left(v + \tfrac{1}{2}\right)$$

$$\omega_{nK} \approx \mu^{-\frac{1}{2}}\left[k_{n0} + \frac{3\hbar^2 K(K+1)}{\mu r_0^4}(1 + 2A_n r_K / k_n)\right]^{1/2}. \tag{11.33}$$

Here v is the harmonic oscillator quantum number, $v = 0, 1, 2, \ldots$

The above vibrational energy levels are those of a harmonic oscillator and result from the approximate expansion of the potential up to quadratic displacements from equilibrium. A more exact treatment clearly must include cubic, quartic, and higher order terms in the potential. Alternatively, the potential $U_n(r)$ may be approximated by a closed analytic expression which is both accurate and simple. An expression of this form has been proposed by Morse (1929):

$$U_n(r) = U_{n0} + B_n\{1 - \exp[-\beta_n(r - r_0)]\}^2, \tag{11.34}$$

where B_n, β_n, and r_0 are three parameters that must be properly chosen to fit the observed potential curve. The energy eigenvalues (relative to the potential minimum U_{n0}) corresponding to this potential may be solved for exactly and are

$$E_{nv} = \hbar\omega_{n0}\left(v + \tfrac{1}{2}\right) - \frac{\hbar^2\omega_0^2}{4B_n}\left(v + \tfrac{1}{2}\right)^2, \tag{11.35a}$$

where

$$\omega_{n0} = \beta_n\left(\frac{2B_n}{\mu}\right)^{1/2}. \tag{11.35b}$$

Note that the first term in E_{nv} corresponds to a simple harmonic oscillator, coming from the first nonconstant term in an expansion of the Morse potential about its minimum. The vibrational quantum number v is an integer lying in the range

$$0 \leqslant v \leqslant \frac{(2\mu B_n)^{1/2}}{\beta_n\hbar} - \tfrac{1}{2}. \tag{11.36}$$

The upper limit corresponds to the condition $\partial E/\partial v = 0$. Two properties of vibrational levels correctly predicted by Eqs. (11.35) and (11.36) of the Morse potential are that there are a finite number of discrete vibrational levels below B_n and that the energy levels are more closely spaced with increasing v.

Selection Rules and Emission Frequencies

The selection rules for vibration-rotation transitions are:

1. $d \neq 0$ (11.37a)
2. $\left.\dfrac{d(d)}{dr}\right|_{r=r_0} \neq 0$ (11.37b)
3. $v = -1$ (emission) or
   $v = +1$ (absorption) (11.37c)
4. $K = \pm 1$ for $\Lambda = 0$
   $K = \pm 1, 0$ for $\Lambda \neq 0$. (11.37d)

Here Λ is the component of electronic orbital angular momentum along the internuclear axis (figure axis). The electronic states $\Lambda=0,1,2,3,\ldots$ are denoted by Σ, Π, Δ, $\Phi,\ldots$, respectively, in analogy with atomic spectroscopic notation.

The second of these rules requires that d change during a change in vibrational state. The third is the familiar rule from quantum theory for harmonic oscillators. The fourth rule is more complicated. Changes in the vibrational state of the molecule do not affect its parity, which must change in a dipole transition (§10.4). For $\Lambda=0$, the parity is determined completely by the rotational quantum number K, and we obtain the same selection rule as in the pure rotational transitions, Eq. (11.30b). For $\Lambda\neq0$,

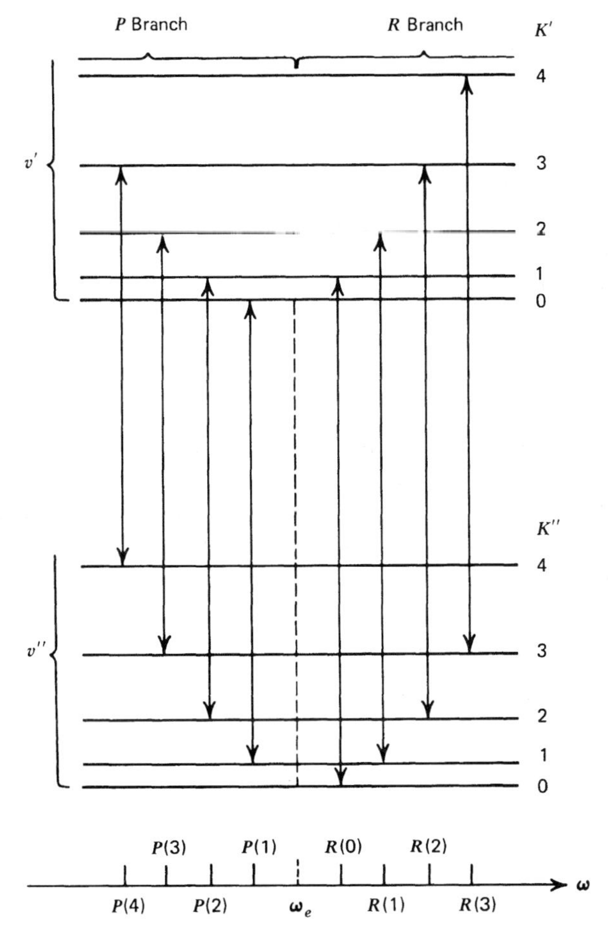

***Figure 11.5** Term diagram for P and R branches in vibrational transitions. Here v and K are vibrational and rotational quantum numbers.*

however, each rotational level splits into two almost degenerate levels, corresponding to the two possible signs of Λ. This is called Λ *doubling*. These two levels have opposite parity, thus allowing an overall change of parity even when $\Delta K=0$.

Note that in either emission or absorption, both $\Delta K=+1$ and $\Delta K=-1$ are allowed, because the majority of the total energy change is in the vibrational transition. This allows a classification of the rotational "fine-structure" according to the change in K as follows

$$\Delta K=-1: R \text{ branch} \tag{11.38a}$$

$$\Delta K=+1: P \text{ branch} \tag{11.38b}$$

$$\Delta K=0: Q \text{ branch (when allowed).} \tag{11.38c}$$

Here $\Delta K=K''-K'$, where K' refers to the upper state, and K'' to the lower state. The P and R branches are illustrated in Fig. 11.5.

11.5 ELECTRONIC-ROTATIONAL-VIBRATIONAL SPECTRA

Energy Levels

An approximate expression for the energy levels of electronic states is

$$E_{nvJ}=V_{n0}+\gamma_n\hbar^2\Lambda^2+\alpha_n\hbar J(J+1)+\left(v+\tfrac{1}{2}\right)\hbar\omega_n, \tag{11.39}$$

where Λ is the component of electron angular momentum $\mathbf{L}$ along the axis separating the two nuclei, $\mathbf{J}=\mathbf{K}+\mathbf{L}$ is the total angular momentum, and V_{n0}, γ_n, α_n, and ω_n are all constant for a given electronic state of quantum number n. The rotational and vibrational energies in Eq. (11.39) are similar to the forms discussed previously and are quite adequate approximations when electronic transitions occur (change in n).

Selection Rules and Emission Frequencies

The selection rules governing *electronic dipole transitions* in a diatomic molecule are:

1. $\Delta\Lambda=-1, 0, +1$ (11.40a)
2. $\Delta J=-1, 0, +1$, but $J=0\rightarrow J=0$ is not allowed and $\Delta J=0$ is not allowed if $\Lambda=0\rightarrow\Lambda=0$. (11.40b)
3. $\Delta v=$any positive or negative integer. (11.40c)

Since the dipole transition is an electronic one, there is no restriction on v.

Again, as in vibrational spectra, we can consider emission frequencies for transitions in which $\Delta J = -1$, the R branch, $\Delta J = 0$, the Q branch, and $\Delta J = +1$, the P branch:

$$\omega_{nvJ} = \frac{E_{nvJ} - E_{n'v'(J+\Delta J)}}{\hbar}$$

$$\equiv \omega_{nn'} + v\omega_n - v'\omega_{n'} + H(J), \tag{11.41}$$

$$H(J) = \begin{cases} (J+1)[J\alpha_n - (J+2)\alpha_{n'}], & P \quad (11.42a) \\ (\alpha_n - \alpha_{n'})J(J+1), & Q \quad (11.42b) \\ J[(J+1)\alpha_n - (J-1)\alpha_{n'}], & R. \quad (11.42c) \end{cases}$$

The dominant term in Eq. (11.41) is $\omega_{nn'}$, a frequency corresponding to the difference in a potential energy of the minima of two curves of the form of Fig. 11.6. The vibrational and rotational terms are successively finer structures on the electronic levels. For given n and n' (and hence, given $\omega_{nn'}$, ω_n and $\omega_{n'}$) Eq. (11.41) indicates that the vibrational fine structure forms a *progression* of uniformly spaced frequencies. For a given n, n', v, and v', the rotational fine structure is superimposed on the vibrational structure to form a *band.*

Several interesting features are apparent from Eq. (11.42) and the selection rules on ΔJ. Selection rule 2. and the requirement that J be positive forbid $J = 0$ in both the Q and R branches, respectively. Furthermore, $\alpha_n / \alpha_{n'}$ is typically not an integer, so that the bracketed expression in

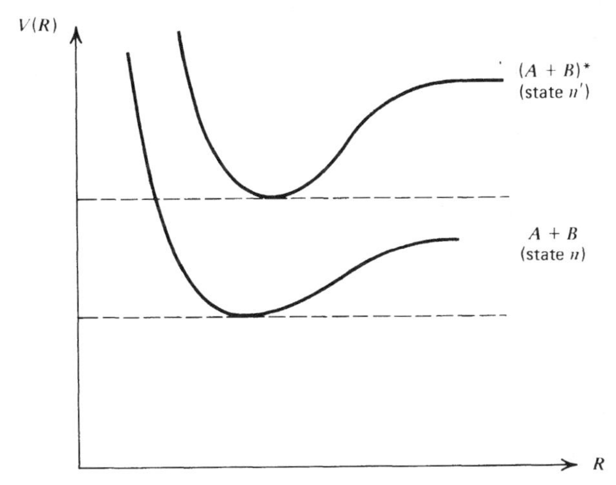

Figure 11.6 Potential energy as a function of nuclear separation of a molecule in its electronic ground state and in an excited electronic state.

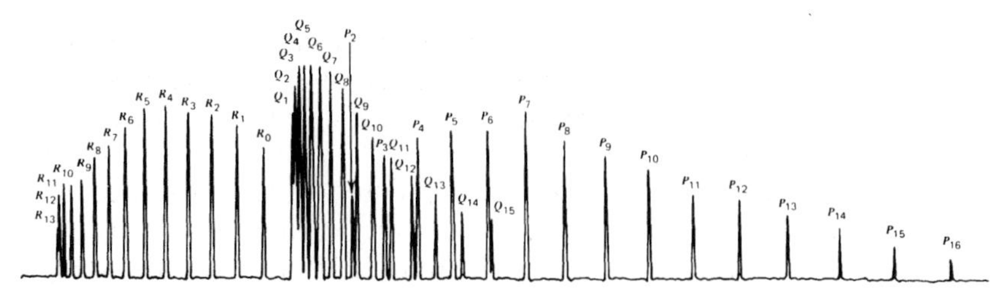

***Figure 11.7** Spectral lines observed in the molecule AlH, illustrating the P, R and Q branches. (Taken from Bingel, W. A. 1969, Theory of Molecular Spectra, Wiley, New York.)*

the P and R branch does not vanish for any value of J. Consequently, $H(J)\neq 0$ for all branches, and there is a missing line in the sequence at frequency $\omega_0 \equiv \omega_{nn'} + v\omega_n - v'\omega_{n'}$ termed, alternatively, the zero gap, null line, or band origin, as in the rotational-vibrational spectra. This is shown as the dotted line in Fig. 11.5. Since the Q and R branches converge on the null line as $J \rightarrow 0$ (the P branch converges on it as J is artificially extrapolated to -1), the null line may be used to identify the origin of J within a band.

Another striking feature of Eq. (11.42) is that the line spacing is quite nonuniform in J, in contrast to rotation-vibration levels. In the P and R

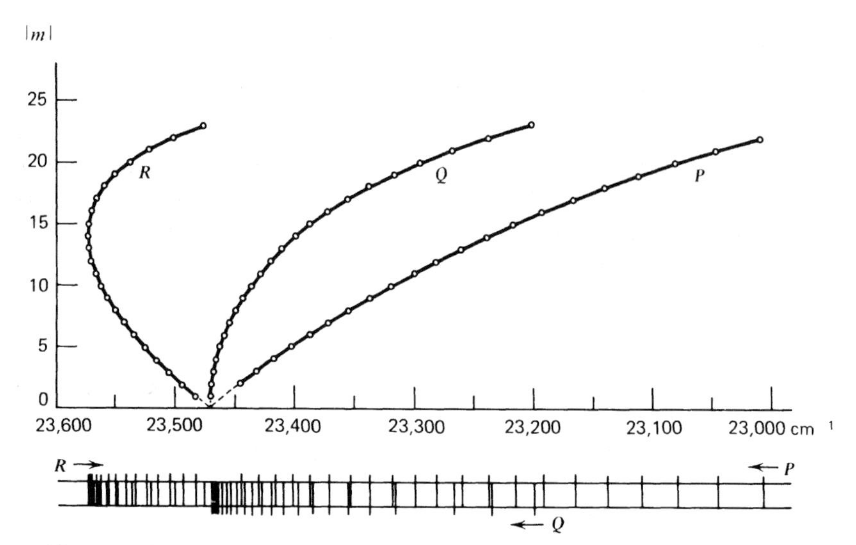

***Figure 11.8** Fortrat diagram for bands shown in Fig. 11.7. Here |m| = J, the total angular momentum quantum number. (Taken from Herzberg, G., 1950, Spectra of Diatomic Molecules, Van Nostrand, New York.)*

branches, the line spacing may not be montonic with J and can reverse at a particular value of $J \equiv J_{\text{head}}$, where $H(J)$ is at an extremum. Near J_{head}, the line spacing is relatively narrow; for J much bigger or smaller than J_{head}, the spacing gets broader. The result is a sharp edge at the boundary of the band, called the *band head.* It is straightforward to determine the location of the band head, and this is developed in Problems 11.3 and 11.4. Figure 11.7 below gives an observed band at 4241 Å in the spectrum of AlH. The lines are labeled according to R, Q and P, with subscripts indicating the J of the branch. Lines corresponding to R_{14}, R_{15}, and so on are not shown.

A theoretical plot of the location of the lines in the J-ω_{nvJ} plane is called a *Fortrat diagram.* Figure 11.8 gives a Fortrat diagram of the same band shown in Fig. 11.7, where $|m| \equiv J$. Note that the R branch reverses at R_{13}, as is observed in the R band head of Fig. 11.7. Reversals do not occur in the Q and P branches for this spectrum, but the lowest frequency lines in the Q branch are similar in form to the band head in the R branch.

PROBLEMS

11.1—Consider an electrically neutral medium of diatomic molecules in thermal equilibrium at temperature T. Each molecule contains a nucleus of mass M_p and a nucleus of mass $2M_p$ at an equilibrium separation r_0.

a. Estimate r_0 in terms of fundamental constants.

b. Estimate the cross section σ_c for collisions between molecules.

c. It is experimentally observed that, as a function of mass density ρ of the medium, the line width of the rotational lines has the form shown in Fig. 11.9. If only Doppler and collisional broadening are present, estimate ρ_0 and show that it may be written completely in terms of fundamental constants, independent of M_p.

11.2—Derive Eq. (11.12).

11.3—Show that both the P and the R branches of the electronic-vibrational-rotational transitions, Eqs. (11.41), (11.42) may be combined into a single formula for the emission frequency of the form

$$\begin{aligned}\omega_{nvJ} &= \omega_{nn'} + v\omega_n - v'\omega_{n'} \\ &\quad + j(\alpha_n + \alpha_{n'}) + j^2(\alpha_n - \alpha_{n'}),\end{aligned}$$

where j ranges over both positive and negative integer values.

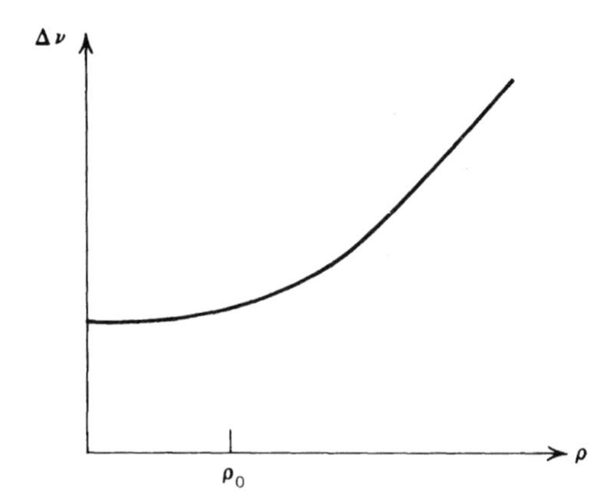

***Figure 11.9** Line width as a function of density for emission from a medium of diatomic molecules.*

11.4—Derive an expression for the J value and frequency of the band head in electronic-vibrational-rotational transitions, give the criteria for whether the band head occurs in the R or P branch, and give the criteria for whether the frequency of the band head lies above or below the band origin.

11.5—Show that the Q branch in electronic-vibrational-rotational transitions does not have a true band head but may have the observed appearance of one under certain conditions.

11.6—For the situation described in Problem 11.1, estimate, in terms of fundamental constants, the range in T over which purely rotational emission lines will be observed from a substantial fraction of the molecules.

REFERENCES

Baym, G. 1969, *Lectures on Quantum Mechanics*, (Benjamin, New York).
Bingel, W. A. 1969, *Theory of Molecular Spectra*, (Wiley, New York).
Leighton, R. B. 1959, *Principles of Modern Physics*, (McGraw-Hill, New York).
Herzberg, G. 1950, *Molecular Spectra and Molecular Structure*, *Vol.* 1, (Van Nostrand, New York).
Morse, P. M. 1929, *Phys. Rev.*, **34**, 57.

SOLUTIONS

1.1—The flux at the film plane is

$$F_\nu = \int I_\nu \cos\theta \, d\Omega \approx I_\nu \cos\theta \, \Delta\Omega,$$

since $\Delta\Omega \ll 1$. Now we have

$$\Delta\Omega = \frac{\Delta A \cos\theta}{r^2}$$

where $\Delta A = \pi(d/2)^2$ and $r = L/\cos\theta$. Thus

$$F_\nu = \frac{\pi \cos^4\theta}{4f^2} I_\nu,$$

where $f \equiv L/d$.

1.2—We first find the energy absorbed in volume dV and time dt due to radiation in solid angle $d\Omega$ and frequency range $d\nu$. Let dA be the cross-sectional area of the volume normal to the radiation. Then the energy absorbed is $-dI_\nu\, dA\, dt\, d\nu\, d\Omega$, where $dI_\nu = -\alpha_\nu I_\nu\, dl$ by the absorption law (1.20) and where dl is the thickness of the volume along the direction of the radiation. Since $dV = dA\, dl$, the energy absorbed is $\alpha_\nu I_\nu\, dV\, dt\, d\Omega\, d\nu$. Integrating over solid angle gives the energy absorbed per unit volume per unit time per unit frequency range to be $4\pi\alpha_\nu J_\nu$, which is also equal to $c\alpha_\nu u_\nu$ by (1.7). Noting that each photoionization requires an amount $h\nu$ of energy and that $\alpha_\nu = n_a \sigma_\nu$ we obtain the number of photoionizations per unit volume per unit time:

$$4\pi n_a \int_{\nu_0}^{\infty} \frac{\sigma_\nu J_\nu}{h\nu}\, d\nu = c n_a \int_{\nu_0}^{\infty} \frac{\sigma_\nu u_\nu\, d\nu}{h\nu}.$$

1.3

a. The transfer equation with no absorption is

$$\frac{dI}{ds} = j = \frac{\Gamma}{4\pi}.$$

Here I is defined in terms of photon numbers rather than energy (photons–cm^{-2}–s^{-1}–ster^{-1}). Integrating along a line through the center gives

$$I_0 = j \cdot 2R = \frac{R\Gamma}{2\pi}.$$

b. The observed average intensity $\bar{I}$ is equal to the total flux divided by the solid angle accepted by the detector, $\Delta\Omega_{\text{Det}} = \pi(\Delta\theta_{\text{Det}})^2$, and $\Delta\theta_{\text{Det}}$ is the detector half angle. Since the total luminosity L of the source (photons–s^{-1}) is simply equal to $(4\pi/3)R^3\Gamma$, the flux is

$$F = \frac{L}{4\pi d^2} = \frac{R^3\Gamma}{3d^2},$$

where d is the distance to the source. Thus

$$\bar{I} = \frac{R^3\Gamma}{3\pi d^2 (\Delta\theta_{\text{Det}})^2}.$$

Noting that $\Delta\theta_s = R/d$ is the angular size of the source, we can write

$$\frac{\bar{I}}{I_0} = \frac{2}{3}\left(\frac{\Delta\theta_s}{\Delta\theta_{\text{Det}}}\right)^2.$$

For a completely unresolved source $\Delta\theta_s \ll \Delta\theta_{\text{Det}}$, so that $\bar{I} \ll I_0$.

1.4

a. Assume that the luminous object has spherical symmetry, so that the flux F at distance r is just $L/(4\pi r^2)$. From Eq. (1.34) the outward radiation force per unit mass on the cloud is

$$f_{\text{rad}} = \frac{\kappa F}{c} = \frac{\kappa L}{4\pi c r^2}.$$

The inward gravitational force per unit mass due to the object is

$$f_{\text{grav}} = \frac{GM}{r^2}.$$

The condition of ejection is that $f_{\text{grav}} < f_{\text{rad}}$, which can be written

$$\frac{M}{L} < \frac{\kappa}{4\pi G c}.$$

b. The cloud experiences an inward force per unit mass $G_{\text{eff}} M/r^2$, where the "effective" gravitational constant is given by

$$G_{\text{eff}} = G - \frac{\kappa L}{4\pi M c}.$$

Thus the effective potential per unit mass is $-G_{\text{eff}} M/r$. Note that G_{eff} is negative under conditions for ejection. Setting up the conservation of energy connecting the state at $r = R$ and $r = \infty$, we obtain

$$-\frac{G_{\text{eff}} M}{R} = \tfrac{1}{2} v^2,$$

which can be written

$$v^2 = \frac{2}{R}\left(\frac{\kappa L}{4\pi c} - GM\right).$$

c. The minimum luminosity occurs when the inequality in part (a) becomes an equality. Substitution of $\kappa = \sigma_T / m_H$ then gives the stated result.

1.5

a. The brightness is $I_\nu = F_\nu / \Delta\Omega$, where $\Delta\Omega = \pi(\Delta\theta)^2$. Here $\Delta\theta = \theta/2 = 2.15$ arc min $= 6.25 \times 10^{-4}$ radian. Thus

$$I_\nu = 1.3 \times 10^{-13} \text{ erg cm}^{-2}\text{s}^{-1}\text{Hz}^{-1}\text{ster}^{-1}$$

$$T_b = \frac{c^2}{2\nu^2 k} I_\nu = 4.2 \times 10^7\ K.$$

Since $h\nu \ll kT_b$, the use of the Rayleigh–Jeans approximation is appropriate.

b. $T_b \propto I_\nu \propto (\Delta\theta)^{-2}$. If the true $\Delta\theta$ is smaller, the true T_b will be larger than stated above.

c. From Eq. (1.56b) we find $\nu_{\max} = 2.5 \times 10^{18}$ Hz.

d. The best that can be said is $T \geqslant T_b$. This follows from Eq. (1.62) with $T_b(0) = 0$. In general, the maximum emission of any thermal emitter at given temperature T will occur when the source is optically thick (see Problem 1.8 *d*).

1.6—Since $u(T) = aT^4$, Eqs. (1.40) can be written

$$\left(\frac{\partial S}{\partial T}\right)_V = 4aVT^2, \qquad \left(\frac{\partial S}{\partial V}\right)_T = \frac{4}{3}aT^3.$$

It follows that $S = (4/3)\, aVT^3 + \text{constant}$. The constant must be chosen to be zero, so that $S \to 0$ as $T \to 0$ (third law of thermodynamics).

1.7

a. The equation of statistical equilibrium [Eq. (1.69)] with $B_{21} = 0$ becomes

$$n_1 B_{12} \bar{J} = n_2 A_{21}.$$

With the Boltzmann law (1.70) this implies

$$\bar{J} = \frac{A_{21}}{B_{12}} \frac{g_2}{g_1} e^{-h\nu_0/kT}.$$

This cannot equal the Planck function with A_{21}/B_{12} independent of temperature. However, the choice $(A_{21}/B_{12})(g_2/g_1) = 2h\nu_0^3/c^2$ does yield $\bar{J} = B_{\nu_0}$ in the Wien limit, $h\nu_0 \gg kT$, [cf. Eq. (1.54)].

b. The main difference between the interactions of neutrinos and photons with the atom is that the former particles are fermions, whereas the latter are bosons. Stimulated emission in a fermion field would place two particles in the same state and thus violate the Pauli exclusion principle. This process is replaced by *inhibited emission*, in which an atom in the excited state is prevented from emitting a fermion when one is already present. The analysis using the Einstein coefficient is identical to the photon case, except $B_{21} \to -B_{21}$. One then has, at equilibrium, for atoms of temperature T,

$$n_1 B_{12} \bar{J} = n_2 A_{21} - n_2 B_{21} \bar{J}$$

$$\bar{J} = \frac{A_{21}/B_{21}}{(n_1/n_2)(B_{12}/B_{21}) + 1}$$

$$\bar{J} = \frac{A_{21}/B_{21}}{(g_1 B_{12}/g_2 B_{21}) e^{h\nu/kT} + 1} = \frac{2h\nu^3}{c^2(e^{h\nu/kT} + 1)},$$

and one obtains the same relationship as before for the Einstein coefficients. These coefficients are properties of the atom alone and clearly must be the same, regardless of the external interactions used to derive them.

1.8

a. Note that $j_\nu = P_\nu / 4\pi$ and that, effectively, $\alpha_\nu = 0$, since the cloud is optically thin. Then, using Eq. (1.24),

$$I_\nu(b) = \int j_\nu(z)\,dz = \frac{P_\nu}{2\pi}\sqrt{R^2 - b^2}\ .$$

b. The total power emitted by the cloud is $L = (4/3)\pi R^3 P$, where $P = \int P_\nu\, d\nu$. Then

$$L = 4\pi R^2 \sigma T_{\text{eff}}^4,$$

by definition of T_{eff}, so that

$$T_{\text{eff}} = \left(\frac{PR}{3\sigma}\right)^{1/4}.$$

c. Let d be the distance from the spherical cloud to the earth. Energy conservation gives a relation between F_ν, the flux at the earth, and P_ν:

$$4\pi d^2 F_\nu = \frac{4}{3}\pi R^3 P_\nu,$$

$$F_\nu = P_\nu \frac{R^3}{3d^2}.$$

d. From Eq. (1.30), with $S_\nu = B_\nu(T)$, $I_\nu(0)=0$, $\tau_\nu \ll 1$,

$$I_\nu = B_\nu(T)(1-e^{-\tau_\nu}) \approx \tau_\nu B_\nu(T) \ll B_\nu(T).$$

With the definition of T_b from Eq. (1.59),

$$B_\nu(T_b) \ll B_\nu(T),$$

and the monotoncity of $B_\nu(T)$ with T, we have $T_b \ll T$.

e. For the optically thick case the results are:

a′. From Eq. (1.30) with $\tau_\nu \gg 1$ and with $S_\nu = B_\nu(T)$ we have $I_\nu = B_\nu(T)$ independent of b.

b′. Since $I_\nu = B_\nu$, the flux at the surface is the blackbody flux, so $T_{\text{eff}} = T$.

c′. The monochromatic flux at the surface is $\pi B_\nu(T)$ [cf. Eq. (1.14)], so using the inverse square law gives

$$F_\nu(d) = \pi\left(\frac{R}{d}\right)^2 B_\nu(T).$$

d′. From (a′) and Eq. (1.59) we have $B_\nu(T_b) = B_\nu(T)$, which implies $T_b = T$.

1.9—Ray A starts on the central object with intensity $B_\nu(T_c)$, and this is essentially the observed intensity at $\nu=\nu_1$, where the absorption in the shell is negligible. The observed intensity at $\nu=\nu_0$ depends on whether the source function in the shell, namely, $B_\nu(T_s)$, is greater or smaller than the incident intensity $B_\nu(T_c)$. (See Eq. (1.30) and subsequent discussion.) When $T_s < T_c$ we have $B_\nu(T_s) < B_\nu(T_c)$, and the intensity is reduced by passing through the shell, so that $I_{\nu_1}^A$ is larger than $I_{\nu_0}^A$. When $T_s > T_c$, $I_{\nu_0}^A$ will be larger than $I_{\nu_1}^A$.

Ray B starts with zero intensity, which is the observed intensity at $\nu=\nu_1: I_{\nu_1}^B = 0$. At $\nu=\nu_0$ the observed intensity will be somewhere between zero and the maximum $B_\nu(T_s)$, depending on the optical depth. In any case, $I_{\nu_0}^B > I_{\nu_1}^B$ always. These cases are illustrated in Fig. S.1.

1.10—The radiative diffusion equation is of the form [cf. Eq. (1.119b)]

$$\frac{1}{3}\frac{\partial^2 J_\nu}{\partial \tau^2} = \epsilon(J_\nu - B_\nu), \tag{1}$$

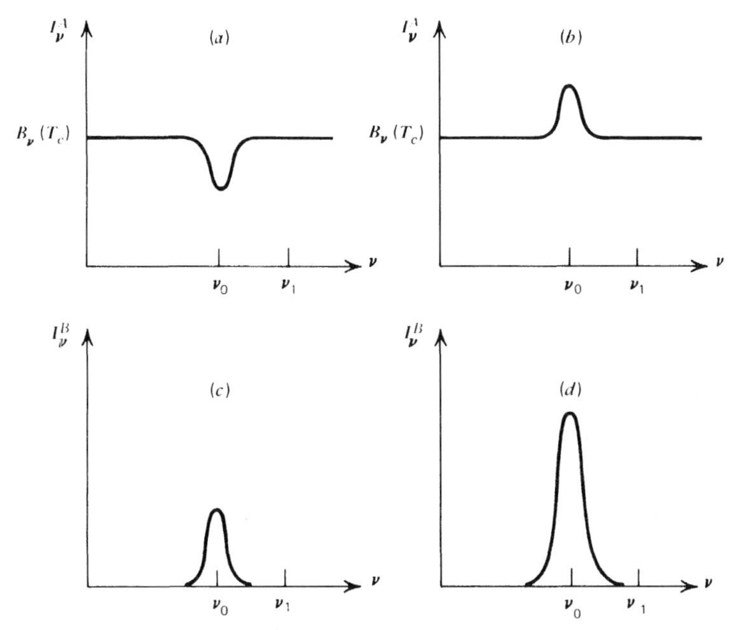

***Figure S.1** Intensity from a blackbody surrounded by a thermal absorbing shell (a) along ray A when $T_s < T_c$, (b) along ray A when $T_s > T_c$, (c) along ray B when $T_s < T_c$, (d) along ray B when $T_s > T_c$.*

where

$$\epsilon \equiv \frac{\alpha_\nu}{\sigma + \alpha_\nu} \tag{2}$$

is the probability per interaction that the photon will be absorbed. The general solution to Eq. (1) is

$$J_\nu - B_\nu = C_1 e^{\tau_*} + C_2 e^{-\tau_*}, \tag{3}$$

where

$$\tau_* \equiv \sqrt{3\epsilon}\ \tau = \sqrt{3\tau_\nu(\tau_\nu + \tau_s)}\ , \tag{4}$$

and C_1 and C_2 are independent of τ and to be determined by boundary conditions. The proper boundary conditions for a semi-infinite half-space are that J_ν remain finite as $\tau \rightarrow \infty$ and that there be no incident intensity at $\tau = 0$: $I_\nu(\tau = 0, \mu < 0) = 0$. The first boundary condition requires $C_1 = 0$, that is,

$$J_\nu(\tau) - B_\nu = C_2 e^{-\tau_*}. \tag{5}$$

Note that J_ν approaches B_ν after an effective optical depth τ_* [answer to part (b)].

Now, using the two-stream approximation to the boundary condition at $\tau=0$, Eq. (1.124a)

$$\frac{1}{\sqrt{3}}\frac{\partial J}{\partial \tau} = J \qquad \text{at } \tau = 0,$$

we obtain

$$C_2 = -B_\nu(1+\sqrt{\epsilon}\,)^{-1},$$

$$J_\nu(\tau) = B_\nu\left(1 - \frac{e^{-\tau_*}}{1+\sqrt{\epsilon}}\right). \tag{6}$$

In the Eddington approximation the flux, $F_\nu(\tau)$, satisfies [cf. Eqs. (1.113b) and (1.118)]

$$F_\nu(\tau) = 2\pi \int \mu I_\nu(\mu,\tau)\,d\mu$$

$$= 4\pi H = \frac{4\pi}{3}\frac{\partial J}{\partial \tau}.$$

Thus, we have the result for the emergent flux:

$$F_\nu(0) = \frac{4\pi}{\sqrt{3}}\frac{\sqrt{\epsilon}}{1+\sqrt{\epsilon}}B_\nu. \tag{7}$$

For small ϵ this differs from Eq. (1.103) only by a factor $1/\sqrt{3}$. (Note $F_\nu(0) = L_\nu/A$). Note, further, that from $J_\nu(\tau)$, one may compute the source function

$$S_\nu(\tau) = (1-\epsilon)J_\nu + \epsilon B_\nu$$

$$= B_\nu\left[1 - (1-\sqrt{\epsilon}\,)e^{-\tau_*}\right] \tag{8}$$

and thus the intensity at any optical depth. In the two-stream approximation the intensity in the outward direction ($\mu = 1/\sqrt{3}$) at $\tau = 0$ is

$$I_\nu^+(0) = \frac{2B_\nu\sqrt{\epsilon}}{1+\sqrt{\epsilon}} \tag{9}$$

2.1—Writing

$$A(t)=\tfrac{1}{2}(\mathcal{A}e^{-i\omega t}+\mathcal{A}^*e^{i\omega t}),$$

$$B(t)=\mathrm{Re}\,\mathcal{B}e^{-i\omega t}=\mathrm{Re}\,\mathcal{B}^*e^{i\omega t},$$

and noting the results $\langle 1\rangle=1$ and $\langle e^{\pm 2i\omega t}\rangle=0$, we obtain

$$\langle AB\rangle=\tfrac{1}{2}\mathrm{Re}\mathcal{A}^*\mathcal{B}=\tfrac{1}{2}\mathrm{Re}\mathcal{A}\mathcal{B}^*.$$

2.2

a. Maxwell's equations (2.6), with $\mathbf{j}=\sigma\mathbf{E}$, are

$$\nabla\cdot\mathbf{E}=4\pi\rho \qquad \nabla\cdot\mathbf{B}=0,$$

$$\nabla\times\mathbf{E}=-\frac{1}{c}\frac{\partial\mathbf{B}}{\partial t} \qquad \nabla\times\mathbf{B}=\frac{4\pi\sigma\mu}{c}\mathbf{E}+\frac{\epsilon\mu}{c}\frac{\partial\mathbf{E}}{\partial t}.$$

We seek plane-wave solutions, so we assume solutions of the form (2.18a, b) and $\rho=\rho_0\exp[i(\mathbf{k}\cdot\mathbf{r}-\omega t)]$. This gives

$$i\mathbf{k}\cdot\hat{\mathbf{a}}_1E_0=4\pi\rho_0 \qquad i\mathbf{k}\cdot\hat{\mathbf{a}}_2B_0=0$$

$$i\mathbf{k}\times\hat{\mathbf{a}}_1E_0=\frac{i\omega}{c}\hat{\mathbf{a}}_2B_0 \qquad i\mathbf{k}\times\hat{\mathbf{a}}_2=\frac{-i\omega m^2}{c}\hat{\mathbf{a}}_1E_0,$$

where $m^2=\mu\epsilon(1+4\pi\sigma i/\omega\epsilon)$. Dotting the vector $\mathbf{k}$ into the last equation, we find $\mathbf{k}\cdot\hat{\mathbf{a}}_1\equiv 0$, which implies $\rho_0=0$ from the first equation. Thus these equations have the same form as Eqs. (2.19), except for the additional m^2 factor. The solution proceeds analogously, leading to the dispersion relation

$$k^2=\frac{\omega^2m^2}{c^2}.$$

b. Take $\mathbf{k}$ along the z-axis, with $\mathbf{E}$ and $\mathbf{B}$ along the x- and y-axes, respectively. Then

$$\mathbf{E}=\hat{\mathbf{x}}E_0e^{-\mathrm{Im}(m)z\omega/c}e^{i[\mathrm{Re}(m)z\omega/c-\omega t]},$$

$$\mathbf{B}=\hat{\mathbf{y}}B_0e^{-\mathrm{Im}(m)z\omega/c}e^{i[\mathrm{Re}(m)z\omega/c-\omega t]},$$

and the Poynting vector is

$$\mathbf{S}=\hat{\mathbf{z}}\frac{c|E_0|^2}{8\pi}e^{-2\mathrm{Im}(m)z\omega/c}=\mathbf{S}(z=0)e^{-\alpha_\nu z},$$

where the absorption coefficient α_ν is given by

$$\alpha_\nu = \frac{2\omega}{c}\mathrm{Im}(m).$$

2.3

a. Substituting $\mathbf{F}_{\text{Lorentz}} = \mathrm{Q\ E}$ and $\mathbf{F}_{\text{visc}} = -\beta\mathbf{v}$ into the force equation gives $\mathrm{v} = \mathrm{QE}/\beta$. The direction of the velocity rotates uniformly in a plane normal to the propagation direction with period $2\pi/\omega$. Thus the radius is found from

$$2\pi r = \oint v\,dt$$

to be $r = QE/\beta\omega$.

b. The power dissipated is $P = -\mathbf{v}\cdot\mathbf{F}_{\text{visc}} = \beta v^2 = Q^2E^2/\beta$. Since the orbit of the charge is constant in time, this is the power transmitted to the fluid.

c. The magnetic force is in the direction of propagation and has magnitude $F_{\text{mag}} = QBv/c = QEv/c = Q^2E^2/\beta c$. Here we have used $|\mathbf{E}| = |\mathbf{B}|$ for a free wave.

d. Using the center of the charge's motion as an origin we find the magnitude of the torque to be $\tau = |\mathbf{F}_{\text{Lorentz}} \times \mathbf{r}| = Q^2E^2/\beta\omega$. For a left-hand circularly polarized wave the E-vector, and thus the charge, rotates counterclockwise as viewed facing the wave. This imparts a torque *along* the direction of propagation. The opposite holds for right-hand polarization. Thus $\tau = \pm Q^2E^2/\beta\omega$.

e. The absorption cross section can be found from $P = \sigma S$ where the Poynting flux is $S = cE^2/4\pi$. Thus $\sigma = 4\pi Q^2/\beta c$.

f. P, F_{mag}, and τ are the rates of energy, momentum, and angular momentum, respectively, given to the fluid. From the results above we have the ratios $F_{\text{mag}}/P = (\text{momentum})/(\text{energy}) = 1/c$ and $\tau/P = (\text{angular momentum})/(\text{energy}) = \pm 1/\omega$. Assuming the quantum relation $E = \hbar\omega$ for a single photon then implies the relations $p = E/c = h/\lambda = \hbar k$ and $J = \pm E/\omega = \pm\hbar$. Since these refer to properties of photons, they are applicable in general, not just to the limited problem considered here.

g. The case of linear polarization leads to the (primed) results:

a′. Again $\mathbf{v}=Q\mathbf{E}/\beta$. Since $\mathbf{E}$ oscillates harmonically along one axis ($E_x=E_0\cos\omega t$), so does the particle. Taking an appropriate origin, we find for the displacement: $x(t)=(QE_0/\omega\beta)\sin\omega t$. The maximum displacement from the origin is $QE_0/\omega\beta$.

b′. The average power dissipated is $\langle P\rangle=-\langle\mathbf{v}\cdot\mathbf{F}_{\text{visc}}\rangle=\beta\langle v^2\rangle=(Q^2E_0^2/\beta)\langle\cos^2\omega t\rangle=Q^2E_0^2/2\beta$.

c′. The average magnetic force is along the direction of propagation and has the magnitude $\langle F_{\text{mag}}\rangle=\langle QvB/c\rangle=(Q/c)\langle\mathbf{v}\cdot\mathbf{E}\rangle=(Q^2E_0^2/\beta c)\langle\cos^2\omega t\rangle=Q^2E_0^2/2\beta c$.

d′. There is no torque on the fluid, since $\mathbf{F}_{\text{visc}}$ always acts along a line through the origin.

e′. The absorption cross section can be found from the relation $\langle P\rangle=\sigma\langle S\rangle$. The average Poynting vector is $cE_0^2/8\pi$, by Eq. (2.24b). Thus $\sigma=4\pi Q^2/\beta c$, as before.

f′. The power and magnetic force are the same as before, with $E^2\to E_0^2/2$. Their ratio is the same, and we conclude that $p=E/c$. The angular momentum is zero, however. Quantum mechanically this comes about because a linearly polarized photon is a superposition of two circularly polarized photons of opposite helicity.

2.4—Suppose that $\nabla\times\mathbf{H}=4\pi c^{-1}\mathbf{j}$. Taking the divergence of both sides yields $\nabla\cdot\mathbf{j}=0$. But the equation of charge conservation is $\nabla\cdot\mathbf{j}=-\partial\rho/\partial t$. Therefore, this form of the field equation applies only to the special case $\dot\rho=0$.

Furthermore, omitting the displacement current form the derivation leading to (2.17) gives:

$$\nabla^2\mathbf{E}=\frac{4\pi\sigma\mu}{c^2}\frac{\partial\mathbf{E}}{\partial t}$$

$$\nabla^2\mathbf{B}=\frac{4\pi\sigma\mu}{c^2}\frac{\partial\mathbf{B}}{\partial t}.$$

That is, the equations for $\mathbf{E}$ and $\mathbf{B}$ become (parabolic) diffusion equations rather than (hyperbolic) wave equations.

3.1

a. By analogy with the Larmor formula for electric dipole radiation, the power radiated by a magnetic dipole is

$$P=\frac{2|\ddot{\mathbf{m}}|^2}{2c^3}. \tag{1}$$

To evaluate this expression, we first note that the component of $\mathbf{m}$ along the rotation axis, $|\mathbf{m}|\cos\alpha$, is a *constant*. Thus

$$|\ddot{\mathbf{m}}| = \omega^2 |\mathbf{m}| \sin\alpha, \tag{2}$$

since $m_x = m\sin\alpha\sin\omega t$ and $m_y = m\sin\alpha\cos\omega t$. Next, we wish to express $|\mathbf{m}|$ in terms of B_0, the magnetic field strength at the star surface. In electrostatics, the electric field due to a dipole of strength $d \equiv el$ is obtained in the following way:

$$\mathbf{E} = \nabla\phi,$$

$$\phi = \frac{e}{|\mathbf{r}|} - \frac{e}{|\mathbf{r}+\mathbf{l}|} \approx \frac{d\cos\theta}{r^2},$$

$$\mathbf{E} = -\frac{2d\cos\theta}{r^3}\hat{\mathbf{r}} - \frac{d\sin\theta}{r^3}\hat{\boldsymbol{\theta}}.$$

At the "pole," $\theta = 0$, the electric field has a magnitude $E = 2d/r^3$. By analogy, the magnetic field B_0 at the magnetic pole has the value

$$B_0 = \frac{2m}{R^3}. \tag{3}$$

Substituting Eqs. (2) and (3) into (1), we obtain

$$P = \frac{\omega^4 R^6 B_0^2 \sin^2\alpha}{6c^3}. \tag{4}$$

b. Assuming the neutron star is a homogeneous body, its rotational energy , E_{rot}, satisfies

$$E_{\mathrm{rot}} = \tfrac{1}{5} M R^2 \omega^2. \tag{5}$$

Now, using $P = -\dot{E}_{\mathrm{rot}} = -2/5 M R^2 \omega\dot{\omega}$ and substituting from Eq. (4), we obtain

$$\tau \equiv \frac{-\omega}{\dot{\omega}} = \frac{12 M c^3}{5 R^4 \omega^2 B_0^2 \sin^2\alpha}. \tag{6}$$

c. To obtain quick quantitative estimates of functions as their arguments assume particular values, astrophysicists frequently write equations with all the quantities normalized to some standard values. Thus, for this problem, we can express Eqs. (4) and (6) in the form

$$P=3.1\times10^{43}\text{erg s}^{-1}\left(\frac{\omega}{10^4\text{s}^{-1}}\right)^4\left(\frac{R}{10^6\text{ cm}}\right)^6\left(\frac{B_0}{10^{12}\text{ gauss}}\right)^2\sin^2\alpha$$

$$\tau=42\text{yr}\left(\frac{M}{M_\odot}\right)\left(\frac{R}{10^6\text{ cm}}\right)^{-4}\left(\frac{\omega}{10^4\text{s}^{-1}}\right)^{-2}\left(\frac{B_0}{10^{12}\text{ gauss}}\right)^{-2}\sin^{-2}\alpha.$$

Thus for $\omega=10^4\ \text{s}^{-1}$, $10^3\ \text{s}^{-1}$, $10^2\ \text{s}^{-1}$, $P/3.1\times10^{43}$ erg s^{-1} has values 1, 10^{-4}, 10^{-8}, and $\tau/42$ yr has values 1, 10^2, and 10^4.

3.2—Since $v_\perp \ll c$ we can compute the radiation field $\mathbf{E}_{\text{rad}}$ from Eq. (3.15a):

$$\mathbf{E}_{\text{rad}}=\frac{e}{c^2r}\mathbf{n}\times(\mathbf{n}\times\dot{\mathbf{u}}).$$

Also, since the system is axially symmetric, and we will eventually average the motion over time, no generality is lost by taking $\mathbf{n}$ in the y-z plane (see Fig. S.2). The plane normal to $\mathbf{n}$ contains the unit vectors $\hat{\mathbf{a}}_1=-\hat{\mathbf{x}}$ and $\hat{\mathbf{a}}_2$,

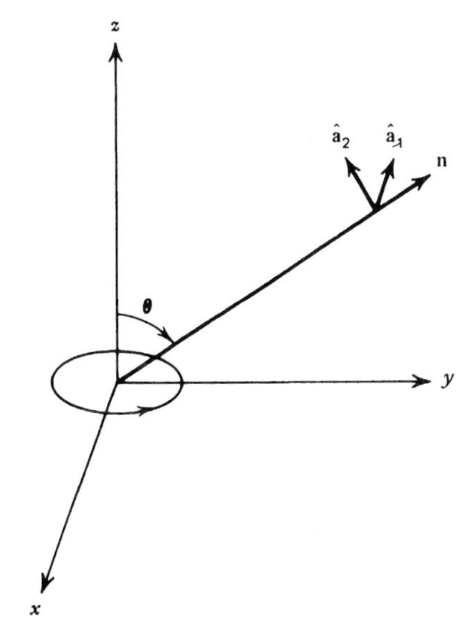

Figure S.2 Geometry for polarization decomposition of radiation emitted by a circulating charge.

which lies in the y-z plane. Letting $\omega = v_\perp/a$, we have for the particle

$$\mathbf{r} = a(\hat{\mathbf{x}}\cos\omega t + \hat{\mathbf{y}}\sin\omega t)$$
$$\mathbf{u} = v_\perp(-\hat{\mathbf{x}}\sin\omega t + \hat{\mathbf{y}}\cos\omega t)$$
$$\dot{\mathbf{u}} = -v_\perp\omega(\hat{\mathbf{x}}\cos\omega t + \hat{\mathbf{y}}\sin\omega t).$$

The components of $\mathbf{n}$ and $\hat{\mathbf{a}}_2$ are given by

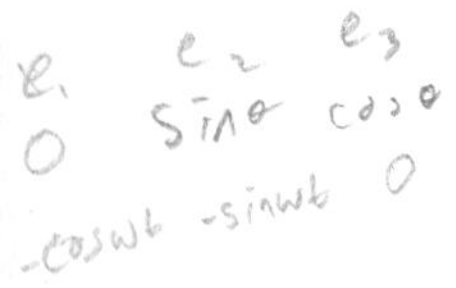

$$\mathbf{n} = \hat{\mathbf{y}}\sin\theta + \hat{\mathbf{z}}\cos\theta,$$
$$\hat{\mathbf{a}}_2 = -\mathbf{y}\cos\theta + \hat{\mathbf{z}}\sin\theta.$$

We now have

$$\mathbf{n}\times\dot{\mathbf{u}} = -v_\perp\omega(\hat{\mathbf{a}}_1\cos\theta\sin\omega t - \hat{\mathbf{a}}_2\cos\omega t),$$
$$\mathbf{n}\times(\mathbf{n}\times\dot{\mathbf{u}}) = -v_\perp\omega(\hat{\mathbf{a}}_1\cos\omega t + \hat{\mathbf{a}}_2\cos\theta\sin\omega t),$$
$$\mathbf{E}_{\text{rad}} = -\frac{ev_\perp\omega}{rc^2}(\hat{\mathbf{a}}_1\cos\omega t + \hat{\mathbf{a}}_2\cos\theta\sin\omega t).$$

a. The power per solid angle is found from

$$\frac{dP}{d\Omega} = \frac{c}{4\pi}|\mathbf{E}_{\text{rad}}|^2 r^2 = \frac{e^2v_\perp^2\omega^2}{4\pi c^3}(\cos^2\omega t + \cos^2\theta\sin^2\omega t).$$

Averaging this over time gives

$$\left\langle\frac{dP}{d\Omega}\right\rangle = \frac{e^2v_\perp^2\omega^2}{8\pi c^3}(1+\cos^2\theta).$$

b. Comparing the formula for $\mathbf{E}_{\text{rad}}$ with Eq. (2.37) (taking the $\hat{\mathbf{x}}$ and $\hat{\mathbf{y}}$ directions of that equation to be now $\hat{\mathbf{a}}_1$ and $\hat{\mathbf{a}}_2$, respectively), we find

$$\mathcal{E}_1 = -\frac{ev_\perp\omega}{rc^2}, \qquad \mathcal{E}_2 = -\frac{ev_\perp\omega}{rc^2}\cos\theta$$

$$\phi_1 = 0, \qquad \phi_2 = \pi/2.$$

The Stokes parameters are, therefore, from Eqs. (2.40),

$$I = A(1+\cos^2\theta)$$
$$Q = A(1-\cos^2\theta)$$
$$U = 0$$
$$V = -2A\cos\theta,$$

where $A \equiv (ev_\perp \omega / rc^2)^2$. The radiation is 100% elliptically polarized ($I^2 = U^2 + Q^2 + V^2$), with $\hat{\mathbf{a}}_1$ and $\hat{\mathbf{a}}_2$ being principal axes. Special cases are

$\theta = 0$	:	left-hand circular polarization.
$\theta = \pi/2$	:	linear polarization along $\hat{\mathbf{a}}_1$
$\theta = \pi$	:	right-hand circular polarization

c. Since $\mathbf{E}_{\text{rad}}$ contains only $\sin\omega t$ and $\cos\omega t$ terms, the radiation is monochromatic at frequency ω. (See, however, Problem 3.7 when radiation of higher order than dipole is included.)

d. Setting the magnetic force equal to the centripetal force gives $\omega : m\omega^2 r = e\omega r B/c$,

$$\omega = \frac{eB}{mc}.$$

Using the result of part (a) gives $\langle P \rangle$:

$$\langle P \rangle = \int \langle \frac{dP}{d\Omega} \rangle d\Omega$$

$$= \frac{e^2 v_\perp^2 \omega^2}{8\pi c^3} \int_0^{2\pi} d\phi \int_0^{\pi} (1 + \cos^2\theta) \sin\theta \, d\theta$$

$$= \frac{2}{3} \frac{e^2 v_\perp^2 \omega^2}{c^3}.$$

Using $r_0 = e^2/mc^2$ and $\beta_\perp = v_\perp / c$, this becomes

$$\langle P \rangle = \tfrac{2}{3} r_0^2 c \beta_\perp^2 B^2.$$

e. The Lorentz force law for an incident electric field E gives $ma\omega^2 = eE$ or $v_\perp = eE/m\omega$. Thus

$$\langle \frac{dP}{d\Omega} \rangle = \frac{r_0^2 c E^2}{4\pi} (1 + \cos^2\theta).$$

Now use the results

$$\langle \frac{dP}{d\Omega} \rangle = \langle S \rangle \frac{d\sigma}{d\Omega}$$

$$\langle S \rangle = \frac{cE^2}{4\pi}$$

to obtain

$$\left(\frac{d\sigma}{d\Omega}\right)_{\text{pol}} = \tfrac{1}{2} r_0^2 (1+\cos^2\theta).$$

To obtain the cross section for unpolarized radiation we should average this cross section with one for circular polarization of the opposite helicity. But since these cross sections do not depend on helicity the unpolarized results are the same:

$$\left(\frac{d\sigma}{d\Omega}\right)_{\text{unpol}} = \tfrac{1}{2} r_0^2 (1+\cos^2\theta).$$

This is the same result obtained previously in Eq. (3.40). The total cross section is just equal to the Thomson cross section, independent of the polarization:

$$\sigma = \int \frac{d\sigma}{d\Omega}\, d\Omega = \frac{8\pi r_0^2}{3}.$$

3.3

a. Use Eq. (3.15a) with $q\dot{u} = -\omega^2 d \cos\omega t$ for each dipole, noting that the retarded times for each differ by $\Delta t = (L/c)\sin\theta$ (see Fig. S.3). Then

$$\begin{aligned}|\mathbf{E}_{\text{rad}}| &= -\frac{\omega^2}{rc^2}\left[d_1 \cos\omega t + d_2 \cos\omega(t-\Delta t)\right]\sin\theta \\ &= -\frac{\omega^2}{rc^2}\left[(d_1 + d_2\cos\delta)\cos\omega t + d_2 \sin\delta \sin\omega t\right]\sin\theta,\end{aligned}$$

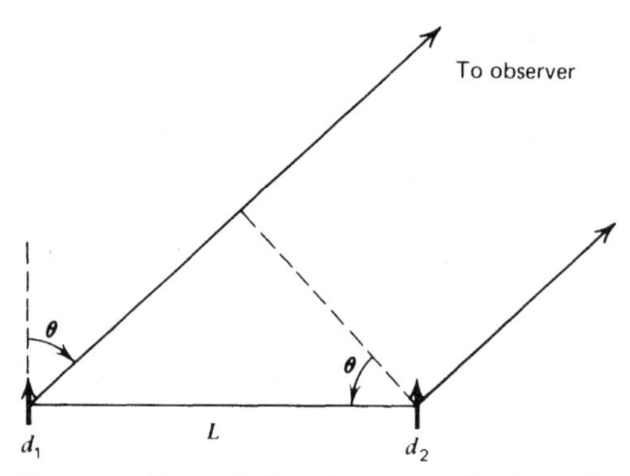

***Figure S.3** Geometry for emission from two dipole radiators separated by distance L.*

where $\delta = \omega \Delta t = \omega L \sin\theta / c$. Squaring and averaging over time, we find

$$\langle |E_{\text{rad}}|^2 \rangle = \frac{\omega^4 \sin^2\theta}{2r^2c^4}\left[(d_1 + d_2\cos\delta)^2 + (d_2 \sin\delta)^2\right]$$

$$= \frac{\omega^4 \sin^2\theta}{2r^2c^4}(d_1^2 + 2d_1d_2\cos\delta + d_2^2).$$

We have finally,

$$\left\langle \frac{dP}{d\Omega} \right\rangle = \frac{cr^2}{4\pi}\langle |E_{\text{rad}}|^2 \rangle$$

$$= \frac{\omega^4 \sin^2\theta}{8\pi c^3}(d_1^2 + 2d_1d_2\cos\delta + d_2^2).$$

b. When $L \ll \lambda$, we have $\delta \equiv 2\pi L \sin\theta / \lambda \ll 1$, and

$$\left\langle \frac{dP}{d\Omega} \right\rangle = \frac{\omega^4 \sin^2\theta}{8\pi c^3}(d_1 + d_2)^2,$$

which is the radiation from an oscillating charge with dipole moment $d_1 + d_2$.

3.4

a. If the cloud is unresolved, then by symmetry there can be no net polarization. Physically, the polarization from different regions of the cloud cancel.

b. Figure S.4*a* shows a typical scattering event in the scattering plane. Radiation from the object can be decomposed into two linearly polarized beams of equal magnitude, one in the plane of scattering and one normal to it. The first produces scattered radiation with polarization direction in the plane of scattering, the second having direction normal to it. These are not of equal magnitude, being in the ratio $\cos^2\theta : 1$ respectively [cf. discussion leading to Eq. (3.41)]. The normal component thus dominates for each value of θ. Integration along the line of sight then gives an observed intensity with dominant component normal to the scattering plane, or, to the observer viewing the plane of the sky, normal to the radial line connecting the object and the point of observation. The plane of the sky with its observed polarization directions is illustrated in Fig. S.4*b*.

c. That the central object can be clearly seen implies that $\tau_e = n_e \sigma_T R \lesssim 1$, and thus $n_e \lesssim (R\sigma_T)^{-1} = 5 \times 10^5\ \text{cm}^{-3}$.

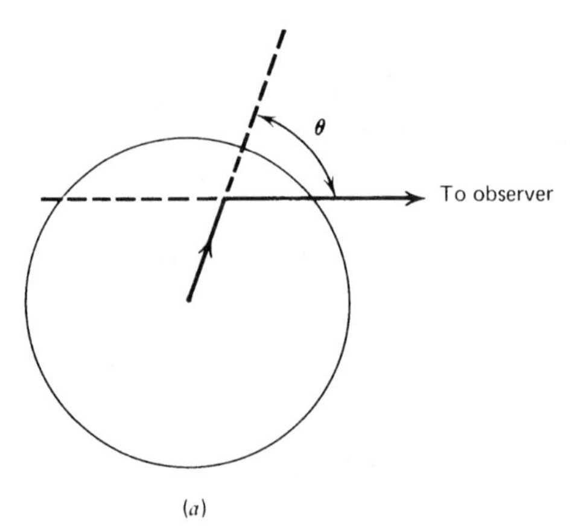

Figure S.4a Scattering event from a spherical cloud.

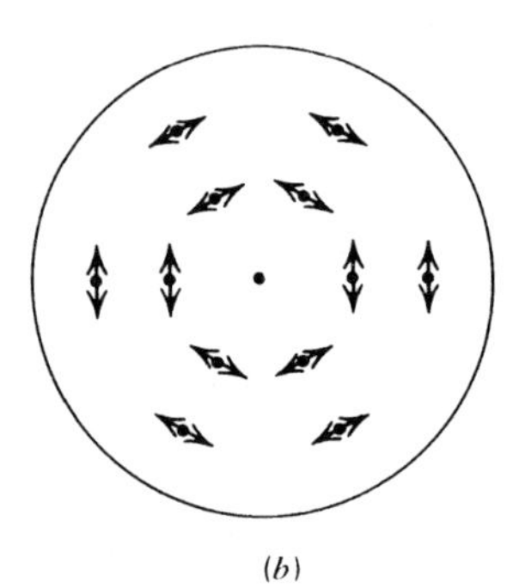

Figure S.4b Observed polarization directions in the plane of the sky.

3.5—Since the field inside the sphere is independent of position, it acts like a single dipole with moment:

$$\mathbf{d} = \frac{4}{3}\pi a^3 \alpha E_0 \left(1 + \frac{4\pi\alpha}{3}\right)^{-1} \cos\omega t\,\hat{\mathbf{x}},$$

where the incident field is $E_0 \cos\omega t\,\hat{\mathbf{x}}$. From Eq. (3.23b) with $k = \omega/c$ we obtain the time averaged power:

$$\langle P \rangle = (4\pi\alpha/3)^2 (1 + 4\pi\alpha/3)^{-2} k^4 a^6 c E_0^2/3.$$

From $\sigma = \langle P \rangle / \langle S \rangle$ with $\langle S \rangle = cE_0^2/8\pi$, we get the required result.

3.6

a. The Fourier transform can be performed explicitly by changing variables in each term of the sum over i:

$$\hat{E}(\omega)=\sum_{i=1}^{N}\frac{1}{2\pi}\int_{-\infty}^{\infty}E_0(u)e^{i\omega u}e^{i\omega t_i}\,du$$

$$=\frac{1}{2\pi}\int E_0(u)e^{i\omega u}du\sum e^{i\omega t_i}$$

$$=\hat{E}_0(\omega)\sum e^{i\omega t_i}.$$

b. Explicitly, we have

$$\left|\sum_{i=1}^{N}e^{i\omega t_i}\right|^2=\sum_{i=1}^{N}\sum_{j=1}^{N}e^{i\omega t_j}e^{-i\omega t_i}$$

$$=\sum_{j=1}^{N}e^{i\omega t_j}e^{-i\omega t_j}+\sum_{j\neq i}\sum e^{i\omega(t_j-t_i)}$$

$$=N+\sum_{j\neq i}\sum e^{i\omega(t_j-t_i)}.$$

Now, since t_i and t_j are randomly distributed, the second term averages to zero.

c. Equation (2.33) gives the spectrum:

$$\frac{dW}{dA\,d\omega}=c|\hat{E}(\omega)|^2=N\left(\frac{dW}{dA\,d\omega}\right)_{\text{single pulse}}.$$

d. In this case we may take each $t_i\approx 0$, because all the pulses have the same arrival time, to order (size of region)/(wavelength). Thus

$$\hat{E}(\omega)\sim N\hat{E}_0(\omega)$$

and

$$\frac{dW}{dA\,d\omega}=N^2\left(\frac{dW}{dA\,d\omega}\right)_{\text{single pulse}}.$$

3.7—Let the charge move in the x-y plane, and its position be denoted by $\mathbf{r}_0(t)$. Then

$$\mathbf{r}_0(t) = r_0(\cos\omega_0 t\,\hat{\mathbf{x}} + \sin\omega_0 t\,\hat{\mathbf{y}})$$
$$\dot{\mathbf{r}}_0(t) = \omega_0 r_0(-\sin\omega_0 t\,\hat{\mathbf{x}} + \cos\omega_0 t\,\hat{\mathbf{y}}).$$

The current is

$$\mathbf{j}(\mathbf{r},t) = e\dot{\mathbf{r}}_0(t)\,\delta(\mathbf{r}-\mathbf{r}_0(t)). \tag{1}$$

Since $\mathbf{j}(\mathbf{r},t)$ is clearly periodic, we write it as a Fourier series:

$$\mathbf{j}(\mathbf{r},t) = \tfrac{1}{2}\mathbf{j}_0^1(\mathbf{r}) + \sum_{n=1}^{\infty}\left[\mathbf{j}_n^1(\mathbf{r})\cos n\omega_0 t + \mathbf{j}_n^2(\mathbf{r})\sin n\omega_0 t\right] \tag{2}$$

with

$$\mathbf{j}_n^1(\mathbf{r}) = \frac{1}{\pi}\int_0^{2\pi}\mathbf{j}(\mathbf{r},t)\cos n\omega_0 t\,d(\omega_0 t) \tag{3a}$$

$$\mathbf{j}_n^2(\mathbf{r}) = \frac{1}{\pi}\int_0^{2\pi}\mathbf{j}(\mathbf{r},t)\sin n\omega_0 t\,d(\omega_0 t). \tag{3b}$$

From Eq. (3.31), we have the Fourier terms for the l-pole contribution to $\mathbf{A}_n^1, \mathbf{A}_n^2$:

$$\left[\mathbf{A}_n^i(\mathbf{r})\right]_l = C(k,r)\int \mathbf{j}_n^i(\mathbf{r}')(\mathbf{n}\cdot\mathbf{r}')^{(l-1)}\,d^3\mathbf{r}', \tag{4}$$

where $i=1,2$ for the coefficient of the cosine or sine term, respectively, in the series. Now, substituting Eq. (3) into Eq. (4) and performing the $d^3\mathbf{r}'$ integral first, we have for the dipole contribution ($l=1$):

$$\left[\mathbf{A}_n^1(\mathbf{r})\right]_1 \propto \int_0^{2\pi}(-\sin\omega_0 t\,\hat{\mathbf{x}} + \cos\omega_0 t\,\hat{\mathbf{y}})\cos n\omega_0 t\,d(\omega_0 t) \propto \delta_{n,1}\hat{\mathbf{y}}, \tag{5a}$$

$$\left[A_n^2(\mathbf{r})\right]_1 \propto \delta_{n,1}\hat{\mathbf{x}}, \tag{5b}$$

where we have used the orthogonality property of sines and cosines and δ is the Kronecker delta. Thus the dipole contribution to the vector potential is nonzero only at $n=1$ ($\omega=\omega_0$) and the cosine and sine coefficients are vectors along the $\hat{\mathbf{y}}$ and $\hat{\mathbf{x}}$ directions, respectively.

For the quadrupole contribution we obtain, using Eqs. (1), (3), and (4) and performing the $\mathbf{r}'$ integration,

$$[\mathbf{A}_n^1(\mathbf{r})]_2 \propto \int_0^{2\pi} (-\sin\omega_0 t\,\hat{\mathbf{x}} + \cos\omega_0 t\,\hat{\mathbf{y}}) \cdot (n_x \cos\omega_0 t + n_y \sin\omega_0 t)\cos n\omega_0 t\, d(\omega_0 t), \tag{6}$$

where n_x and n_y are the x and y components of the unit vector $\mathbf{n}$. Now, using standard trigonometric identities, we can write this as

$$\begin{aligned}[\mathbf{A}_n^1(\mathbf{r})]_2 &\propto \frac{\hat{\mathbf{x}}}{2}\int_0^{2\pi}[-n_x \sin 2\omega_0 t - n_y(1-\cos 2\omega_0 t)]\cdot \cos n\omega_0 t\, d(\omega_0 t) \\ &\quad + \frac{\hat{\mathbf{y}}}{2}\int_0^{2\pi}[n_x(1+\cos 2\omega_0 t) + n_y \sin 2\omega_0 t]\cdot \cos n\omega_0 t\, d(\omega_0 t) \\ &\propto \frac{\hat{\mathbf{x}}}{2} n_y \delta_{n,2} + \frac{\hat{\mathbf{y}}}{2} n_x \delta_{n,2}.\end{aligned} \tag{7a}$$

Analogously, we have the result,

$$[\mathbf{A}_n^2(\mathbf{r})]_2 \propto \frac{\hat{\mathbf{x}}}{2} n_x \delta_{n,2} + \frac{\hat{\mathbf{y}}}{2} n_y \delta_{n,2}. \tag{7b}$$

Thus the quadrupole contribution is nonzero only at $n=2$ ($\omega = 2\omega_0$). It should be clear that, in the general case, the l-pole contribution is solely at the harmonic $\omega = l\omega_0$.

4.1—The key idea in this problem is that, because of relativistic beaming of radiation, portions of the surface that are at sufficiently large angles from the observer's line of sight never communicate with him. Because the sphere is optically thick, only surface elements can be observed. Referring to Fig. S.5 we see that the cones of emission at any surface point (half-angle of order γ^{-1}) include the observer's direction only for a limited region of the sphere, between points B and B'. Emission from points such as C will not reach the observer.

The observed duration of any pulse has as a lower bound the time delay between the observed radiation from A and B:

$$\Delta t \gtrsim \frac{R}{c}(1-\cos\theta_c) \approx \frac{R\theta_c^2}{2c}.$$

But $\theta_c = \gamma^{-1}$ from the geometry, so $R \lesssim 2\gamma^2 c\,\Delta t$.

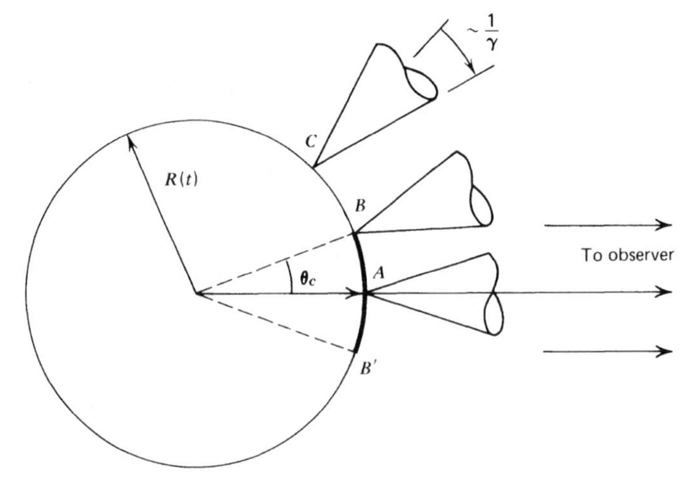

***Figure S.5** Geometry of emission cones from points on the surface of a rapidly expanding shell.*

4.2—Suppose that K and K' count the same set of M stars. Since M is a scalar, they must agree on the count:

$$P(\theta,\phi)\,d\Omega = P(\theta',\phi')\,d\Omega',$$

$$P(\theta',\phi') = P(\theta,\phi)\frac{d\Omega}{d\Omega'}\,.$$

Now, if we take θ as the angle of the incident light ray with the velocity axis (i.e., $\theta=\pi$ corresponds to the forward direction), we may use Eq. (4.95) for $d\Omega/d\Omega'$ to obtain

$$P(\theta',\phi') = \frac{P(\theta,\phi)(1-\beta^2)}{(1+\beta\cos\theta')^2} = \frac{N}{4\pi}\,\frac{(1-\beta^2)}{(1+\beta\cos\theta')^2}\,.$$

Note that $P(\theta',\phi') = P(\theta,\phi)$ if $\beta=0$, and that

$$\int P(\theta',\phi')\,d\Omega' = \frac{N}{4\pi}\cdot 2\pi\int_0^{\pi}\frac{(1-\beta^2)}{(1+\beta\cos\theta')^2}\sin\theta'\,d\theta'$$

$$= N.$$

Finally, since $P(\theta',\phi')$ has a maximum at $\theta'=\pi$, the stars bunch up in the forward direction.

4.3

a. Use Eqs. (4.2) and (4.5) to compute changes in times and velocities measured in different frames:

$$dt=\gamma\left(dt'+\frac{v}{c^2}\,dx'\right)=\gamma\sigma\,dt',$$

$$du_x=\gamma^{-2}\sigma^{-2}\,du'_x,$$

$$du_y=\gamma^{-1}\sigma^{-2}\left(\sigma\,du'_y-\frac{vu'_y}{c^2}\,du'_x\right).$$

Hence

$$a_x=\frac{du_x}{dt}=\gamma^{-3}\sigma^{-3}\frac{du'_x}{dt'}=\gamma^{-3}\sigma^{-3}a'_x,$$

$$a_y=\frac{du_y}{dt}=\gamma^{-2}\sigma^{-3}\left(\sigma\frac{du'_y}{dt'}-\frac{vu'_y}{c^2}\frac{du'_x}{dt'}\right),$$

$$=\gamma^{-2}\sigma^{-3}\left(\sigma a'_y-\frac{vu'_y}{c^2}\,a'_x\right)$$

A similar result holds for a_z.

b. If the particle is at rest instantaneously in K', then $u'_x=u'_y=u'_z=0$. Then $\sigma=1$, and from part (a),

$$a'_{\parallel}=\gamma^3 a_{\parallel},$$

$$a'_{\perp}=\gamma^2 a_{\perp}.$$

4.4

a. An inertial frame instantaneously at rest with respect to the rocket measures its acceleration as g. Transforming from this frame (the "primed frame") to the earth frame with problem 4.3 gives

$$a=\frac{d^2x}{dt^2}=\gamma^{-3}a'=\gamma^{-3}g.$$

Note that the choice of which frame is "primed" is not arbitrary, because in only one is the rocket instantaneously at rest.

b. Since $\gamma=(1-\beta^2)^{-1/2}$, the variables are immediately separable:

$$\int \frac{d\beta}{(1-\beta^2)^{3/2}} = \frac{g}{c}\int dt.$$

Let $\beta=\sin u$. Then the integral becomes

$$\int \frac{du}{\cos^2 u} = \tan u = \frac{\beta}{\sqrt{1-\beta^2}},$$

so that

$$\frac{\beta}{\sqrt{1-\beta^2}} = \frac{gt}{c} + \text{constant}.$$

Since $\beta=0$ at $t=0$, the constant vanishes. Inverting the last expression gives

$$\beta(t) = \frac{gt/c}{\sqrt{(gt/c)^2+1}}.$$

c. Set $\beta=c^{-1}dx/dt$ and separate variables again:

$$\int dx = \int \frac{gt\,dt}{\sqrt{(gt/c)^2+1}}.$$

Substituting $u=(gt/c)^2+1$ to transform the integral,

$$x = \frac{c^2}{2g}\int \frac{du}{u^{1/2}},$$

$$x = \frac{c^2}{g}\sqrt{\left(\frac{gt}{c}\right)^2+1} + \text{constant}.$$

Since $x=0$ at $t=0$, we obtain

$$x = \frac{c^2}{g}\left[\sqrt{\left(\frac{gt}{c}\right)^2+1} - 1\right].$$

d. To find the proper time of the rocket we set

$$dt = \gamma(t)d\tau,$$

$$\gamma(t) = (1-\beta^2)^{-1/2} = \sqrt{(gt/c)^2+1}\ ,$$

$$\int \frac{du}{\sqrt{u^2+1}} = \frac{g\tau}{c} + \text{constant},$$

$$\sinh^{-1}\left(\frac{gt}{c}\right) = \frac{g\tau}{c} + \text{constant}.$$

Since $t=0$ when $\tau=0$,

$$\frac{gt}{c} = \sinh\left(\frac{g\tau}{c}\right).$$

e. By symmetry, the journey consists of four segments of equal distance and time. The maximum distance away from earth is twice the result of part (a), which with part (d) can be written

$$d = \frac{2c^2}{g}\left[\cosh\left(\frac{g\tau}{c}\right) - 1\right].$$

With $\tau = 10$ yr, $g = 980$ cm s^{-2}, this yields $d = 2.8\times 10^{22}$ cm. (One could visit the center of our galaxy).

f. Four times the time given in part (d) is

$$T = \frac{4c}{g}\sinh\left(\frac{g\tau}{c}\right) = 5800 \text{ yr}$$

Unless their friends have also been exploring, the answer is "no."

g. Changing g, to $2g$ gives $T = 8.8\times 10^8$ yr and $d = 4.2\times 10^{26}$ cm. (Were it not for energetic and shielding considerations, round-trip intergalactic travel within one's lifetime would be possible.)

4.5—We demonstrate a simple counterexample: $A^\alpha = B^\alpha = (1,0,0,0)$. Using the boost (4.20) in the transformation law (4.30), we obtain $A'^\alpha = B'^\alpha = \gamma(1,-\beta,0,0)$. Now $A^\alpha B^\alpha = 1$, which does not equal $A'^\alpha B'^\alpha = (1+\beta^2)/(1-\beta^2)$, unless $\beta = 0$.

4.6—Make an arbitrary boost in a direction that lies in the y-z plane. The photon now may have nonvanishing p^y and p^z as well as p^x. A pure rotation lines up the coordinate frame again so that only p^x is nonvanishing, but p^x does not now have its original magnitude. So, make a final boost along p^x either to redshift it or to blueshift it to the original value. Since $E^2-p^2=0$, E also has its original value. You can easily convince yourself that the product of these transformations is not a pure rotation; there is in general a net boost left over. An example is:

$$\underset{x\text{-boost}}{\begin{bmatrix} \gamma' & \gamma' v' & 0 & 0 \\ \gamma' v' & \gamma' & 0 & 0 \\ 0 & 0 & 1 & 0 \\ 0 & 0 & 0 & 1 \end{bmatrix}} \underset{x\text{-}y\text{ rotation}}{\begin{bmatrix} 1 & 0 & 0 & 0 \\ 0 & (1-v^2)^{1/2} & v & 0 \\ 0 & -v & (1-v^2)^{1/2} & 0 \\ 0 & 0 & 0 & 1 \end{bmatrix}}$$

$$\times \underset{y\text{-boost}}{\begin{bmatrix} \gamma & 0 & \gamma v & 0 \\ 0 & 1 & 0 & 0 \\ \gamma v & 0 & \gamma & 0 \\ 0 & 0 & 0 & 1 \end{bmatrix}} \begin{bmatrix} E \\ E \\ 0 \\ 0 \end{bmatrix} = \begin{bmatrix} E \\ E \\ 0 \\ 0 \end{bmatrix}$$

where v' is chosen to satisfy the equality $\gamma\gamma'(1+v')=1$, which gives $v'=v^2/(v^2-2)$.

4.7

a. Suppose that the blob moves from points 1 to 2 in a time Δt (see Fig. S.6). Because 2 is closer to the observer than 1, the apparent time difference between light received by him $(\Delta t)_{\text{app}}$ is

$$(\Delta t)_{\text{app}} = \Delta t\left(1 - \frac{v}{c}\cos\theta\right),$$

(c.f. discussion of Doppler effect, §4.4). The apparent velocity on the sky is

$$v_{\text{app}} = \frac{v\Delta t \sin\theta}{(\Delta t)_{\text{app}}} = \frac{v \sin\theta}{1 - \dfrac{v}{c}\cos\theta}.$$

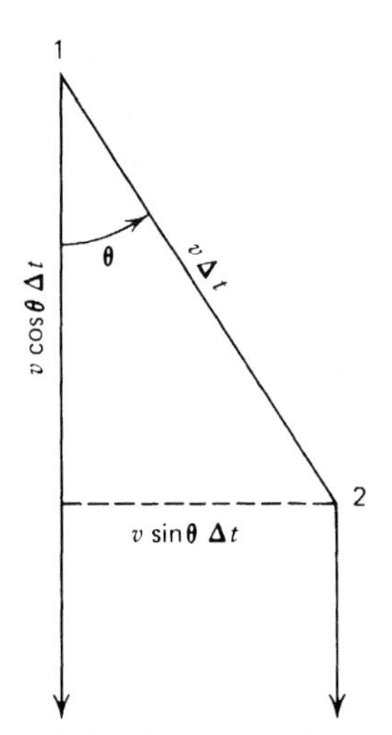

Figure S.6 Geometry of emission from a moving source.

b. Differentiation with respect to θ and setting to zero yields the critical angle θ_c:

$$\cos\theta_c = \frac{v}{c} \equiv \beta; \qquad \sin\theta_c = \sqrt{1-\beta^2} = \gamma^{-1}.$$

The maximum apparent v is thus

$$v_{\max} = \frac{v\sqrt{1-\beta^2}}{1-\beta^2} = \gamma v.$$

This clearly exceeds c when $\gamma \gg 1$.

4.8—Let K be a frame in which the two velocities are $\mathbf{v}_1$ and $\mathbf{v}_2$. The four-velocities are

$$\vec{U}_1 = \gamma(v_1)(1, \mathbf{v}_1), \qquad \vec{U}_2 = \gamma(v_2)(1, \mathbf{v}_2).$$

Let K' be observer 1's rest frame. The four-velocities are:

$$\vec{U}_1' = (1, \mathbf{0}), \qquad \vec{U}_2' = \gamma(v)(1, \mathbf{v}),$$

and $\vec{U}_1' \cdot \vec{U}_2' = \gamma(v)$, so that $v^2 = 1 - (\vec{U}_1' \cdot \vec{U}_2')^{-2}$. But $\vec{U}_1 \cdot \vec{U}_2 = \vec{U}_1' \cdot \vec{U}_2' = \text{scalar}$, so that $v^2 = 1 - (\vec{U}_1 \cdot \vec{U}_2)^{-2}$. Using the above expressions for $\vec{U}_1$ and $\vec{U}_2$, we obtain the desired result.

4.9—We must find a tensor expression that reduces to $\mathbf{j}=\sigma\mathbf{E}$ in the fluid rest frame. From Eq. (4.59) for $F_{\mu\nu}$, the "time components" F_{0k}, contain the electric field. We also know that the fluid four-velocity, U^{μ}, has only time components in the fluid rest frame. Thus we try

$$j^{\mu}=\sigma F^{\mu\nu}U_{\nu}. \tag{1}$$

(We use units where $c=1$.) This equation has the right space components in the rest frame

$$j^{k}=\sigma F^{k\nu}U_{\nu}=\sigma E^{k},$$

but the time component ($\mu=0$) in the rest frame gives $\rho=0$, an unacceptable constraint. Thus we want to subtract out of Eq. (1) its time component in the rest frame, that is, we need to project out only that part which is orthogonal to $\vec{U}$:

$$j^{\alpha}-j^{\beta}U_{\beta}U^{\alpha}=\sigma F^{\alpha\nu}U_{\nu}, \tag{2}$$

where we have used $F^{\alpha\nu}U_{\alpha}U_{\nu}=0$. Now, Eq. (2) is correct. It is manifestly a tensor equation; its space components give $\mathbf{j}=\sigma\mathbf{E}$ in the rest frame (where $U^{k}=0$) and its time component in the rest frame gives $0=0$, that is, no constraint on ρ.

4.10

a. If the radiation is isotropic in K', there is no preferred direction in that frame, so by symmetry the particle must remain at rest in K', that is, $a'^{\mu}=0$. Since $\vec{a}$ is a four-vector that vanishes in K', it also vanishes in K.

b. In K':

$$\vec{P}'_{\text{tot}}=(W',\mathbf{0}),$$

in units where $c=1$. Transforming to K, one obtains

$$\vec{P}_{\text{tot}}=\gamma W'(1,\boldsymbol{\beta}),$$

giving a *spatial* momentum, $\mathbf{P}_{\text{tot}}=\gamma W'\boldsymbol{\beta}$.

c. As shown in (a), the acceleration vanishes in both frames. Momentum is conserved because the particle loses mass, even though its speed in K is unchanged. The mass loss $\Delta m'=W'$ in K' is measured as $\Delta m=\gamma W'$

by an observer in K. Since the particle has speed β, the associated momentum change is $-\beta\Delta m = -\beta\gamma W'$, which just balances the momentum of the radiation.

4.11

a. Let $\gamma' \equiv (1-v'^2)^{-1/2}$ and $\vec{P}$ and $\vec{P}'$ be the total four-momentum vectors before and after absorption. Then $(c=1)$,

$$\vec{P} = (m + h\nu, h\nu, 0, 0),$$

$$\vec{P}' = \gamma' m'(1, v', 0, 0).$$

Conservation of energy and momentum gives $\vec{P} = \vec{P}'$, or $m + h\nu = \gamma' m'$ and $h\nu = \gamma' v' m'$. Thus

$$\frac{m'}{m} = \left(1 + \frac{2h\nu}{mc^2}\right)^{1/2}.$$

b. Suppose that in the lab frame K the particle now initially has velocity v. In the frame K' in which the particle is at rest the photon has frequency

$$\nu' = \gamma\nu(1 - \beta\cos\theta).$$

In frame K' we now perform the same computation as in part (a), and we obtain

$$\frac{m}{m'} = \left(1 + \frac{2h\nu'}{mc^2}\right)^{-1/2}$$

Because m/m' is a scalar quantity (the ratio of rest masses), this equation now holds in any frame, including K.

4.12

a. Since the photons carry no angular momentum with respect to the star (unpolarized radiation), none can be given to the particle.

b. By consideration of a Lorentz frame instantaneously at rest with respect to the particle, plus the result of problem 4.10(a), we have immediately that the particle's v and direction cannot change at the instant of emission. Then

$$\frac{l}{l_0}=\frac{(mvr)_{\text{after}}}{(mvr)_{\text{before}}}=\frac{m}{m'}.$$

Thus from Problem 4.11(b) we have

$$\frac{l}{l_0}=\frac{m}{m'}=\left(1+\frac{2\gamma h\nu}{mc^2}\right)^{-1/2},$$

where $\theta=\pi/2$ in the Doppler formula.

c. Expanding to first order in $h\nu/mc^2$,

$$\frac{\Delta l}{l_0}\approx-\frac{h\nu}{mc^2}.$$

d. The net effect of absorbing and reemitting many photons is for the particle to slowly spiral in towards the sun (assuming $\Delta l/l_0 \ll 1$ per orbit), with no change in its mass; the entire effect, in the lab frame, comes from a nonradial redirection of the incident photons. Letting $\mathfrak{L}$ be the sun's luminosity and $\dot{N}$ be the rate of photon absorption,

$$\dot{N}=\frac{\mathfrak{L}\sigma}{h\nu 4\pi r^2}.$$

Now, using the fact that $l\propto r^{1/2}$ for circular orbits,

$$\frac{1}{l}\frac{dl}{dt}=\frac{1}{2}\frac{1}{r}\frac{dr}{dt}=\left(\frac{\Delta l}{l}\right)_{\text{per photon}}\dot{N}.$$

Combining the above equations with part (c) we have an equation for dr/dt, whose solution is

$$r^2(t)=r_0^2-\left(\frac{\mathfrak{L}\sigma}{\pi mc^2}\right)t.$$

Substituting $\mathfrak{L}\approx 4\times10^{33}$ erg s^{-1}, $m\approx10^{-11}$ g, $\sigma\sim10^{-8}$ cm^{-2}, $R_\odot\sim7\times10^{10}$ cm, 1 AU$\sim1.5\times10^{13}$ cm, we find that the time for r to decrease from 1 AU to $R_\odot$ is

$$t\sim5\times10^4 \text{ yr}$$

4.13

a. From Eq. (4.110) we have

$$\frac{I_\nu}{\nu^3} = \frac{I'_{\nu'}}{\nu'^3}.$$

Substituting the Planck function $I_\nu = B_\nu(T)$ from Eq. (1.51) we obtain

$$\begin{aligned} I'_{\nu'} &= \frac{2h\nu'^3}{c^2}\left(e^{h\nu/kT} - 1\right)^{-1}, \\ &= \frac{2h\nu'^3}{c^2}\left(e^{h\nu'\gamma(1-\beta\cos\theta')/kT} - 1\right)^{-1}, \end{aligned}$$

using the Doppler formula (4.12b) with $\theta \to \theta + \pi$, since radiation propagates in the direction opposite to the viewing angle. If we define

$$T' \equiv \frac{T}{\gamma(1-\beta\cos\theta')} = T\frac{\sqrt{1-v^2/c^2}}{1-(v/c)\cos\theta'},$$

then $I'_{\nu'} = B_{\nu'}(T')$, and for each direction the observed radiation is blackbody.

b. Expanding for $\beta \ll 1$, we obtain $T' \approx T(1+\beta\cos\theta')$, so that $T'_{\max} \approx T(1+\beta)$ and $T'_{\min} \approx T(1-\beta)$. Then

$$\begin{aligned} I_{\max} &\propto \left(e^{h\nu'/kT'_{\max}} - 1\right)^{-1} \propto 1+\beta, \\ I_{\min} &\propto \left(e^{h\nu'/kT'_{\min}} - 1\right)^{-1} \propto 1-\beta, \end{aligned}$$

for $h\nu/kT \sim 0.18 \ll 1$. Thus

$$\frac{I_{\max} - I_{\min}}{I_{\max} + I_{\min}} \approx \beta \lesssim 10^{-3},$$

so that $v \lesssim 300$ km s^{-1}.

4.14—Let the primed frame be the instantaneous rest frame of the particle. Then

$$F_\parallel = \frac{dp_\parallel}{dt} = \frac{\gamma(dp'_\parallel + \beta\, dE')}{\gamma(dt' + \beta/c\, dx')} = \frac{dp'_\parallel}{dt'} = F'_\parallel,$$

because the particle is instantaneously at rest in this frame. Similarly,

$$F_\perp = \frac{dp_\perp}{dt} = \frac{dp'_\perp}{\gamma\, dt'} = \gamma^{-1} F'_\perp .$$

Then, from (4.92),

$$\begin{aligned} P &= \frac{2e^2}{3c^3}\left(a_\parallel'^2 + a_\perp'^2\right) \\ &= \frac{2e^2}{3m^2c^3}\left(F_\parallel^2 + \gamma^2 F_\perp^2\right). \end{aligned}$$

4.15—The two scalar invariants of the electromagnetic field strengths are $\mathbf{E}\cdot\mathbf{B}$ and $\mathrm{E}^2 - \mathrm{B}^2$. If we can show that $W_{em}^2 - |\mathbf{S}|^2/c^2$ can be written solely in terms of these two invariants, then it must be an invariant. Since $W_{em} = (8\pi)^{-1}(E^2 + B^2)$ and $\mathbf{S} = (c/4\pi)\mathbf{E}\times\mathbf{B}$, we have

$$\begin{aligned} 64\pi^2\left(W_{em}^2 - \frac{1}{c^2}|\mathbf{S}|^2\right) &= (E^2 + B^2)^2 - 4|\mathbf{E}\times\mathbf{B}|^2 \\ &= E^4 + 2E^2B^2 + B^4 - 4E^2B^2\sin^2\theta \\ &= E^4 + 2E^2B^2 + B^4 - 4E^2B^2\left(1 - \frac{(\mathbf{E}\cdot\mathbf{B})^2}{E^2B^2}\right) \\ &= (E^2 - B^2)^2 + 4(\mathbf{E}\cdot\mathbf{B})^2. \end{aligned}$$

4.16—The solution to this problem requires some tensor index manipulation plus use of Maxwell's equations in tensor form.

a.

$$\begin{aligned} 4\pi T^{\mu\nu}\eta_{\mu\nu} &= \eta_{\mu\nu} F^{\mu\alpha}F^{\nu}{}_{\alpha} - \tfrac{1}{4}\eta^{\mu\nu}\eta_{\mu\nu}F^{\alpha\beta}F_{\alpha\beta} \\ &= F^{\mu\alpha}F_{\mu\alpha} - \tfrac{1}{4}\cdot 4F^{\alpha\beta}F_{\alpha\beta} \\ &= 0. \end{aligned}$$

b.

$$\begin{aligned} 4\pi T^{\mu\nu}{}_{,\nu} &= F^{\mu\alpha}{}_{,\nu}F^{\nu}{}_{\alpha} + F^{\mu\alpha}F^{\nu}{}_{\alpha,\nu} - \tfrac{1}{4}\eta^{\mu\nu}\left(F^{\alpha\beta}{}_{,\nu}F_{\alpha\beta} + F^{\alpha\beta}F_{\alpha\beta,\nu}\right) \\ &= -F^{\mu}{}_{\alpha}F^{\alpha\nu}{}_{,\nu} + F^{\mu\alpha,\nu}F_{\nu\alpha} - \tfrac{1}{2}F^{\alpha\beta,\mu}F_{\alpha\beta} \\ &= -F^{\mu}{}_{\alpha}F^{\alpha\nu}{}_{,\nu} - F_{\alpha\beta}\left(F^{\mu\alpha,\beta} + \tfrac{1}{2}F^{\alpha\beta,\mu}\right). \end{aligned}$$

Now,

$$F_{\alpha\beta}F^{\mu\alpha,\beta}=F_{\beta\alpha}F^{\mu\beta,\alpha}=F_{\alpha\beta}F^{\beta\mu,\alpha},$$

relabeling indices and using antisymmetry of F. Thus

$$4\pi T^{\mu\nu}{}_{,\nu}=-F^{\mu}{}_{\alpha}(F^{\alpha\nu}{}_{,\nu})-\tfrac{1}{2}F_{\alpha\beta}(F^{\mu\alpha,\beta}+F^{\beta\mu,\alpha}+F^{\alpha\beta,\mu})=0,$$

because the two quantities in parentheses vanish according to Maxwell's equations in free space. [See Eqs. (4.60) and (4.61).]

5.1

a. The optically thin luminosity is equal to the volume $V=(4/3)\pi R^3(t)$ times the power radiated per unit volume, Eq. (5.15b):

$$\mathfrak{L}_{\text{thin}}=1.7\times10^{-27}n_e n_p T_0^{1/2}V,$$

where we have taken $\bar{g}_B=1.2$. Now, $n_e=n_p=M_0/m_p V$, where $m_p=$ hydrogen mass. Thus

$$\mathfrak{L}_{\text{thin}}=1.6\times10^{20}M_0^2T_0^{1/2}R^{-3}(t).$$

b. The optically thick luminosity is equal to the surface area $4\pi R^2(t)$ times the blackbody flux, Eq. (1.43):

$$\mathfrak{L}_{\text{thick}}=7.1\times10^{-4}T_0^4R^2(t).$$

c. The transition between thick and thin cases occurs roughly when $\mathfrak{L}_{\text{thin}}\approx\mathfrak{L}_{\text{thick}}$. Setting the above expressions equal for $t=t_0$ we obtain

$$R(t_0)\approx4.7\times10^4M_0^{2/5}T_0^{-7/10}.$$

[An alternate solution follows by setting $\alpha_R^{ff}R(t_0)\approx1$, using Eq. (5.20). This yields a result of the same form, but with coefficient 2.0×10^4.]

d. See Fig. S.7.

5.2—The knee in the spectrum gives T:

$$T=\frac{E_{\max}}{k}\approx10^9\text{ K}.$$

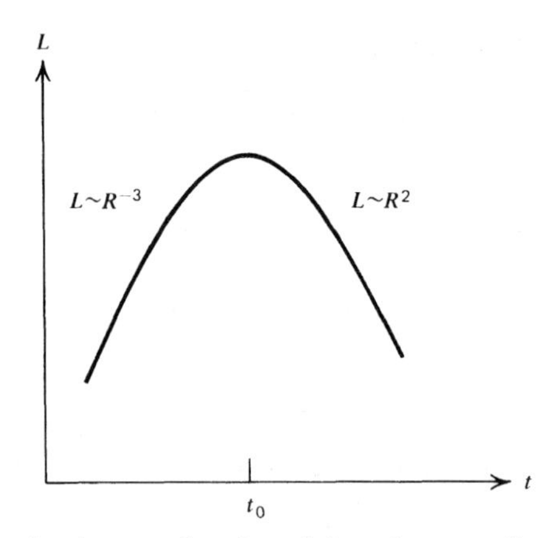

***Figure S.7** Luminosity as a function of time, from a collapsing sphere.*

From Eq. (5.15b) we obtain

$$F = \frac{1}{4\pi L^2}\frac{4\pi R^3}{3}\left(1.4\times 10^{-27} T^{1/2} n_e n_i Z^2 \bar{g}_B\right).$$

At $T = 10^9$ K the gas is completely ionized. If we can assume it is pure hydrogen, $n_i = n_e$. (Including a typical helium abundance makes only a negligible difference.) Then

$$n_i n_e \approx n_H^2 = \left(\frac{\rho}{m_H}\right)^2 = 3.6\times 10^{47}\rho^2.$$

Taking $Z = 1$ and $\bar{g}_B = 1.2$ gives

$$F = 2.0\times 10^{20}\rho^2 T^{1/2} R^3 L^{-2}. \tag{1}$$

Hydrostatic equilibrium gives another constraint on ρ and R. From the virial theorem we know that $2\times$ (kinetic energy/particle) $=$ $-$(gravitational energy/particle) or

$$3kT \sim \frac{GMm_H}{R}.$$

For $T = 10^9$ K this implies

$$R \approx 5\times 10^8 \left(\frac{M}{M_\odot}\right) \text{ cm}, \tag{2}$$

where $M_\odot$ = mass of sun $\approx 2\times 10^{33}$ g.

Combining Eqs. (1) and (2) gives

$$\rho \approx 4\times 10^{-26} L F^{1/2}\left(\frac{M}{M_\odot}\right)^{-3/2}.$$

Substituting in the measured values of F and L we obtain

$$\rho \approx 1.2\times 10^{-7}\text{g cm}^{-3}\left(\frac{M}{M_\odot}\right)^{-3/2}. \tag{3}$$

Now, to get an optical depth we must first determine the dominant opacity source, free-free (bremsstrahlung) κ_{ff} or scattering κ_{es}. Using Eq. (5.20) for the Rosseland mean of κ_{ff} we have

$$\frac{\kappa_R^{ff}}{\kappa_{es}} \approx \frac{0.7\times 10^{23}\rho T^{-7/2}}{0.4} \approx 10^{-15}\left(\frac{M}{M_\odot}\right)^{-3/2}.$$

Thus, for $M/M_\odot \gg 10^{-10}$, $\kappa_R^{ff} \ll \kappa_{es}$ and the "effective" opacity coefficient is [cf. Eq. (1.97)],

$$\kappa_* \sim \sqrt{\kappa_R^{ff}\kappa_{es}} \sim 10^{-8}\left(\frac{M}{M_\odot}\right)^{-3/4}\ \text{cm}^2\ \text{g}^{-1}.$$

The effective optical depth τ_* is

$$\tau_* \sim R\rho\kappa_* \sim 6\times 10^{-7}\left(\frac{M}{M_\odot}\right)^{-5/4}.$$

Thus, for $M/M_\odot \gg 10^{-5}$, the source is effectively thin, $\tau_* \ll 1$, and the assumption of bremsstrahlung emission is justified. For complete consistency, however, one must also check to see whether inverse Compton cooling (Chapter 7) is important. [See Problem 7.2.]

6.1—By conservation of energy, $d/dt(\gamma mc^2) = -$power radiated. Therefore, using Eq. (6.5) and $B_\perp = B\sin\alpha$ we have

$$\dot{\gamma} = \frac{-P}{mc^2} = -A\beta^2\gamma^2 \approx -A\gamma^2,$$

since $\beta \approx 1$. This equation is easily integrated to yield

$$-\gamma^{-1} = -At + \text{constant}.$$

The boundary condition implies constant $= -\gamma_0^{-1}$, so that

$$\gamma = \frac{\gamma_0}{1 + A\gamma_0 t}.$$

At $t = t_{1/2}$ we have $\gamma = \gamma_0/2$. Thus

$$t_{1/2} = (A\gamma_0)^{-1} = \left(\frac{2e^4}{3m^3c^5}\gamma_0 B_\perp^2\right)^{-1}.$$

To correctly account for the radiation reaction (and decrease of γ) in the particle equation of motion, the electric field of the self-radiation must be added to Eq. (6.1).

6.2

a. We assume that the magnetic field is frozen into the gas. [This is almost always a good approximation for problems on a cosmic scale. See, e.g., Alfven, H., *Cosmical Electrodynamics* (Clarendon, Oxford, 1963)]. The magnetic flux through a loop moving with the gas is then a constant, and since area scales as l^2, the magnetic field is proportional to l^{-2}.

b. The action integral for a particle in a periodic orbit is defined as

$$\mathcal{Q} = \oint \mathbf{p} \cdot d\mathbf{r},$$

where $\mathbf{p}$ = momentum and where the integral is taken over one period. This quantity is an *adiabatic* invariant, that is, it is approximately constant for slow changes in the external parameters, such as the magnetic field [see, e.g., Landau and Lifshitz, *Mechanics* 3rd ed., (Pergamon, New York, 1976)].

The motion of the electron separates into uniform motion along the field and circular motion around the field. We may apply the adiabatic invariant to the circular motion alone. From Eq. (6.3), we find

$$p_\perp c = eBa,$$

where a is the radius of the projected motion on the normal plane. From the adiabatic invariant we have $p_\perp a$ = constant, and from part (a) we have $B \propto l^{-2}$. Thus $p_\perp \propto l^{-1}$ and $a \propto l$. Since the orbit contracts at the same rate as the contraction of the gas, the flux through it remains constant.

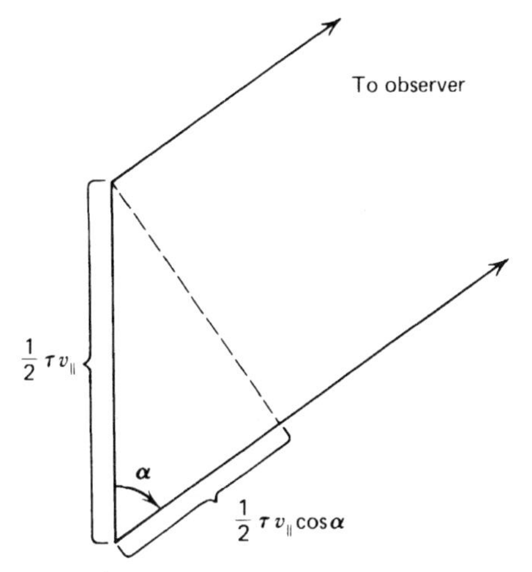

Figure S.8 Geometry of Doppler shift for a particle spiraling in a magnetic field with pitch angle α.

c. For relativistic particles $p\propto\gamma$ so that $\gamma\propto l^{-1}$. From Eq. (6.7b) and $B\propto l^{-2}$ we have $P\propto l^{-6}$. From Eq. (6.17) we have $\omega_c\propto l^{-4}$. From Problem 6.1 we have $t_{1/2}\propto l^5$.

6.3—Suppose that the time for the particle to go from 1 to 2 and back from 2 to 1 is τ. For the second half of each cycle, of duration $\tau/2$, *no* radiation from the particle reaches the observer, since the radiation beam, of halfwidth $\gamma^{-1}\ll 1$, is directed at an angle $\pi-2\alpha$ or greater away from the observer. Now, from simple time delay, the apparent duration of the first half of the cycle is (Fig. S.8)

$$\tau_{\text{app}}=\frac{1}{2}\tau\left(1-\frac{v_{\|}}{c}\cos\theta\right)$$
$$=\frac{1}{2}\tau\left(1-\frac{v}{c}\cos^2\alpha\right)\approx\frac{1}{2}\tau\sin^2\alpha.$$

Thus the fraction of each cycle in which the particle appears to radiate is

$$\tfrac{1}{2}\sin^2\alpha.$$

6.4

a. When the source is optically thin the observed flux $F_\nu\propto j_\nu=\alpha_\nu S_\nu\propto\nu^{-(p-1)/2}$; when it is optically thick $F_\nu\propto S_\nu\propto\nu^{5/2}$. From the given spectrum, $(p-1)/2=1/2$ or $p=2$. At the critical frequency ν_2, where the synchrotron source becomes optically thick, we have two pieces of information:

$$\int \alpha^s_{\nu_2} ds \approx \alpha^s_{\nu_2} R \approx 1,$$

and

$$F_0 = S_{\nu_2} \Omega = \frac{\pi S_{\nu_2} R^2}{d^2}.$$

Therefore,

$$C\left(\frac{B}{B_0}\right)^2 \left(\frac{\nu_2}{\nu_0}\right)^{-3} R = 1,$$

$$A\left(\frac{B}{B_0}\right)^{-1/2} (\nu_2/\nu_0)^{5/2} \Omega = F_0,$$

and we have

$$\frac{B}{B_0} = \left[A(\nu_2/\nu_0)^{5/2} \Omega F_0^{-1}\right]^2,$$

$$R = C^{-1}\left(\frac{\nu_2}{\nu_0}\right)^{-7} (A \Omega F_0^{-1})^{-4}.$$

b. The spectrum below ν_1 seems to be dominated by bound-free absorption, exhibiting the clear ν^2 dependence of an optically thick thermal emitter. The frequency ν_1 is thus the frequency at which the hydrogen gas becomes optically thick, so that we have an additional relation for the distance to the source

$$\int \alpha^{bf}_{\nu_1} ds \approx \alpha^{bf}_{\nu_1} d \approx 1,$$

or

$$d = D^{-1}\left(\frac{\nu_1}{\nu_0}\right)^3.$$

Using the expressions for R, d, and $\Omega = \pi R^2/d^2$ we have

$$\Omega^9 = \pi A^{-8} C^{-2} D^2 \left(\frac{\nu_1}{\nu_0}\right)^{-6} \left(\frac{\nu_2}{\nu_0}\right)^{-14} F_0^8.$$

6.5

a. Integrating the two intensities of Eqs. (6.32) over the electron distribution, we have for the linear polarization

$$\Pi = \frac{\int G(x)\gamma^{-p}\,d\gamma}{\int F(x)\gamma^{-p}\,d\gamma}.$$

We now write γ in terms of x, $\gamma \propto x^{-1/2}$, obtaining

$$\Pi = \frac{\int G(x)x^{(p-3)/2}\,dx}{\int F(x)x^{(p-3)/2}\,dx}. \tag{1}$$

Now, using Eq. (6.35), and the property of the Γ function, $\Gamma(q+1) = q\Gamma(q)$, we obtain

$$\Pi = \frac{p+1}{p+7/3}. \tag{2}$$

b. The polarization of the frequency-integrated emission is, using Eqs. (6.32),

$$\Pi = \frac{\int G(x)\,dx}{\int F(x)\,dx},$$

where we have used the fact that $\omega \propto x$. Comparing this integral with that of Eq. (1) above, we see that they are equal for $p=3$. Thus substituting this value into Eq. (2) above, we obtain

$$\Pi = \frac{4}{3+\frac{7}{3}} = 75\%.$$

7.1—The energy transfer to a photon of energy ϵ, in a single scattering, has the form [cf. Eq. (7.36)]

$$\Delta\epsilon = \epsilon\left(\frac{4kT}{mc^2} - \frac{\epsilon}{mc^2}\right).$$

Thus, for photons of energy $\epsilon \ll 4kT$, the energy gain per scattering can be put into the approximate differential form

$$\frac{d\epsilon}{dN} \sim \epsilon \frac{4kT}{mc^2},$$

where dN is the differential number of scatterings. After N scatterings, the energy of a photon of initial energy ϵ_i is thus

$$\frac{\epsilon_N}{\epsilon_i} \sim e^{(4kT/mc^2)N} \qquad \text{for } \epsilon_N \ll 4kT.$$

The exponential nature of the energy gain is apparent from the initial equation. In a medium of optical depth $\tau_{es} \gg 1$, the characteristic photon scatters $\sim \tau_{es}^2$ times before escaping, because of the random walk nature of the scattering process. Thus setting $N = \tau_{es}^2$ gives

a.

$$\frac{\epsilon_f}{\epsilon_i} \sim e^y, \qquad y \equiv \frac{4kT}{mc^2}\tau_{es}^2.$$

b. When $\epsilon_f \sim 4kT$, the initial equation shows that photons stop gaining energy from the electrons; the process has saturated. Thus, to obtain τ_{crit},

$$\frac{4kT}{\epsilon_i} \sim e^{y_{\text{crit}}},$$

$$(4kT/mc^2)\tau_{\text{crit}}^2 = \ln\left(\frac{4kT}{\epsilon_i}\right)$$

$$\tau_{\text{crit}} = \left[\frac{mc^2}{4kT}\ln\left(\frac{4kT}{\epsilon_i}\right)\right]^{1/2}.$$

c. From part (a) the parameter is $y \equiv (4kT/mc^2)\tau_{es}^2$.

7.2—From the solution to Problem 5.2, the size R, density ρ, and temperature T, of the emitting region satisfy

$$R = 5 \times 10^8 \text{ cm} \frac{M}{M_\odot}$$

$$\rho = 1.2 \times 10^{-7} \text{ g cm}^{-3}\left(\frac{M}{M_\odot}\right)^{-3/2}$$

$$T = 10^9 \text{ K}.$$

From Problem 7.1, inverse Compton is important if the "Comptonization parameter" $y \equiv (4kT/mc^2)\tau^2$ exceeds unity. Now,

$$\tau_{es} \sim \kappa_{es}\rho R,$$

so that

$$y \sim 400\left(\frac{M}{M_\odot}\right)^{-1}.$$

Thus if $M \gg 400\ M_\odot$, inverse Compton can be ignored, and the determination of T, ρ, and R on the assumption of pure bremsstrahlung cooling is self-consistent. On the other hand, if $M \lesssim 400\ M_\odot$, then the model is self-inconsistent, because inverse Compton cooling was ignored in determining the energy balance.

7.3

a. From Eq. (6.17c) for the characteristic synchrotron frequency, we have, in normalized units, (taking $\sin\alpha = 3^{-1/2}$)

$$h\nu_c \approx 0.10\ \text{eV}\left(\frac{\gamma}{10^4}\right)^2\left(\frac{B}{0.1G}\right).$$

The ratio of the photon's energy to the electron rest mass energy, in the electron rest frame, is then given approximately by

$$\frac{\gamma h\nu_c}{mc^2} \approx 2.0\times 10^{-3}\left(\frac{\gamma}{10^4}\right)^3\left(\frac{B}{0.1G}\right).$$

b. The energy associated with a temperature of 1 K is $\sim 0.86\times 10^{-4}$ eV. The blackbody spectrum peaks at $\sim 2.8\ kT$. Thus the characteristic photon in a blackbody spectrum of temperature T has an energy $\sim 2.4\times 10^{-4}T$ eV. The ratio of a microwave photon energy to electron rest mass in the latter's rest frame is, therefore,

$$\frac{\gamma h\nu}{mc^2} \approx 1.4\times 10^{-5}\left(\frac{\gamma}{10^4}\right).$$

Note that in both (a) and (b), for the *second scattering*, the relevant ratio is a factor γ^2 higher and no longer less than unity for $\gamma \sim 10^4$!

7.4

a. Computation of Δ, Eq. (7.53):
First we set $c = 1$ in our computations. Let the initial photon four-momentum be $\vec{P}_\gamma = \hbar\omega(1, \mathbf{n})$, final photon four-momentum be $\vec{P}_{\gamma_1} =$

$\hbar\omega_1(1,\mathbf{n}_1)$, initial electron four-momentum be $\vec{P}_e=(E,\mathbf{p})$, and final electron four-momentum be $\vec{P}_{e_1}$. Then, expanding out the expression

$$|\vec{P}_{e_1}|^2=|\vec{P}_e+\vec{P}_\gamma-\vec{P}_{\gamma_1}|^2$$

gives

$$E\hbar\omega-\hbar\omega\mathbf{p}\cdot\mathbf{n}=\hbar^2\omega\omega_1(1-\mathbf{n}\cdot\mathbf{n}_1)+\hbar\omega_1 E-\hbar\omega_1\mathbf{p}\cdot\mathbf{n}_1. \tag{1}$$

Here $\mathbf{n}$, $\mathbf{n}_1$, and $\mathbf{p}$ are the initial and final photon directions and the initial electron momentum, respectively. From Eq. (1) we obtain

$$\Delta\equiv\frac{\hbar(\omega_1-\omega)}{kT}=\frac{x\mathbf{p}\cdot(\mathbf{n}_1-\mathbf{n})-x^2kT(1-\mathbf{n}\cdot\mathbf{n}_1)}{E-\mathbf{p}\cdot\mathbf{n}_1+xkT(1-\mathbf{n}\cdot\mathbf{n}_1)} \tag{2}$$

where $x\equiv\hbar\omega/kT$. Now, since $\mathbf{p}/m$ is of order $\alpha\equiv(kT/m)^{1/2}$, to lowest order in α we may replace the denominator of Eq. (2) by $E=m$, where m is the electron mass, and neglect the second term in the numerator, $O(\alpha^2)$, in comparison with the first, thus obtaining (putting back factors of c)

$$\Delta=x\mathbf{p}\cdot\frac{(\mathbf{n}_1-\mathbf{n})}{mc}+O\left(\frac{kT}{mc^2}\right). \tag{3}$$

b. Computation of I_2, Eq. (7.54):
Let χ be the angle between the vector $\mathbf{p}$ and the vector $(\mathbf{n}_1-\mathbf{n})$. Then, using Eq. (7.53) for Δ^2, we obtain

$$\begin{aligned} I_2&\equiv\int\int d^3p f_e\Delta^2\frac{d\sigma}{d\Omega}d\Omega \\ &=\left(\frac{x}{mc}\right)^2\int d^3p p^2\cos^2\chi f_e\int|\mathbf{n}_1-\mathbf{n}|^2\frac{d\sigma}{d\Omega}d\Omega. \end{aligned} \tag{4}$$

Now, since $d\sigma/d\Omega$ does not depend on $\mathbf{p}$, to lowest order in $v/c\sim\alpha$, the integral over p may be done independently of the integral over photon directions. Next, substitute in Eq. (7.49) for the Maxwellian electron distribution, f_e, and let χ be the polar angle for the d^3p integration, that is, $d^3p=p^2\,dp\,d\cos\chi\,d\phi$. The integration over d^3p then gives

$$I_2=x^2n_e\frac{kT}{mc^2}\int\frac{d\sigma}{d\Omega}|\mathbf{n}_1-\mathbf{n}|^2d\Omega. \tag{5}$$

Finally, let $\mathbf{n}_1$ lie along the polar axis for the $d\Omega = d\cos\theta\, d\phi$ integration, so that $|\mathbf{n}_1 - \mathbf{n}|^2 = 2(1-\cos\theta)$. Substituting Eq. (7.1b) for $d\sigma/d\Omega$, we obtain the desired result,

$$I_2 = \tfrac{3}{4}x^2 n_e \sigma_T \frac{kT}{mc^2}\int_{-1}^{1}(1-x+x^2-x^3)\,dx$$

$$= 2x^2 n_e \sigma_T \left(\frac{kT}{mc^2}\right).$$

c. Computation of $\partial n/\partial t$, Eqs. (7.55):
To conserve the total number of photons, $\partial n/\partial t$ must have the functional form of Eq. (7.55a), since

$$\frac{d}{dt}\int nx^2\,dx = -\int_0^\infty \frac{\partial}{\partial x}\left[x^2 j(x)\right]dx = -x^2 j(x)|_0^\infty, \tag{7}$$

that is, the change in total photon number arises only from a flux through the boundaries of energy space. Next, write Eq. (7.55a) in the form

$$\frac{\partial n}{\partial t} = -\frac{2}{x}j - \frac{\partial j}{\partial x}, \tag{8}$$

and Eq. (7.52) in the form

$$\frac{\partial n}{\partial t} = C_1(x)n'' + C_2(n,x)n' + C_3(n,x). \tag{9}$$

Equations (8) and (9) must be *functionally identical.* Comparing the highest x derivatives in these two equations, we see that j must contain a term linear in n', with coefficient independent of n and no terms in n''. Thus j must be of the functional form

$$j = g(x)\left[n' + h(n,x)\right]. \tag{10}$$

8.1—Consider two media, of refractive indices n_r and n_r'. Let θ and θ' be the angles of incidence and refraction of a beam of radiation incident on an area $d\sigma$ of the surface of separation of the two media. Let I_ν and I_ν' be the intensities of the incident and refracted beam, respectively. Then, assuming that no energy is lost by reflection at the interface, we have

$$I_\nu \cos\theta\, d\sigma\, d\Omega = I_\nu' \cos\theta'\, d\sigma\, d\Omega', \tag{1}$$

where $d\Omega = d\cos\theta\, d\phi$, $d\Omega' = d\cos\theta'\, d\phi'$. Now, $d\phi = d\phi'$, and by Snell's law

$$n_r \sin\theta = n_r' \sin\theta'. \tag{2}$$

Squaring and differentiating Eq. (2) leads to

$$n_r^2 \cos\theta\, d\cos\theta = n_r' \cos\theta'\, d\cos\theta'. \tag{3}$$

Now, combining Eqs. (1) and (3) gives

$$\frac{I_\nu}{n_r^2} = \frac{I_\nu'}{n_r'^2}.$$

In a medium in which the refractive index changes continuously and slowly on the scale of a wavelength, we can imagine that the photon path is made up of a number of short segments in regions of constant refractive index, to which the above result applies. Thus I_ν / n_r^2 is seen to be an invariant over general paths. Note that the assumption of no reflection loss at the interface between media becomes completely valid in the continuous limit.

8.2—The Fourier transform of $\psi(r,t)$ with respect to r is simply $A(k)\exp[-i\omega(k)t]$ from its definition. Therefore, from Parseval's theorem we have

$$\int_{-\infty}^{\infty} |\psi|^2 dr = (2\pi)^{-1} \int_{-\infty}^{\infty} |A\exp(-i\omega t)|^2 dk = (2\pi)^{-1} \int_{-\infty}^{\infty} |A(k)|^2 dk, \tag{1}$$

since ω is real. Thus the normalization of the packet remains constant in time (no absorption). Now consider the result

$$r\psi(r,t) = \int_{-\infty}^{\infty} A(k) e^{-i\omega t} \frac{1}{i} \frac{\partial}{\partial k} e^{ikr}\, dk$$

$$= \int_{-\infty}^{\infty} e^{ikr} i \frac{\partial}{\partial k} \left[A(k) e^{-i\omega t} \right] dk$$

where we have integrated by parts. This shows that the Fourier transform of $r\psi$ is $i\partial/\partial k(Ae^{-i\omega t})$. Using the generalized Parseval's theorem [cf. Eq. (2.31)],

$$\int A^*(r) B(r)\, dr = (2\pi)^{-1} \int \hat{A}^*(k) \hat{B}(k)\, dk,$$

with $A=\psi$ and $B=r\psi$, we obtain

$$\int_{-\infty}^{\infty} r|\psi|^2\,dr = \frac{1}{2\pi}\int_{-\infty}^{\infty} A^* e^{i\omega t} i\frac{\partial}{\partial k}(Ae^{-i\omega t})\,dk,$$

$$= \frac{1}{2\pi}\int_{-\infty}^{\infty}\left(|A|^2\frac{\partial\omega}{\partial k}t + iA^*\frac{\partial A}{\partial k}\right)dk,$$

which depends on time linearly. Therefore,

$$\frac{d}{dt}\int_{-\infty}^{\infty} r|\psi|^2\,dr = \frac{1}{2\pi}\int_{-\infty}^{\infty}|A(k)|^2\frac{\partial\omega}{\partial k}\,dk, \tag{2}$$

and dividing by Eq. (1), which is independent of time, yields the desired result:

$$\frac{d}{dt}\langle r(t)\rangle = \left\langle\frac{\partial\omega}{\partial k}\right\rangle. \tag{3}$$

Suppose that the wave packet is localized in both space and wave number, within the restrictions of the uncertainty relation $\Delta k\,\Delta r \gtrsim 1$, of course. If $\partial\omega/\partial k$ changes slowly over the scale Δk about the central wave number k_0, then the packet will move with the group velocity $(\partial\omega/\partial k)_{k=k_0}$. This is the usual statement of the group velocity property, but Eq. (3) also holds when the packet is spread arbitrarily in space and wave number.

8.3—Taking the derivative of Eq. (8.31) with respect to ω and dividing the resulting equation by Eq. (8.20), we obtain

$$\frac{d\Delta\theta/d\omega}{dt_p/d\omega} = 1.7\times10^7\ \mathrm{s}^{-1}\langle B_\parallel\rangle,$$

where $\langle B_\parallel\rangle$ is measured in Gauss. Note the interesting result that the frequency dependence cancels out of the above expression. Substituting $d\Delta\theta/d\omega = 1.9\times10^{-4}$ s and $dt_p/d\omega = 1.1\times10^{-5}\ \mathrm{s}^2$, we obtain the result

$$\langle B_\parallel\rangle = 1.0\times10^{-6}$$

without needing to know the frequency at which the measurements were made!

9.1

a. The properly normalized *antisymmetric* wave function is:

$$\psi(\mathbf{r}_1,\mathbf{r}_2,s_1,s_2)=\frac{1}{\sqrt{2}}\left[u_a(1)u_b(2)-u_b(1)u_a(2)\right],$$

where 1 and 2 include space and spin coordinates. The operator whose expectation value we want is

$$R^2=(\mathbf{r}_1-\mathbf{r}_2)^2=r_1^2-2\mathbf{r}_1\cdot\mathbf{r}_2+r_2^2.$$

Now, use Dirac notation for integrals,

$$\int u_a^*(\mathbf{r}_1,s_1)\mathbf{r}_1 u_b(\mathbf{r}_1,s_1)d^3r_1\equiv\langle a|\mathbf{r}|b\rangle,$$

use the fact that $\mathbf{r}_1$ only operates on functions of $\mathbf{r}_1$,

$$\int u_a^*(\mathbf{r}_1)\mathbf{r}_2 u_a(\mathbf{r}_1)d^3r_1=\mathbf{r}_2,$$

and use the orthogonality of orbitals

$$\langle a|b\rangle=0 \qquad \text{for } a\neq b,$$

to obtain

$$\begin{aligned}\langle R^2\rangle&=\int\psi^*(\mathbf{r}_1,\mathbf{r}_2)\left[r_1^2-2\mathbf{r}_1\cdot\mathbf{r}_2+r_2^2\right]\psi(\mathbf{r}_1,\mathbf{r}_2)d^3r_1\,d^3r_2\\&=\langle a|\mathbf{r}^2|a\rangle+\langle b|\mathbf{r}^2|b\rangle\\&\quad-2\langle a|\mathbf{r}|a\rangle\cdot\langle b|\mathbf{r}|b\rangle+2|\langle a|\mathbf{r}|b\rangle|^2.\\&=(\mathbf{r}^2)_a+(\mathbf{r}^2)_b-2|\mathbf{r}_a|\,|\mathbf{r}_b|+2|\mathbf{r}_{ab}|^2.\end{aligned}$$

b. Note that the dipole operator vanishes between states of the same parity (see §10.4)

c. Separating space and spin parts,

$$u_a(\mathbf{r}_1,s_1)=u_a(\mathbf{r}_1)|s_1\rangle,$$

we have

$$\langle a|\mathbf{r}|b\rangle=\int u_a^*(r)\mathbf{r}u_b(r)d^3r\langle s_a|s_b\rangle,$$

since **r** does not operate on $|s\rangle$. Since different spin states are orthogonal, for example, $|s_a\rangle=(1,0)$, $|s_b\rangle=(0,1)$, electrons of different spins give $\langle s_a|s_b\rangle=0=\langle a|\mathbf{r}|b\rangle$.

d. Since $|\langle a|r|b\rangle|^2$ is a positive definitive quantity, comparison of parts (a) and (c) shows that $\langle R^2\rangle$ is larger for same spin electrons by a term $2|\langle a|r|b\rangle|^2$.

9.2

a. $2s^2$: These are equivalent electrons and must be in opposite spin states, so $S=0$, and only singlets occur. Each orbital has zero orbital angular momentum, so $L=0$, and only S states occur. The parity is even: $(-1)^{0+0}=1$. There is only one possible value for $J=L+S$, that is, zero. The one possible term and level is, therefore, $^1S_0$.

b. $2p3s$: These are nonequivalent electrons, so that all combinations of spins are allowed. Therefore both $S=0$ and $S=1$ are possible, and singlets and triplets occur. The orbital angular momenta of the orbitals are $l=1$ and $l=0$, so that $L=1$ and only P states can occur. The parity is odd: $(-1)^{0+1}=-1$. The angular momentum of the singlet state $(S=0, L=1)$ can be only $J=1$. The triplet state $(S=1)$ can combine with $L=1$ to yield $J=0,1,2$. Therefore, the terms and levels are: $^1P_1^O$ and $^3P_{2,1,0}^O$.

c. $3p4p$: These are nonequivalent electrons, allowing $S=0$ and 1, so that singlets and triplets occur. The values of L, found from combining $l=1$ and $l=1$, are $L=0$, 1 and 2, so that S, P, and D states can occur. The parity is even: $(-1)^{1+1}=1$. There are six ways to pick L and S which lead to the following terms and levels: $^3D_{3,2,1}$, $^3P_{2,1,0}$, $^3S_1$, $^1D_2$, $^1P_1$, $^1S_0$.

d. $2p^43p$: There are four equivalent $2p$ electrons and one $3p$ electron. In cases such as this, find the terms of the equivalent electrons first, then combine each in turn with the remaining nonequivalent electron. The terms of p^4 are the same as for p^2, which are ^1S, ^1D, ^3P. (See §9.4.) The ^1S term plus the $3p$ electron gives rise to $^2P_{3/2,1/2}^O$. The ^1D term plus $3p$ yields $^2F_{7/2,5/2}^O$, $^2D_{5/2,3/2}^O$, and $^2P_{3/2,1/2}^O$. The ^3P term plus $3p$ yields $^4D_{7/2,5/2,3/2,1/2}^O$, $^4P_{5/2,3/2,1/2}^O$, $^4S_{3/2}^O$; and $^2D_{5/2,3/2}^O$, $^2P_{3/2,1/2}^O$ and $^2S_{1/2}^O$.

9.3—Recall the statistical weights: $2(2l+1)$ for each nonequivalent electron in a configuration; $(2L+1)$ $(2S+1)$ for a term; and $(2J+1)$ for a level.

a. From $l_1 = l_2 = 0$, we have $2(2l_1+1)\ 2(2l_2+1) = 4$, but two states violate the Pauli principle, and the remaining two are indistinguishable. Thus $N_{\text{conf}} = 1$. From $L=0$ and $S=0$ we obtain $N_{\text{term}} = (2L+1)(2S+1) = 1$. From $J=0$ we obtain $N_{\text{level}} = (2J+1) = 1$.

b. From $l_1 = 0$, $l_2 = 1$ we have $N_{\text{conf}} = 2(2\cdot 0+1)\ 2(2\cdot 1+1) = 12$. The $(L=1, S=0)$ and $(L=1, S=1)$ terms give $N_{\text{term}} = (2\cdot 1+1)(2\cdot 0+1) + (2\cdot 1+1)(2\cdot 1+1) = 3+9 = 12$. The levels have $J = 1, 2, 1, 0$ so that $N_{\text{level}} = 3+5+3+1 = 12$.

c. From $l_1 = 1$, $l_2 = 1$ we have $N_{\text{conf}} = 2(2\cdot 1+1)\ 2(2\cdot 1+1) = 36$. The terms ^3D, ^3P, ^3S, ^1D, ^1P, ^1S yield $N_{\text{term}} = 15+9+3+5+3+1 = 36$. The levels have $J = 3, 2, 1, 2, 1, 0, 1, 2, 1, 0$, so that $N_{\text{level}} = 7+5+3+5+3+1+3+5+3+1 = 36$.

9.4—Using the definitions of λ, ξ, and γ, the Saha equation can be written

$$\frac{\chi_j}{kT} + \ln\left(\frac{N_{j+1}}{N_j}\right) = \ln\left(\frac{2U_{j+1}}{U_j}\right) + \gamma \approx \gamma, \tag{1}$$

since γ is large compared to $\ln(2U_{j+1}/U_j)$.

a. The transition from stage j to $j+1$ is defined by $N_j \approx N_{j+1}$. From Eq. (1) we have $kT \sim \chi_j/\gamma$.

b. From Eq. (1) we have

$$\frac{d\ln(N_{j+1}/N_j)}{d\ln T} = \frac{\chi_j}{kT} + \frac{3}{2},$$

since $\xi \propto T^{-3/2}$. From part (a) we know $\chi/kT \approx \gamma \gg 3/2$, so that

$$\frac{\Delta T}{T} = \left[\frac{d\ln(N_{j+1}/N_j)}{d\ln T}\right]^{-1} \approx \gamma^{-1}.$$

c. The ratio of excited to ground state populations in state j is given by the Boltzmann law:

$$\frac{N_{i,j}}{N_{0,j}} = \frac{g_i}{g_0} e^{-\chi_{i,j}/kT},$$

where g_i and g_0 are statistical weights and $\chi_{i,j}$ is the excitation potential. Using part (a) we have

$$\frac{\chi_{i,j}}{kT} \approx \gamma \frac{\chi_{i,j}}{\chi_j}.$$

Except for very low-lying states, $\chi_{i,j}$ is of order χ_j, so that the exponential term is very small and $N_{0,j} \gg N_{i,j}$.

9.5

a. The Saha equation has the following form:

$$\frac{N_p N_e}{N_H} \equiv \delta N_e = \left(\frac{2\pi m_e kT}{h^2}\right)^{3/2} \exp\left(-\frac{\chi_H}{kT}\right).$$

The statistical factor $2U_p/U_H$ is unity. Eliminate N_e by writing it in terms of ρ and δ, using $N_e = N_p$ (neglect H^- and H_2):

$$\begin{aligned}\rho &= N_H m_H + N_e(m_e + m_p) \approx (N_H + N_e)\\ &= m_H N_e(1+\delta^{-1})\\ N_e &= \frac{\rho\delta}{m_H(\delta+1)}.\end{aligned}$$

Thus

$$\frac{\delta^2}{\delta+1} \equiv \Delta(\rho, T) = \frac{m_H}{\rho}\lambda^{-3} \exp\left(-\frac{\chi_H}{kT}\right),$$

where λ is the thermal de Broglie wavelength of Problem 9.4.

b. Solving the quadratic equation in δ we obtain

$$\delta = \tfrac{1}{2}\left[\Delta + (\Delta^2 + 4\Delta)^{1/2}\right].$$

10.1—The selection rules are:

1. Configuration changes by exactly one orbital, for which $\Delta l = \pm 1$.
2. $\Delta S = 0$.
3. $\Delta L = \pm 1, 0$.
4. $\Delta J = \pm 1, 0$, except $J = 0$ to $J = 0$.

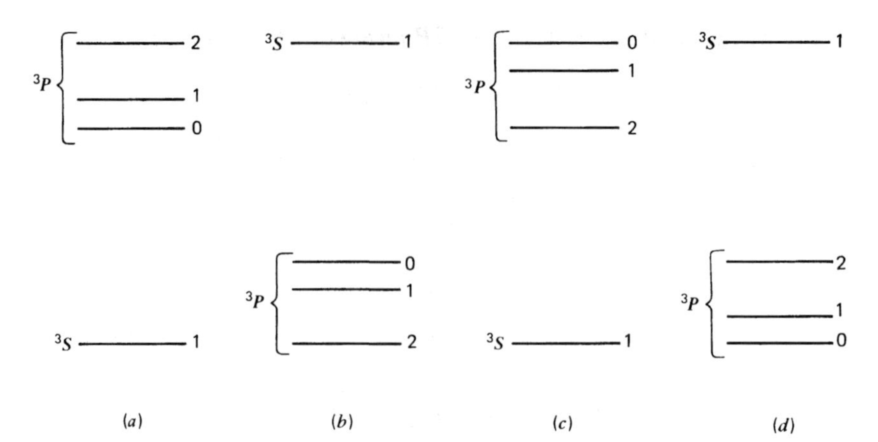

Figure S.9 Energy spacings for ^3S and ^3P levels when ^3P term is (a) normal, upper, (b) inverted, lower, (c) inverted, upper, (d) normal, lower.

Without knowing the configurations we can say nothing about (1) because, for example, $1s2s\ ^3S_1 \rightarrow 1s2p\ ^3P_{0,1,2}$ is allowed while $1s2s\ ^3S \rightarrow 2p3d\ ^3P_{0,1,2}$ is not. Rules (2) and (3) are satisfied, both terms being triplets and $\Delta L = \pm 1$. Finally, all the transitions

$$^3S_1 \rightarrow {}^3P_0$$
$$^3S_1 \rightarrow {}^3P_1$$
$$^3S_1 \rightarrow {}^3P_2$$

satisfy (4) and are therefore allowed.

The four possible arrangements of the two terms with both normal and inverted ^3P term are illustrated in Fig. S.9. Let C be the energy difference between the closest ^3P levels ($J=0$ and $J=1$ levels). Then the energy difference between the other two adjacent levels ($J=1$ and $J=2$) is $2C$, by the Lande interval rule. Also, let ΔE_0 be the energy difference between the $^3S_1$ level and the closest ^3P level, corresponding to the least energetic spectral line.

The energies of the three spectral lines are given in terms of two cases:

1. Normal upper or inverted lower ^3P term (Fig. S.9a and b):

$$^3S_1 \leftrightarrow {}^3P_0, \quad \Delta E = \Delta E_0,$$
$$^3S_1 \leftrightarrow {}^3P_1, \quad \Delta E = \Delta E_0 + C,$$
$$^3S_1 \leftrightarrow {}^3P_2, \quad \Delta E = \Delta E_0 + 3C.$$

2. Normal lower or inverted upper $3P$ term (Fig. S.9*c* and *d*):

$$^3S_1 \leftrightarrow {}^3P_2, \quad \Delta E = \Delta E_0,$$

$$^3S_1 \leftrightarrow {}^3P_1, \quad \Delta E = \Delta E_0 + 2C,$$

$$^3S_1 \leftrightarrow {}^3P_0, \quad \Delta E = \Delta E_0 + 3C.$$

10.2

a. $3s\ ^2S_{1/2} \leftrightarrow 4s\ ^2S_{1/2}$. Not allowed. Parity does not change; jumping electron has $\Delta l = 0$.

b. $2p\ ^2P_{1/2} \leftrightarrow 3d\ ^2D_{5/2}$. Not allowed. $\Delta J = 2$.

c. $3s3p\ ^3P_1 \leftrightarrow 3p^2\ ^1D_2$. Not allowed. $\Delta S = 1$.

d. $2p3p\ ^3D_1 \leftrightarrow 3p4d\ ^3F_2$. Allowed.

e. $2p^2\ ^3P_0 \leftrightarrow 2p3s\ ^3P_0$. Not allowed. $J=0 \rightarrow J=0$.

f. $3s2p\ ^1P_1 \leftrightarrow 2p3p\ ^1P_1$. Allowed.

g. $2s3p\ ^3P_0 \leftrightarrow 3p4d\ ^3P_1$. Not allowed. Parity does not change; jumping electron has $\Delta l = 2$.

h. $1s^2\ ^1S_0 \leftrightarrow 2s2p\ ^1P_1$. Not allowed. Two electrons jump.

i. $2p3p\ ^3S_1 \leftrightarrow 2p4d\ ^3D_2$. Not allowed. $\Delta L = 2$.

j. $2p^3\ ^2D_{3/2} \leftrightarrow 2p^3\ ^2D_{1/2}$. Not allowed. Parity does not change; configuration does not change.

10.3—Comparison of Eqs. (10.10c), (10.23) and (10.29b) shows that the oscillator strength may be written as

$$g_i f_{if} = \sum \frac{2}{3} \frac{m}{\hbar} \omega_{if} |\mathbf{r}_{fi}|^2, \tag{1}$$

where g_i is the statistical weight of the initial state, the sum is over degenerate levels of the initial and final states, ω_{if} is the frequency of the transition, and

$$|\mathbf{r}_{fi}|^2 \equiv \left| \int \psi_f^* \mathbf{r} \psi_i \, d^3r \right|^2. \tag{2}$$

The initial wave function is the $(n,l,m) = (1,0,0)$ ground state of hydrogen, which has the form [cf. Eqs. (9.10) and (9.16)]

$$\psi_{100} = \pi^{-1/2} a_0^{-3/2} e^{-r/a_0}. \tag{3}$$

The final wave function may be any of the three states $(n,l,m)=(2,1,-1)$, $(2,1,0)$, $(2,1,1)$, corresponding to wave functions

$$\begin{aligned}\psi_{21-1}&=r^{-1}R_{21}Y_{1-1},\\ \psi_{210}&=r^{-1}R_{21}Y_{10},\\ \psi_{211}&=r^{-1}R_{21}Y_{11},\end{aligned}\tag{4}$$

where

$$R_{21}\equiv 2^{-3/2}a_0^{-5/2}3^{-1/2}r^2e^{-r/2a_0},\tag{5}$$

and Y_{lm} are the spherical harmonics.

Now, it is easily shown that the operator $|\mathbf{r}_{fi}|^2$ may be written as

$$|\mathbf{r}_{fi}|^2=\tfrac{1}{2}|(x+iy)_{fi}|^2+\tfrac{1}{2}|(x-iy)_{fi}|^2+|z_{fi}|^2,\tag{6}$$

where the latter "spherical operators" are conveniently expressible in terms of spherical harmonics:

$$x\pm iy=r(8\pi/3)^{1/2}Y_{1\pm1},\tag{7a}$$

$$z=r(4\pi/3)^{1/2}Y_{10}.\tag{7b}$$

Using Eqs. (3) to (7) the position matrix element may then be written as

$$|\mathbf{r}_{fi}|^2=\frac{1}{18}a_0^{-8}\mathfrak{R}^2|\mathcal{Q}|^2,\tag{8}$$

where

$$\mathfrak{R}\equiv\int_0^\infty r^4e^{-3r/2a_0}dr=\left(\tfrac{2}{3}\right)^5 4!\,a_0^5,$$

$$|\mathcal{Q}|^2\equiv\left|\int Y^*{}_{1m}Y_{11}d\Omega\right|^2+\left|\int Y^*{}_{1m}Y_{10}d\Omega\right|^2+\left|\int \psi^*{}_{1m}\psi_{1-1}d\Omega\right|^2.$$

Now, by the orthogonality relations of the Y_{lm}, only one term in $|\mathcal{Q}|^2$ contributes for each m, and this contribution is unity.

Finally, performing the sum indicated in Eq. (1), that is, summing over $m=0,\ \pm1$ and multiplying by 2 for the two possible spin states of the initial electron we obtain

$$g_i f_{if} = \frac{2^{17}}{3^{10}} \frac{a_0^2 m\omega_{if}}{\hbar} = \frac{2^{17}}{3^{10}} \frac{\hbar^3 \omega_{if}}{me^4}. \tag{9}$$

Now, ω_{if} is the frequency of transition between the $n=2$ and $n=1$ levels:

$$\omega_{if} = \frac{E_{if}}{\hbar} = \frac{e^2}{2a_0\hbar}\left(1 - \frac{1}{4}\right) = \frac{3}{8}\frac{me^4}{\hbar^3}. \tag{10}$$

Substituting Eq. (10) into (9), we obtain the desired result:

$$g_i f_{if} = \frac{2^{14}}{3^9}.$$

10.4—The initial wave function ψ_i for the $1s$ state of a hydrogen like ion is

$$\psi_i = \left(\frac{Z^3}{\pi a_0^3}\right)^{1/2} e^{-Zr/a_0}.$$

The final wave function ψ_f for the continuum electron in the Born approximation is taken to be the free electron state

$$\psi_f = V^{-1/2} e^{i\mathbf{q}\cdot\mathbf{r}},$$

where $\mathbf{q} = \mathbf{p}/\hbar$ is the wave number of the electron. The normalization $V^{-1/2}$ is consistent with the derivation of Eq. (10.52), the final electron being localized to a volume V.

In the nonrelativistic regime we can neglect the retardation factor $\exp(i\mathbf{k}\cdot\mathbf{r})$ in the matrix element (dipole approximation), so we must evaluate $|\langle f|\mathbf{l}\cdot\nabla|i\rangle|^2$. By the Hermitian property this is equivalent to evaluating $|\langle i|\mathbf{l}\cdot\nabla|f\rangle|^2$. Now

$$\langle i|\mathbf{l}\cdot\nabla|f\rangle = \int \psi_i^* \mathbf{l}\cdot\nabla\psi_f \, d^3\mathbf{r}$$

$$= i(\mathbf{l}\cdot\mathbf{q})\int \psi_i^* \psi_f \, d^3\mathbf{r}.$$

This latter integral can be evaluated by choosing a polar coordinate system with respect to the direction of $\mathbf{q}$. Let θ be the polar angle between $\mathbf{r}$ and $\mathbf{q}$, and let $\mu = \cos\theta$. Then

$$\int \psi_i^* \psi_f d^3x = \left(\frac{Z^3}{\pi V a_0^3}\right)^{1/2} 2\pi \int_0^\infty dr\, r^2 \int_{-1}^1 d\mu\, e^{-Zr/a_0} e^{iqr\mu},$$

$$= \left(\frac{Z^3}{\pi V a_0^3}\right)^{1/2} \frac{4\pi}{q} \int_0^\infty r\, dr\, e^{-Zr/a_0} \sin qr,$$

$$= 8\pi \left(\frac{Z^5}{\pi V a_0^5}\right)^{1/2} \left[\left(\frac{Z}{a_0}\right)^2 + q^2\right]^{-2},$$

the integrals being elementary. Using these results in the cross-section formula (10.52) we obtain

$$\frac{d\sigma_{bf}}{d\Omega} = \frac{32\alpha\hbar}{m\omega}\left(\frac{Z}{a_0}\right)^5 \frac{q(\mathbf{l}\cdot\mathbf{q})^2}{[(Z/a_0)^2 + q^2]^4}.$$

When the energy of the electron is large compared to the ionization energy, $\hbar\omega \approx \hbar^2 q^2/2m \gg Z^2 e^2/2a_0$, it follows that $q^2 \gg Z^2/a_0^2$, so that we may write

$$\frac{d\sigma_{bf}}{d\Omega} \to \frac{32\alpha\hbar}{m\omega}\left(\frac{Z}{a_0}\right)^5 \frac{(\mathbf{l}\cdot\mathbf{q})^2}{q^7}.$$

We now integrate this over solid angles to obtain the total cross section. It is convenient to use now polar coordinates with respect to the direction $\mathbf{l}$. Note that

$$\int d\Omega \frac{(\mathbf{l}\cdot\mathbf{q})^2}{q^7} = \frac{2\pi}{q^5}\int_{-1}^1 d\mu\, \mu^2 = \frac{4\pi}{3q^5}.$$

Therefore, we obtain for the integrated cross section,

$$\sigma_{bf} = \frac{32\alpha\hbar}{m\omega}\left(\frac{Z}{a_0}\right)^5 \frac{4\pi}{3q^5}.$$

Using the relation $\hbar\omega \approx \hbar^2 q^2/2m$ to eliminate q, we obtain finally Eq. (10.53) of the text.

10.5—Since the source is optically thin the spectrum of the emitted radiation is proportional to the emission function (1.73), and thus has the same shape as the profile function $\phi(\nu)$. For this case $\phi(\nu)$ is given by Eq. (10.78), where $\Gamma=\gamma$ is the natural width of Eq. (10.73). In the limiting case of $\Delta\nu_D \ll \gamma/4\pi$ the profile is essentially given by Eq. (10.73), and the observed half-width is independent of temperature. In the limiting case of $\Delta\nu_D \gg \gamma/4\pi$ the profile is essentially given by Eq. (10.68), and the observed half-width grows as the square root of temperature. The critical temperature T_c separating these two cases is found by setting $\Delta\nu_D(T_c)=\gamma/4\pi$. Using Eq. (10.69) we obtain

$$kT_c = \frac{1}{8}(\gamma/\omega_0)^2 m_H c^2, \tag{1}$$

if we assume that hydrogen atoms, $m_a = m_H$, are emitting. For Lyman-α we have $\gamma = A_{21}$ and using Eq. (10.34) we obtain

$$\gamma = \frac{2e^2\omega_0^2}{m_e c^3}\left(\frac{g_1 f_{12}}{g_2}\right)$$

Now $g_1=2$, $g_2=2(2l+1)=6$ for the $2p$ state and $gf=2^{14}/3^9$ from Problem 10.3. The value of ω_0 is found from Eq. (10.42a), with $n=1$ and $n'=2$,

$$\hbar\omega_0 = \frac{3}{4}\frac{e^2}{2a_0} = \frac{3}{8}\frac{m_e e^4}{\hbar^2}.$$

With these values Eq. (1) becomes

$$kT_c = \frac{2^{21}}{3^{18}}\alpha^6 m_H c^2,$$

where $\alpha \equiv e^2/\hbar c$ is the fine structure constant. Numerically,

$$T_c = 8.5\times 10^{-3}\ \text{K}.$$

For most cases of astrophysical interest $T \gg T_c$, and the Doppler broadening dominates, at least near line center. It should be noted, however, that far from line center the Lorentz part of the broadening, which falls off as $(\nu-\nu_0)^{-2}$, will eventually dominate the Doppler part, which falls off extremely rapidly as $\exp[-(\nu-\nu_0)^2/\Delta\nu_D^2]$ [cf. Eqs. (10.77) to (10.79)].

10.6—The dipole matrix elements of $\mathbf{r}\equiv(x,y,z)$ can be conveniently expressed in terms of the matrix elements z_{if} and $(x\pm iy)_{if}$. The matrix element of $z=r\cos\theta$ between states with (l,m) and (l',m') is proportional to

$$\int_{-1}^{+1} P_{l'}^{m'}(\mu)\mu P_l^m(\mu)\,d\mu \int_0^{2\pi} e^{i(m-m')\phi}\,d\phi$$

where $\mu=\cos\theta$ and $P_l^m(\mu)$ is the associated Legendre function. Since the second integral vanishes unless $m'=m$ we need consider only

$$\int_{-1}^{1} P_{l'}^{m}(\mu)\mu P_l^m(\mu)\,d\mu.$$

The recurrence relation

$$(2l+1)\mu P_l^m=(l-m+1)P_{l+1}^m+(l+m)P_{l-1}^m$$

and the orthogonality relations for the P_l^m (see, e.g., Arfken, G. 1970, *Mathematical Methods for Physicists*, Academic, New York) imply that

$$z_{if}=0, \text{ unless } m'=m \text{ and } l'=l+1 \text{ or } l'=l-1.$$

The matrix element of $(x\pm iy)=r\sin\theta\, e^{\pm i\phi}$ is proportional to

$$\int_{-1}^{+1} P_{l'}^{m'}(\mu)\sqrt{1-\mu^2}\; P_l^m\,d\mu \int_0^{2\pi} e^{i(m-m'\pm 1)\phi}\,d\phi,$$

which vanishes unless $m'=m+1$ or $m'=m-1$. If $m'=m+1$ we use the recurrence relation $(2l+1)\sqrt{1-\mu^2}\; P_l^{m-1}=P_{l-1}^m-P_{l+1}^m$ and the orthogonality relations to show that

$$(x+iy)_{if}=0, \text{ unless } m'=m+1 \text{ and } l'=l+1 \text{ or } l'=l-1.$$

Similarly, we show that

$$(x-iy)_{if}=0, \text{ unless } m'=m-1 \text{ and } l'=l+1 \text{ or } l'=l-1.$$

Taken together, these results imply the electric dipole selection rules: $\Delta m=0, \pm 1$ and $\Delta l=\pm 1$.

10.7—Let $\phi_c(t)$ be the collision-induced random phase at time t. Then the electric field will be

$$E(t)=Ae^{i\omega_0 t-\gamma t/2+i\phi_c(t)}, \qquad (1)$$

where A is a constant, ω_0 is the fundamental frequency, and γ is the rate of spontaneous decay. We wish to compute the averaged power spectrum

$$\langle|\hat{E}(\omega)|\rangle^2 = \left\langle \left| \int E(t) e^{i\omega t}\, dt \right|^2 \right\rangle. \tag{2}$$

From Eq. (1), we have

$$\begin{aligned} |\hat{E}(\omega)|^2 &= |A|^2 \int_0^\infty \int_0^\infty dt_1\, dt_2\, e^{i(\omega-\omega_0)(t_1-t_2) - \frac{\gamma}{2}(t_1+t_2)} \\ &\cdot e^{i[\phi_c(t_1)-\phi_c(t_2)]} \\ &\equiv |A|^2 \int_0^\infty \int_0^\infty dt_1\, dt_2\, G(t_1,t_2) e^{i[\phi_c(t_1)-\phi_c(t_2)]}. \end{aligned} \tag{3}$$

Now, the only random function in $|\hat{E}(\omega)|^2$ is $\phi_c(t)$. Thus we obtain

$$\langle|\hat{E}(\omega)|^2\rangle = |A|^2 \int \int dt_1\, dt_2\, G(t_1,t_2) \langle e^{i[\phi_c(t_1)-\phi_c(t_2)]} \rangle. \tag{4}$$

Now, we can write

$$\phi_c(t_1) - \phi_c(t_2) = \Delta\phi_c(t_1 - t_2)$$

where $\Delta\phi_c(t_1-t_2)$ is the change of phase during the time interval $t_1 - t_2$, and we wish to compute

$$\langle e^{i\Delta\phi_c(t_1-t_2)} \rangle.$$

Since changes in phase are random, this average vanishes if one or more collisions occurs during $\Delta t \equiv |t_1 - t_2|$ and is unity if no collisions occur during this interval. We are given the mean rate of collisions is ν_{col}, thus implying that the probability for no collisions to occur during Δt is $e^{-|t_1-t_2|\nu_{col}}$ (assuming a Poisson distribution for the collisions). Thus

$$\langle e^{i\Delta\phi_c(t_1-t_2)} \rangle = e^{-|t_1-t_2|\nu_{col}}, \tag{5}$$

and Eq. (4) becomes

$$\langle|\hat{E}(\omega)|\rangle^2 = |A|^2 \int_0^\infty dt_1 \int_0^\infty dt_2\, G(t_1,t_2) e^{-|t_1-t_2|\nu_{col}}. \tag{6}$$

Equation (6), using Eq. (3) for $G(t_1 t_2)$, can be integrated in terms of elementary functions and yields Eq. (10.75).

11.1

a. In order of magnitude the equilibrium separation is the Bohr radius of an atom, since the electron binding is what holds the two positive charges together. So

$$\mathbf{r}_0 \sim a_0 \equiv \frac{\hbar^2}{me^2}.$$

b. Since the molecules will be electrically neutral in the temperature range considered, they can be treated approximately as hard spheres of size $\sim r_0 \sim a_0$. The cross section is thus the simple geometrical form for the area

$$\sigma \sim \pi r_0^2 \sim \pi a_0^2.$$

c. For Doppler broadening, the line width, $\Delta\nu_D$, is

$$\Delta\nu_D = \frac{\nu_0}{c}\sqrt{\frac{2kT}{3M_p}} \sim \frac{E_{\rm rot}}{hc}\sqrt{\frac{2kT}{3M_p}}.$$

For collisional broadening, the line width, $\Delta\nu_c$, is the collision frequency $\nu_{\rm col}$. From part (b), we estimate

$$\begin{aligned}\nu_{\rm col} &= n\sigma_{\rm col}\langle v\rangle \\ &\sim (\rho/M_p)a_0^2\sqrt{kT/M_p}.\end{aligned}$$

For low ρ, the line width is dominated by $\Delta\nu_D$ and is thus independent of ρ. At high ρ, $\nu_{\rm col} \gg \Delta\nu_D$ and the line width increases in proportion to ρ. The transition occurs at a ρ_0 such that

$$\nu_{\rm col}(\rho_0) \sim \Delta\nu_D$$

or

$$\frac{\rho_0}{M_p}a_0^2 \sim \frac{E_{\rm rot}}{hc} \sim \left(\frac{\hbar^2}{M_p a_0^2}\right)\frac{1}{hc}$$

or

$$\rho_0 \sim \frac{\hbar}{ca_0^4} = \alpha\frac{m_e}{a_0^3} \sim 5\times10^{-5}\ \text{g cm}^{-3}.$$

11.2

$$S(R) \equiv \pi^{-1} \int e^{-|\mathbf{r}-\mathbf{R}_A|} e^{-|\mathbf{r}-\mathbf{R}_B|} r^2 \, dr \, d\cos\theta \, d\phi.$$

Make the changes of variables $\mathbf{R} \equiv \mathbf{R}_A - \mathbf{R}_B$, $y \equiv |\mathbf{r} - \mathbf{R}_A|/R$, $x \equiv \cos\theta$, where θ is the angle between $(\mathbf{r} - \mathbf{R}_A)$ and $\mathbf{R}$. Then, after doing the trivial ϕ integral to obtain a factor 2π, one gets

$$S(R) = 2R^3 \int_{-1}^{1} dx \int_0^{\infty} y^2 \, dy \, e^{-R(y + \sqrt{y^2 + 2yx + 1}\,)}.$$

Now, reverse the order of integration, doing the x integral first, and make the change of variables $a^2 \equiv y^2 + 2yx + 1$, $dx = y^{-1}\, a da$, to obtain

$$S(R) = 2R^3 \int_0^{\infty} e^{-Ry} y \, dy \int_{|y-1|}^{y+1} e^{-Ra} a \, da.$$

This elementary double integral is easily done by (e.g., integration by parts) to yield Eq. (11.12).

11.3—Regard the $H(J)$ term in the expression for ω_{nvJ} as a function of α_n and $\alpha_{n'}$, written in the form

$$H(J) = j_1(J)(\alpha_n + \alpha_{n'}) + j_2(J)(\alpha_n - \alpha_{n'}),$$

where the "coefficients" $j_1(J)$ and $j_2(J)$ are to be determined by equating the above expression to Eq. (11.42) for $H(J)$. For the P branch, the two resulting equations for $j_1(J)$ and $j_2(J)$ yield

$$j_1(J) = -(J+1) \equiv j$$
$$j_2(J) = j^2.$$

As J ranges over its allowed positive values, j ranges over negative integer values, $j \leqslant -1$. For the R branch the two resulting equations for $j_1(J)$ and $j_2(J)$ yield

$$j_1(J) = J \equiv j,$$
$$j_2(J) = j^2$$

and j ranges over positive integer values $j \geqslant 1$. Combining the two formulae yields

$$H(J) = j(\alpha_n + \alpha_{n'}) + j^2(\alpha_n - \alpha_{n'}),$$

where j ranges over negative and positive integers, excluding zero, the "band origin."

11.4—Regarding j as a continuous variable in the result of Problem 11.3, we find the band head (i.e., the reversal of line spacing with increasing j) by setting the derivative of $H(J)$ equal to 0, $\partial H/\partial j|_{j_{\text{head}}}=0$, giving

$$j_{\text{head}}=\frac{1}{2}\left(\frac{\alpha_n+\alpha_{n'}}{\alpha_{n'}-\alpha_n}\right).$$

If $j_{\text{head}}<0$ (i.e., $\alpha_{n'}<\alpha_n$), the band head clearly falls in the P branch. If $j_{\text{head}}>0$, (i.e., $\alpha_n<\alpha_{n'}$), the band head falls in the R branch. Generally, j_{head}, as deduced above, is not an integer, so that the true value of the band head corresponds to the nearest integer value to j_{head}. From j_{head}, the value of J_{head} may be deduced from the solution to Problem 11.3, that is,

$$\begin{aligned} J_{\text{head}}&\sim j_{\text{head}} && \text{for } j_{\text{head}}>0, \\ & && (R \text{ branch}) \\ J_{\text{head}}&\sim-(1+j_{\text{head}}) && \text{for } j_{\text{head}}<0 \\ & && (P \text{ branch}) \end{aligned}$$

The frequency of the band head is found by substituting j_{head} into the expression for $H(j)$:

$$\omega_{n\upsilon J}|_{\text{head}}=\omega_0-\frac{1}{4}\frac{(\alpha_n+\alpha_{n'})^2}{(\alpha_n-\alpha_{n'})}.$$

The band head frequency is below or above ω_0, depending on whether $\alpha_n>\alpha_{n'}$ or $\alpha_n<\alpha_{n'}$.

11.5—Using the same arguments as in Problem 11.4, we see that the Q branch would have a band head at the J value satisfying

$$\frac{\partial H(J)}{\partial J}=0,$$

or, using Eq. (11.42b) for the Q branch form of $H(J)$,

$$J_{\text{head}}=-1/2.$$

Since J only has positive integer values, this band head is never actually realized. However, this value is sufficiently near $J=1$ that the Q branch (at low resolution), at $J=1$, resembles a band head.

11.6—Rotational energy levels have energies

$$E_{\text{rot}}=\frac{\hbar^2}{2I}J(J+1),$$

where $I\sim m_p r_0^2\sim m_p a_0^2$, and vibrational and electronic transitions have energies

$$E_{\text{vib}}\sim\left(\frac{m_p}{m_e}\right)^{\frac{1}{2}}E_{\text{rot}},$$

$$E_{\text{elec}}\sim\left(\frac{m_p}{m_e}\right)E_{\text{rot}}.$$

The probability of a given energy level being occupied is proportional to $\exp(-E/kT)$. Thus if $kT\lesssim E_{\text{rot}}(J-2)$, most molecules will be in the $J=1$ rotational ground states, and few rotational transitions can occur. On the other hand, if $kT\gtrsim E_{\text{vib}}$, vibrational levels will be excited, and rotational-vibrational spectra will be produced. To have pure rotation spectra produced, then, one requires

$$\frac{\hbar^2}{m_p a_0^2}\ll kT\ll\frac{\hbar^2}{m_p a_0^2}\left(\frac{m_p}{m_e}\right)^{\frac{1}{2}}.$$

INDEX